Pitiless Bronze:
A Postpatriarchal Examination of Prepatriarchal Cultures

By

Ruth J. Heflin, PhD

Second Edition Paperback

Choeofpleirn Press, LLC

2023

ISBN 979-8-9877852-6-3 (paperback)

Appreciations

For all the feminist scholars who came before me
trying to pinpoint the rise of the patriarchy.
I hear you and honor your work.

ΩΩΩ

For my ever-patient husband who supported me and chivvied me
forward, reminding me that, while doing the research is exciting,
the writing must be done. I love you for your patience and your
careful editing assistance, too, despite the fact you wanted me to
use "et al" instead of listing every scholar's name in order
to give everyone the credit s/he/xi deserves.

ΔΔΔ

For my son who listens every time I tell a story
as though he is hearing it for the first time.

ΥΥΥ

Special thanks to Maria Kvilhaug for introducing me to the
symbolism of bucrania and how horned animal skulls became the
symbol of female reproductive organs.

ααα

Finally, for my mother, who championed my goals
and supported my efforts no matter what I decided
to do. She inspired me to read everything
and to write down what I know.

Table of Contents

---

[1] Previously published in the *Glacial Hills Review*, Summer 2020.

Illustrations

Section One

What Readers Need to Know

Bronze does not just gleam like a merciless sun in a hot desert during the combat on the Trojan plains that Homer wrote about circa 750 BCE in his epic poem, *The Iliad*. The poet seems to purposely describe the bronze weaponry, which was a relatively new technology circa 1250 BCE when the siege of Troy supposedly took place, as "pitiless" because it helped the fighters kill with abandon, without regret, and definitely without caring much about who the metal weapons injured and killed. Thus, Homer sets up for his listeners of his time and for his readers now an important warning contrast between the "pitiless bronze" weapons and the flesh and blood warriors who wielded them, reminding us that young males allowed to, even encouraged to, feed their anger with frenzied blood baths spell doom for others and themselves. Might, after all, does not make murder, mayhem, or war right.

Homer warns us from 2800 years in the past that impersonal technology used ruthlessly can overtake our sense of humanity by giving us vivid, detailed examples from his imagined Troy.

But there is so much more to the story and to the Bronze Age which gave it birth than that apparent warning.

Because not everyone is familiar with the ancient world circa 3000-750 BCE or with the scholarly studies written by historians, archeologists, anthropologists, or sociologists, this first section of the book explains many concepts in ways that both lay citizens and scientists can understand in order to embrace the arguments in the more scholarly examinations of the prepatriarchal cultures discussed in Section Two.

Many of these concepts are important to understand because they reintroduce certain symbols used by ancient humans and reinterpret them through gynocentric eyes, rather than the androcentric view usually applied to them. This step is necessary to demonstrate how our ancient ancestors used symbolism to communicate, and why so many scholars who use blinders based on their patriarchal cultures miss these important ideas.

The chapters do not have to be read in order because some ideas are overlapped in order to increase comprehension.

My ultimate goal in this book is to urge scholars to remove their androcentric glasses in order to put on gynocentric ones that will allow them to better interpret the signs and symbols left by our ancestors prior to the discovery of procreation circa 2400 BCE, which led to the cultural elevation of males as dominant over females in a cultural system that we now call the patriarchy. Otherwise, their confirmation bias will continue to discolor the real ancient history of humanity.

Chapter One
Why Postpatriarchal?

The "patriarchy," which has not reigned as long as some people want us to believe, is in its death throes.

Small groups of men around the world, enabled by some women who find advantages to living in patriarchal cultures, are attempting, right this minute, to re-assert male control over female rights under the law, laws that have, traditionally, been written almost exclusively by men. The fact that so many of these androcentrists are using fear and brute force to attempt to dominate women, again, demonstrates that they are desperate, giving their last gasps before finally being thrown over for a more egalitarian form of governance. Hence, this examination is postpatriarchal.

Whether we are discussing the Islamic State, the Taliban, Southern Baptists, the Texas or Kansas state legislatures, or the conservatively stacked United States Supreme Court, there are concerted groups of men, often assisted by women, attempting to remove women's fundamental right to make our own decisions, again.

For many of the people who support the "patriarchy," the belief that men are superior to women, so should make decisions for them, seems natural, mostly because they have learned to turn a blind eye to contrary evidence. For many male-supremacists, having a penis and testicles automatically makes one human superior to those who possess the more generatively powerful vagina, vulva, and uterus trio, which is why they also oppose anyone who questions the binary male-female gender stereotypes still largely purveyed in so many cultures around the world. The big problem is that these ultra conservatives never question their own skewed logic.[2]

---

[2] If they cannot even determine how twisted their religious views are, bigoted views that sanction their ideas of male supremacy, how will they ever recognize their skewed logic about gender? We might be asking too

What is especially tragic are the number of published scholars who also appear to believe the idea that the "patriarchy" existed long before history (written history) began, despite rampant evidence to the contrary. Feminist scholars have long experienced backwash against the idea that women are equal to men in everything, so it is not particularly surprising when female androcentrists are recruited to argue against the idea of prepatriarchal societies that were undoubtedly more gynocentric than androcentric simply because of three powerful, magical, and spiritual facts: 1) women's ability to bleed monthly and not die, 2) the menstrual cycle's seeming synchrony with the moon's phases, and 3) the fact that women were the only gender creating new human beings seemingly out of the power of their own bodies. How could any ancient human being note these three facts and not be awed by women's magical and spiritual (not to mention physical) superiority to men? To understand this point of view better, we have to peel off our androcentric lenses to see the ancient world anew.

Androcentric scholars like Cynthia Eller in her book, *The Myth of the Matriarchal Prehistory*, get vicious when it comes to mocking feminist talking points about the logic of there having been a matriarchal organization to humanity long before there was a patriarchal one. Remember, when someone resorts to denigrating the person instead of discussing her theory, the negater is admitting s/he/xi does not have any real evidence to share. In discussing the magic of childbirth, Eller mockingly claims that "many feminist matriarchalists—probably more than the national average—are childless" (57). Even women who are not mothers know what motherhood looks, sounds, and feels like, so her comment is completely unnecessary and unkind. Ironically, by the end of her book, Eller has actually convinced herself

---

much of them, but most of the world has had enough of religious groups trying to dictate laws and social mores that clearly favor men over women and impose strict gender identities.

that the "matriarchal myth" provides "an emotionally compelling, inspiring" answer to the question of how modern humans can overcome male domination in most cultures (185). She makes the error of assuming, however, that just because most matriarchies were overcome by patriarchies means humans will never see value in thus never establish matriarchies again (185). Ironically, Eller spends hundreds of pages dismissing valid theories without providing any counter evidence, only personal hearsay. Like so many androcentrists, Eller's real underlying fears seem to be that no one will find value in women as leaders of whole cultures or, possibly, that women will treat men the way women have been treated by men for the last two thousand years.

This androcentric bias carried into scholarly research is especially shocking among the scholars who consider themselves feminists, some of whose biased scholarship I will question in this book. While there are more scholars now who admit that societies during the Eurasian Neolithic period, circa 10,000 BCE to 3000 BCE, were more egalitarian than they were in the Bronze Age, circa 3000 BCE to 1500 BCE, few scholars question when the "patriarchy" actually began, assuming that male-led cultures had to be full-on patriarchies, even though there is ample evidence--such as matrilineal descent and matrifocal marriage (pair bonding) and burial practices, a heavy reliance on Female Creative Magic and moon knowledge for calculating time and seasons, and the presence of powerful goddesses--that they were not.

Unfortunately, for most scholars, who are used to determining approximate time periods and specific cultural identities, there is little room for prepatriarchal cultures that were probably more remnant matriarchal than they were patriarchal (aka gynocentric rather than androcentric). In fact, most scholars raised in androcentric cultures, especially those that lean toward patriarchy, especially through the lens of the Abrahamic religions, rarely question much regarding anti-female views exhibited by scholars discussing cultures that precede the Abrahamic religions, believing fully that any woman who has sex with a man not her husband had to be committing adultery because they assume legal heterosexual pair bonding or monogamy existed as soon as writing occurred or centuries prior, and because the Abrahamic religions are fueled, first, by Hebrew scripture wherein anyone who disregards El/Yahweh/Jehovah as the primary god must be whores, harlots, or prostitutes of some sort—which was apparently the most common, meanest denouncement the ancient Hebrews could think of.[3] This sort of anti-female rhetoric demonstrates how so many scholars who mimic the bias fail to reflect on their own biases that they carry into their research, constantly skewing their "research" in favor of male dominance.

Therefore, let us begin by clearly defining what a **patriarchy** is, or at least how the idea, as a social given, has operated for nearly two and half thousand years. <u>A patriarchy typically asserts that males are superior to women in everything</u>; therefore, women are denied access to things males enjoy the most, such as education, business, socializing, participating in sports, and sex. Even though heterosexual men need women to have sex with in patriarchies, most look down on women who choose to have sex, especially if they enjoy it. Women who enjoy sex are held in contempt, called whores, because women have to be denigrated in patriarchies in order to control what men father children. Because there is no light that comes on after a woman has sex with a man to let her know she is pregnant with that man's child, men in patriarchies are desperate to control women's sex lives,

---

[3] Gwendolyn Leick mentions the Whore of Babylon, who was also "the mother of harlots" and how "Luther attacked the Roman Church for her low moral standards" by comparing her to said Whore (1). If you have read the Christian bible, you might have noticed how many times, especially in the Old Testament, the Judeans are called by various names for prostitutes by various prophets. Women who have sex with men not their husbands were considered the lowest of the low, despite the fact that some believers in El/Yahweh/Jehovah routinely sacrificed children to appease their god.

including their choices about being pregnant, in order to determine paternity. However, as those popular DNA tests have shown us, even today, some of the men that children grow up thinking are their biological fathers are not, so the issue of controlling women's sexuality is still a huge problem for some people, even women who desire to punish other women for obviously having had sex.

There are, however, degrees of patriarchy. So, when I say a culture is patriarchal, I mean it is a full blown patriarchy that actively suppresses women's rights in favor of men's. There are cultures, though, that are merely **androcentric-heavy**, meaning they lean toward being patriarchal by favoring men in many ways, but still allow women some basic human rights, such as the right to own property, to have custody of their children, and to hold elected office. The majority of cultures around the world today are merely androcentric-heavy, although the conservative movements around the world would like to bring back full on patriarchies.

As more **androcentric**[4] ideas took hold in early societies, men sought to exclude women from religious activities, in addition to the other privileges men wanted solely for themselves, primarily because, for many millennia preceding the Bronze Age (circa 3000 BCE), women were seen as the people most closely tied to god-like powers, such as the creation of other human beings, menstrual cycles that mirror the moon's cycle (an indication of cosmic connection), the creation of ceramics, weaving, and knowledge of plants and animals. Frankly, males had to begin replacing, then excluding, women from religious and spiritual roles in order to maintain the façade that men are superior to women because long before men ever had an inkling that they contributed to conception, women were the creators—seemingly able to produce children out of their own bodies through their own magical or spiritual means, which is why so many early divinities or gods were female. Because of

____

4 Androcentric means male-centered.

women's discovery and refinement of ceramics, many early creation myths also involve shaping humans from clay.

In fact, cultures like the Aztecs believed women had to fight a spiritual battle in order to bring another human being successfully into the world, so that they regarded women who died in childbirth as the same kind of hero a man was who died in battle. Male warriors wanted to carry Female Protective Magic into battle, so sought to carry skin, bones, ears, or other body parts of women who died in childbirth with them as amulets in order to draw on another form of Female Magic—menstruation, the power to bleed, but not die, for three to five days. To justify their role as leaders and to feed the gods who they said demanded blood daily, male leaders also pierced their penises to draw blood to feed the gods, a direct imitation of menses as a form of sympathetic magic.

We know that early Eurasian cultures, like the Egyptians (who are actually African but were so advanced that Victorian Europeans could not imagine them as anything other than pre-European) and the Sumerians, believed that women's creative and regenerative spiritual powers were great because they imitated the female pudenda, the Mound of Creation, with their mastabas5, ziggurats, and pyramids. Yet no scholar discusses these important Female Magical Power buildings as such, even though many will admit that these "mounds of creation" are "machines" built to help the dead pass over into the next realm, and despite the fact that many of the myths, including Christian ones, that a person has to be "reborn" into the afterlife. Such bigotry means androcentric scholars have to dance around crediting women with superior spiritual powers.

Even though ground-breaking Marija Gimbutas focused her research on clarifying how the female figures she

____

5 Online etymology sources would have us believe that this obviously Greek word, deriving from "masto," meaning a woman's breast, is actually from Arabic, even though the Arabic term is nothing like "mastaba."

studied were emblematic of women's creative and regenerative powers, she never made the connection between the many pyramids scattered around the world, and the natural human reverence for Female Magical, Protective, and Creative Powers.

ΩΩΩ

Here's a teaser to keep you reading *Pitiless Bronze.*

If a pyramid is a stylized "mons Venus," what kind of ritual were the Aztecs really enacting by cascading bloody human sacrifices down their pyramids?

ΔΔΔ

Before we move on, however, we must clarify a few other definitions.

First, let us address a fake concern among some modern concerns. What is a **woman**? According to the *Oxford English Dictionary*, the word "woman" derives from the Old English word "wifman"[6] (484), but the editors of that esteemed tome are wrong. Instead, the word is from the Old German word "wommen," which also gives us "womb." In fact, the word "men" is also Old German, surviving as the Yiddish "mensch," but also giving us words related to menstruation like menses, menopause, and menarche. The term refers to women's ties to the moon, and "men" refers as a general term to the human beings who worship or at least follow the moon's cycles to determine the most sacred moments of time—those moments tied to actions of the moon, from waxing and waning to the full blood moon of a lunar

---

[6] The word "wife" does arise from "wifman," but not the word "woman." A gender bias exists in the OED entries for this word. I am certain I will hear lots of fuss about this distinction.

9

eclipse. Therefore, human beings in general are "men"[7] because we still follow the moon for reckoning time. We call this organization of time "months." Women still refer to their menstrual cycles as "monthlies," so not all ties with the origins of the word are lost.

A woman, then, is a human being with a womb. However, that basic definition should not be used exclusively because there are cis women born without wombs from a condition called Mayer-Rokitansky-Kuster-Hauser (MRKH) Syndrome. People who fret over the binary distinction between females and males should overcome their prejudices because even the most basic definition has exceptions. Therefore, a trans woman is still a woman.[8] In fact, in the Inanna Temple, we know that there were many such people, including some who were believed to hold the spirits of both genders in themselves, who distinguished themselves from other less sacred people by wearing the clothing of both genders at the same time.

Most frightened androcentrists assume a matriarchy is the exact mirror opposite of a patriarchy with women dominating men in the same way men have oppressed women for two and half millennia. Except for tales of the Amazon warriors told by the Greeks, there is little[9] to no evidence that there were ever cultures of women who actively oppressed the men amongst them in such a brutal manner.

So, when I occasionally use the term **matriarchy**, I am not referring to a culture that oppressed men. Instead, I am

---

[7] While many feminists take umbrage when the word "men" is used to refer to humans in general, they should not. They should, however, still point out when the word is purposefully used in attempts to exclude women.

[8] Telling a trans woman that she is not a woman is akin to telling a cis man who has lost his genitals in an accident that he is no longer a male. The act of opposing trans women's rights is still an act of egynovition, a forceful removal of feminine power. Note the lack of furor over trans men.

[9] Ishtar's reputation for killing her consorts might speak to such a matriarchy.

referring to a culture that revered women, had women leaders, and who practiced both matrilineal descent and matrilocal marriage and burial practices, and whose spiritual ideas probably revolved mainly around aspects of Female Creative Magic. Most of these cultures were more **egalitarian**, but all still had **hierarchies**. Note that an egalitarian culture is one in which people are treated as equals, but we humans, like most mammals, have always had hierarchies, which in its simplest form is the idea that one person or a small group of people leads the rest of a community in some way. Learn to separate "domination" from "leading," because good leaders do not have to actively oppress their people.

Most early human hierarchies were undoubtedly **democratic**, meaning they elected by some sort of vote who would lead when. As androcentrism rose, however, the common people began to have less say in who would lead, with many "kings" proclaiming themselves as such, often claiming they have the divine blessings of specific female gods[10] in order to rule, not merely lead, the people. I will discuss these ideas in more depth in other parts of this book, such as in the chapter on an important myth that must be debunked about gynocentric cultures.

There are many forms of seemingly egalitarian cultures that are more **gynocentric**[11] than androcentric, but there are a few forms of **androcentric-leaning** cultures that also held community elections, such as in early Sumeria. We often think of democracy as a relatively new concept, but most ancient cultures probably practiced some form of voting in order to make community-wide decisions that satisfied most people. After all, human beings are more cooperative than we are competitive; otherwise, we would be constantly killing each

---

[10] Such as Inanna in Sumeria and Ma at in Egypt.

[11] Gynocentric means female-centered. I realize that some people prefer the multiple voweled spelling of the word, but I do not hold with such pretentiousness, since my goal is to reach as many people as possible.

other. Despite international news and the internet seeming to indicate otherwise, most human interactions are peaceful, rather than violent, demonstrating how much we have matured as a species.

In early Sumeria, we know that there were councils of elders elected to help run each city, which are often viewed as individual "states" by many scholars. We also suspect that the small town *capullis*, which was also the designation used for specific parts of the larger Aztecan cities, what we might think of as wards or neighborhoods, were led by elected councils of elders.

Because the tendency has been to view war and conflict as "natural" outcomes among humans, few scholars question why one culture does something one way, while another handles the situation a different way, simply because most do very little comparative study. Far too many androcentric scholars are fascinated by wars and conflict, rather than on cooperation and community building because they still, like some of their ancestors, still find brutality and blood spilling fascinating.

The reality is that most of the monumental structures that still exist 6000+ years later were created with cooperative effort, not because some guy stood over everyone with a whip. While some of these monuments, such as Stonehenge, might have been built by competing villages or clans to see who could bring the most impressive stones or skills to the construction, these groups had to cooperate, first, among themselves, then, second, with the other villages to build these immense structures.

Yet the idea that **"might makes right" is at the very core of every patriarchy**. Some extremist patriarchal men often use brute force to compel people to do their bidding, whether it is shooting girls who want an education on their way to school or kidnapping children to force girls to bear the brutes' children and to force boys to become their next group of soldiers or stoning women who speak out.

By foisting the belief that men are superior to women spiritually, Southern Baptists, Catholics, and most Islamic sects have kept women from achieving leadership roles in their religions.

Sadly, if so many people did not believe these skewed, sexist views, we might have a female pope now.

Will the day a woman is ordained pope[12] be the official end of the patriarchy?

ΩΩΩ

A more important question this book attempts to tackle is: How did the "patriarchy" become a thing?

While Gerda Lerner attempted to tackle this question in her ovular book, *The Creation of Patriarchy* in 1987, and while she did an excellent job at pinpointing the Bronze Age as the one in which the patriarchy began its rise, she did not carry her findings far enough, choosing, instead, to focus her arguments mostly on Judeo-Christian beliefs, which actually did not arise until the Iron Age.

While I could spend time uncovering how the early Israelites and Judeans morphed from more egalitarian cultures that worshipped both Yahweh and Astarte into a one-god/one-king culture during their enforced exile in Babylon when they "discovered" "missing" books from their early religious works (and probably wrote everything down for the first time), I will not dedicate much discussion to what many consider the "original" patriarchal religion[13] that fomented two other patriarchal religions (aka the Abrahamic religions) from its mishmash of collected cultural ideas because, frankly,

---

[12] "Pope" is a Latinate form of "father."

[13] I will not be addressing Hinduism, which is such a complex religion with a very complex history, except to occasionally mention some of its earliest cultures as identified by archaeology. Note that Hinduism has been, according to some, both matriarchal and patriarchal, but the question remains about when patriarchal views came to the fore.

their religious texts come late in history, compared to most other cultures, and other scholars, like William G. Dever in his book, *Did God Have a Wife?* have already covered the topic.

Instead, this book will critically examine androcentric (prepatriarchal) cultures like the Sumerians, the Egyptians, the Mycenaeans, the Minoans, the Hittites, and the Aztecs in order to outline how they transformed from gynocentric cultures into androcentric cultures. I will even prove that human males probably did not know, definitively, of their role in procreation until much later in history than most people realize because **without actual verifiable biological fatherhood** there would be no patriarchy at all.

I will also examine how some scholars' skewed logic because of their androcentric biases has resulted in some rather bigoted scholarship. As Starr Goode put it, "the sway of entrenched patriarchal bias can result in misinterpretations of the obvious" (133).

To be fair and consistent, I will use Before the Common Era (**BCE**) and After the Common Era (**ACE**) as abbreviations indicating time periods. While some scholars might argue that we are living in the Common Era (CE), there are still many cultures that use different calendar systems, so, to be as fair and as unbiased as possible, I am assuming that year Zero is a common enough point for most scholars that it will foster better understanding and extrapolation into whatever dating system each individual scholar uses, but remain familiar enough to lay readers that they will understand what I mean, as well.

To avoid being sexist, I will also use the term **foragers**, instead of hunters-and-gatherers, since there is increasing evidence that pre-literate cultures had fewer divisions along gender lines than modern cultures do. In other words, everyone pitched in, with men learning how to gather and cook important food stuffs, and women learning how to hunt and fight more often than not.

Other chapters in this first portion of the book will explain these and other concepts in fuller context, so I have left them separate from this chapter in order to facilitate readers who are already familiar with those concepts and feel they do not need to read those chapters.

I recognize that many of these ideas will appear foreign and highly questionable to almost everyone, so have painstakingly researched and explained these concepts using a variety of sources, many of which are available on the internet, so that even a lay person can access them to verify my claims.

Making assumptions about individuals based solely on their gender is a very outdated and bigoted way to interact with others, so I hope this book helps many recognize that some modern views on gender can be not only confirmed as similar to what the ancients believed, but also learn that some of their personal views might be incredibly biased because of the cultural influences now imposed on them.

Chapter Two
A Few Definitions and Clarifications

This book is a postpatriarchal examination of prepatriarchal cultures. Many terms about those cultures will be used throughout this book, so this chapter is my attempt to help guide you with definitions and clarifications about what I mean when I use certain words.

First, we cannot escape discussions about gender and sexuality when discussing **gynocentric** or **androcentric** ideas. Gyno- means female, so gynocentric is **female-centered**, which includes anything having to do with cis born women and anything associated with the feminine, which can include transgendered women, who, yes, existed in ancient times. Andro- means male, so androcentric is **male-centered**, which includes anything having to do with cis born men and anything associated with masculinity, especially in its extremes, including **toxic masculinity**, or **the tendency to use aggression, threats of violence, and actual violence to intimidate or to oppress others**.

Because we must discuss sexuality in all its permutations, I will use the term **gender** when referring to ideas about females, males, the genderless, the gender-combined, or gender-changed. I will refrain from using the term **sex** to refer to a person's gender, however, even though many scholars conflate the two ideas. In a discussion of sexual activity, using the term sex to refer to gender just gets too confusing. So, even though some scholars, especially linguists, prefer not to use the term gender to refer to people's sexual identity because they believe it only pertains to language, I will do so anyway. It is better to err on the side of clarity than to follow unnecessary dictates.

Second, realize that dates are vital to understanding when particular ideas or skills were created, as well as when particular cultures flourished because the When of time often influences the Why of events. While I would prefer to use my own dating system, which would begin roughly around what we now call 3000 **BCE**, or **Before the Common Era**, I will continue to use the BCE/CE system with one modification. Instead of using CE, which stands for Common Era, meaning the time we share now, I will use **ACE** to stand for **After the Common Era** because I want to emphasize how erroneous it is to believe that switching from BC, which stood for Before Christ, and AD, which stood for the Latinate term *Anno Domini*, which meant "the year of our lord," simply by changing to BCE/CE abbreviations, makes amends for the Abrahamic world assuming that everything timewise revolved around the mythical birth of Jesus Christ, which is in and of itself an extremely biased mythological view. While I could use BP (Before Present), I want to emphasize just how much older than the Abrahamic religions so many religious ideas around the world were and want to avoid forcing nonscholars to do the math required to figure out the relative BCE/ACE time periods, so I will just pretend that year zero was "the Common Era," even though so many cultures have completely different dating systems.

Many scholars are used to reading and writing **androcentric** articles that use terminology that promotes **the idea that men were better than women** at almost everything in antiquity. Nothing could be further from the truth, of course, so this chapter is an effort to counter such biases that come from specific words that are often used to mean generalized ideas.

For instance, the word **hunter-gatherers** is often used to denote **foraging people** who lived off the land by looking for and harvesting specific food stuffs—from plants to animals, and who created shelters and clothing from those same resources, plus rock and soil, which they used to create shelters and to make tools, including, eventually, making ceramics. Foragers had many skills, from hunting in simple ways, such as using traps, as well as hunting larger game in cooperation with other foragers, to learning how to recognize various plants and soils that could be used in cooking and healing and for manufacturing useful items like rope, bags, nets, and pottery containers. We have no evidence whatsoever that really indicates who did which jobs,[14] but we should know, beyond the shadow of a doubt, that when everyone in each small, mobile group knows what to look for, everyone benefits, so it is most likely that most people in ancient cultures knew how to do a little bit of everything that was needed for everyday living, which is why it is more logical to assume that such early societies were *egalitarian*, treating everyone as equal in importance to the survival of the community as best they could.

One important reason we must get away from using "hunter-gatherers" as a term for foragers is because it carries with it <u>a sexist assumption—that men where the hunters and women were the gatherers</u>, when, in fact, everyone in a community unit would have been on the lookout as they traveled through an area or temporarily set up camp in a particular area for the kinds of resources they needed and wanted. While some scholars have proven that women were capable hunters, far too many **Paleolithic (prior to 12,000 years ago**, generally speaking) and **Neolithic (generally between 12,0000 and 3000 years ago**, depending on the culture) scholars assume that gender-based <u>work roles</u> were

---

[14] Some archaeologists like to argue that fingerprint sizes and hand sizes in some of the works created indicate either gender or age, but we have to realize they are only guessing. Human beings were not as large in most of the eras prior to our own simply because they did not have consistent access to a healthier variety of nutritious foods. A good example to illustrate what I mean can be seen at Fort Riley, Kansas' museum called the Custer House, which has preserved some of Elizabeth Bacon Custer's clothing. Livy, as she was called, lived from 1842-1933 ACE, and, like my mother who was born in 1921, was a small woman, not even five feet tall. My siblings were all taller than my mother, and their children tended to outgrow them, as well, all because we grew up with access to better nutrition and were better educated about what our bodies needed to grow than our ancestors were.

already entrenched and well-defined in the time periods they study, when, in fact, *human beings were probably far more flexible in their social roles as foragers* than they will be later on in sedentary communities.

Even in tribal cultures that existed in the fifteenth century ACE in the Americas, for instance, rarely were tasks so decisively split in any of the cultures. Women needed to know how to defend their homes and their families just as much as the men did, whether they had to defend against animal predators or human ones, so limiting women's abilities to make and to use weapons would not have been a very smart survival tactic. Similarly, even if a group of men went out hunting, seeking raw materials, or trading, they had to know how to cook what they caught along the way, because there was no way they could carry enough food to sustain them on months long excursions. While such groups probably included women, especially when they went to gather sacred items, such as pipestone with which to make ceremonial pipes, the men would have been expected to participate in foraging and preparing the group's food.

Another important reason to use the word foragers instead of hunter-gatherers is because so many cultures that began or returned to herding livestock often combined foraging for other materials, such as plant food stuffs for themselves and their herds as well as clays and minerals for pottery, instead of living in one settled area for longer than three to six months. Because the term foragers still best describes these combination lifestyles,[15] I will use the simpler and non-gendered terms **foragers** and **herders.**

In fact, the assumption many scholars have made in the last several centuries have been mostly viewed only through androcentric eyes, so that many scholars expect to see males as leaders, becoming clearly uneasy when they discover

---

[15] David Wengrow and David Graeber call such flexible movement between farming and foraging "play farming," clearly not taking such survival techniques seriously (70).

women were leaders, especially respected leaders, in many cultures. I lost count of the number of scholarly articles that discussed gynocentric cultures which completely avoided pointing out that the probable leaders organizing those cultures were women, so that the scholarship they present ends up never reflecting on the archeological finds from women's points of view, if the scholars could help it. This kind of androcentric bias is, unfortunately, quite pervasive.

When most people hear the words "**patriarchy**" and "**matriarchy**," they assume that a matriarchy must be the polar opposite of a patriarchy. In other words, they assume that a matriarchy is led by women, organized by women, and actively suppresses men in the same way that the self-declared patriarchy has suppressed women for most of the last two and a half millennia.

People who assume a matriarchy is the polar opposite of patriarchy seem to assume that every group has to exhibit a **hierarchy, where one person or one small group is in control of everyone else**, despite what studies of living matriarchies show us. Those with gynophobia assume the worst, that women, once we "gain the upper hand," will oppress men the same way women have been oppressed—terrified of being diminished in importance.

One myth this book examines is the idea that a patriarchal organization is natural[16] simply because some men are often physically stronger than many women. So it is important to realize that women can be and often are as strong as men physically, but many women have been physically restricted as they grow for the past two and half millennia. When people are not allowed to exercise and demonstrate their physical abilities, it is only natural that those abilities become diminished. In the less than 50 years since women's sports has been more actively encouraged in America, women

---

[16] An important point I won't explore, which should be stated, is the fact that androcentric cultures feed boys more food than they feed girls, so that girls end up, naturally, being smaller and weaker than boys.

have demonstrated their ability to be as strong, some even stronger, pound for pound, than men, and are quickly gaining speed, too, in many sports. The fastest recorded woman in 1922, the first year women were allowed to compete in Western countries, was still only two seconds slower than the fastest man of the time. Since then, both men and women have become faster, but now women are less than a second slower than men. Remember that few men, however, cannot take birth pains because they do not experience menstrual cramps monthly as preparation, so the question of their overall physical "strength" is fluid, as well.

Instead, it took men several millennia to begin to overcome humanity's more gynocentric views of the world in order to get to the point where more people accepted the idea of patriarchal hierarchies as the inevitable result of human evolution. Ample evidence exists, as well, that during those millennia of purposely touting men's "superiority," there were many groups who rebelled against the idea. Such a rebellion could have been the cause of the great Greek Dark Age toward the middle of the Bronze Age.

Another example is a close reading of the Book of Ruth in the Judeo-Christian bible, which reveals that the Judean women still inherited the land, and that the primary purpose of that book in the bible is to convince the women of Judea to allow a different way of calculating inheritance, which is why the writers of Ruth emphasize how much the women of Judea endorse the ideas put forward at the end of that book. The writers of the bible had to go to great lengths, such as in the Book of Judges, to convince women that men alone should rule, despite also demonstrating that women can be judges, prophets, and warriors. They had to admit such facts because the women of the culture remembered these female leaders, spoke about them, and expected them to be honored, so the men had to manipulate their female characters in order to create stories that both lauded male superiority, as they saw it, and still honored the influential women from their history.

Most cultures, like the Judeans, were not originally patriarchal, let alone **patrilineal,**[17] meaning **to trace ancestry or a family name through the father**. However, like most cultures influenced by the rise of androcentrism, the idea of a patriarchy as put forward by the Hebrew bible with the men doing all the "begetting" has been, by now, socially programmed into most people, scholars included.

Ergo, a "**patriarchy**" in its **most extreme form is a culture that believes men are the only people capable of being "the heads of households," political leaders, teachers, and prophets; in short, they believe men are so superior to women and children that they designate anyone else a "minority," meaning they are treated like minors, or as though they are <u>children</u> incapable of making their own decisions**. A full-on patriarchal culture practices **patrilineal inheritance**, where the children, especially the males, inherit names, titles, land, and goods from their fathers, practice **patrilocal marriage**, where the women move to live with the husband's family often taking her husband's family's land location (village or identifying geographical feature) or profession as a surname, as well as recognize **patrilocal funeral processes**, where the women have moved from the places of their birth to be buried with their husbands' families or communities.[18] A full on patriarchy also limits women's ability to move about freely in their communities, with extremists keeping women virtually locked up in their homes or fully shrouded from sight by clothing when in public, a disguise

---

[17] Realize that patrilineality could **only** happen **after** males learned they play a biological role in procreation, circa 2400 BCE.

[18] A conundrum exists, however, because women can and do migrate willingly in many cultures around the world, so the mere fact that a woman is buried near her husband and not her mother does not necessarily indicate patrilocal practices.

that has proven useful for men wanting to pass as women in order to enter secluded areas.

However, there are no full-on patriarchies[19] in the world today, although many political and religious groups like to believe there are. The closest are run with bloody fists by men, who often have to beat, stone, imprison or murder outspoken women to keep their female populations oppressed. Even in the androcentric heavy West, many women keep their surnames or hyphenate them with their husband's name, also controlling how their goods and wealth are distributed to their heirs. Some women even choose to be buried near their own parents rather than with their husbands' parents, if near any parents at all.

Cultures that are not full-on patriarchies are often called **heterarchies** by scholars, with some fully meaning to imply a lack of organization by using the word. However, the origin of the word indicates it is meant to indicate cultures **where both genders are considered equal**, with **no particular gender being considered more important than the other**, which we also call **egalitarianism**. Some scholars have skewed the primary definition for heterarchy, however, and use it to mean organized systems that have a rotating ability, so that whichever group is needed to tackle a particular issue has the right and the opportunity to use their particular skills to mitigate that problem for the group as a whole. For instance, for the more gynocentric Cherokee or Mohawk cultures, there was an "inner chief" or a leader looked to for inner tribal concerns and an "outer chief" who was the leader sought out when outsiders visited. Either could be any gender,[20] but was usually the person, often elected, best able to handle either domestic or foreign issues. Some

---

[19] The closest is Saudi Arabia, but even they now allow women to drive. Sort of.

[20] Many American Indian nations acknowledged and, often, treated as sacred people who were both genders, which is where the idea of being "two-spirited" comes from.

scholars seem to attempt to disparage this more egalitarian way of organizing groups by calling it circular logic, when, in fact, many corporations employ such systems, often called Matrix, Mixed Model, or Circular Platform models, quite efficiently. This book does not use the term heterarchies much, but I believe you should be aware of how some scholars use the term.

Similarly, there are no full-on matriarchal cultures in the world today, either. In fact, other than the mythical Amazons and perhaps the early Ubadians, there possibly has never been a completely matriarchal culture among humans, at least ***not using the polar-opposite definition of matriarchy with women suppressing or oppressing men.*** There are, however, many cultures that consider themselves matriarchal, such as the Mosuo of Tibet and the Hopi of the United States. These cultures often bow to the demands of androcentrism, however, by calling their inheritance systems "uncle" systems because it is the woman's brother, not her husband, who teaches the woman's male children how to be a man in those societies. It is important to recognize that this is not a male-oriented system, but a female-oriented one with the mother's kin taking precedence over the father's relatives.

There are many **matriarchal mammals**, though, with elephants being one of the best examples. The females, led by a matriarch or matron, stay in small female-focused groups, usually a grandmother, her daughters, and their daughters and sons, traveling together by following the matriarch's or matron's, as they are often called, lead. Her long-term memory is vital for the herd's survival because she has lived long enough to know where water and food can be found, even in the toughest of times. Males generally move out of the group when they are old enough, usually joining **bachelor herds**. The matriarch among elephants decides which male gets breeding rights when females come into season, and often oversees the first mating for young females in order to ensure their safety.

Similarly, American bison are also matriarchal. Like elephants, there is a seniority-based hierarchical system within a herd, which is composed of females--grandmothers, mothers, and their children. Males can move with the herd, but only with the females' tolerance, and almost always on the periphery of the herd. While some younger male bison form small bachelor herds, most males prefer to stick closer to the herd, usually along its edges, looking for opportunities to mate.

African lions are also matriarchal, despite the misinformation often touted about the role of the male lions in a pride as being "kings" of the jungle or savannah. Lions, like other mammals, know nothing about genetics, as far as we can tell, so probably never think about the progeny they might produce when they mate. The males sometimes kill the infant cubs of a female if she has been unwilling to mate with him—not because he is obsessed with passing on his genes, but because he knows who he has mated with and who he has not, so eliminates the children of the females with whom he knows he has not mated to bring her into heat, which will make her more willing to have sex with him.

While few ancient humans would have been able to view dolphin sex lives, we also know that these members of the whale family also have highly developed family and bachelor pods, and that the females' clitorises are highly sensitive, which explains why they like to have sex so often, using sex like bonobos and humans often do for pleasure and for social bonding, with some scientists speculating that dolphins need the sexual stimulation in order to ovulate (Ogden).

The list goes on and includes domesticated as well as wild animals. Some dairy farmers, for instance, call what they witness among dairy cows a "pecking order." The senior matron of the herd will lead the way into the barns for milking, with the lower level females jostling for entry behind her, most following a clear hierarchical order behind the older cows.

While many anthropology and archeology scholars love to point to **hierarchies** as a sign of the "advancement" of human societies, the tendency to develop hierarchies is actually a mammalian one, at the very least, so should not, in itself, necessarily be regarded as a marker of social development. In fact, it would be strange if humans did not develop similar social hierarchies long established among other gregarious mammals.

Even though hierarchies exist and are easily identifiable in most mammalian cultures, it becomes more challenging to pinpoint when groups of women have dominated a village, town, or city at any point in human history. I discuss the importance of more fully understanding hierarchies among groups of females in more detail in the chapter on the common myth about female-centered, or gynocentric, communities.

We can, however, accept the fact that, prior to the discovery of procreation, **children traced their connections to a group through their mothers,** which is **matrilineal descent**—an inheritance system which depends wholly on knowing who the child's mother was. Because the ancients had no idea that sperm,[21] which will not be discovered until well After the Common Era, floated around in semen, they only knew that **women were the ancestors of everyone**, so it only makes sense that women were not only the homemakers, literally, who established where a home would be built, how it would be built, and who could live in it, but also became the revered ancestor whose

---

[21] Beware of any and every interpretation of Sumerian and Egyptian literature where the idea that sperm, which was unknown to the ancients and to most people until the 1800s as actual living cells contained within semen, is asserted by the interpreter because that interpreter (e.g. Kramer, Jacobsen, Maier, etc.) is using his androcentric bias to assert that these ancient people knew about sperm's existence.

remains were buried beneath the home, thereby establishing a family's ties to particular lands. Besides burying the family matron beneath a home, loving mothers also buried their infant children beneath their homes, most likely in the hope that the child's spirit would reinhabit her womb, so she could successfully give birth again.

While there are many ancient tales of mothers sending their children down rivers in reed baskets, we almost always know who the mother of a child is, even if she has died during childbirth, so elevating matrilineal connections as "ancestor" spiritual connections made the most sense for ancient people, with some cultures, like the Egyptians, burying their mothers near them in order to help them be more easily "reborn" into the afterlife.

We also know from DNA studies done on archeological remains that, prior to the rise of androcentrism, most human societies practiced **matrilocal funerals and marriages**, where the males moved in with his wife's family (the matron to whom he was bonded or pledged), and he and his children were buried near his wife's family. In biblical terms, the man "cleaved" unto his wife, and her family.

ΔΔΔ

The mistake many scholars make when discussing the differences in types of ancient societies is to assume that all communities with male leaders are full-on patriarchies.

We know, for instance, that patriarchal cultures can be led by women, such as Queen Elizabeth of England and Indira Gandhi of India.

Conversely, it is fully possible for communities that are gynocentric to be led by men, so we should avoid the bigoted trap of assuming that the presence of male leaders signals a patriarchal society.

The reality is that most cultures around the world are a mix of different cultural views about the importance of gender, with some finding it unnecessary to identify leaders by gender at all. In other words, a leader is a leader, not a female or male leader, not a queen or a king, which tends to the be the words favored by those with a clear patriarchal bias because "king" implies a higher status than "queen." In fact, scholars need to learn to differentiate between when terms for leaders change meaning. For instance, the word "lugal," which was a general Sumerian term for a city leader, possibly originating from a common idea of the "big belly" although most scholars prefer to think of the term as "big man," despite the evidence pointing out women as city leaders.

The term, however, is more often than not interpreted as "king," largely because of an inflated gender status the scholars are assuming the male leaders held. However, there is little to no evidence that any one city leader considered herself/himself/ximself a "king" in the same way the term came to be defined in later periods—as a person sanctified to lead because of a particular divine ritual that was successfully performed by that person which demonstrated a deity's blessings, so we need to learn to differentiate between city leaders (what we often call mayors in America) and those leaders who either used effective skills of persuasion (such as Narmer of Egypt?) or used violent oppression to subdue more than one city (such as Sargon of Akkad). Later cultures embraced the idea that "royalty" had divine rights, thus were more spiritually powerful than other people; in this way, the cultures of the late Bronze Age bought into and often actively helped to support (although some did rebel, frequently) the idea of whole geographic areas being "kingdoms," which were greater than the original idea of Houses (being bonded by blood or by pledge to one lineage) or towns/cities.

Humans are complex creatures, so our societies are often more complex than simplistic labels often indicate they are. As complex beings, we also gravitate to charismatic leaders, even when we should know they do not have our best interests at heart.

For the purposes of this book, I will be referring to two main ancient cultural tendencies: **gynocentrism** and **androcentrism**. Even though these labels indicate a binary view of humanity, many cultures recognized at least one other gender, sometimes called by terms such as two-spirited or even genderless. Because almost all of these cultures that recognized a third gender did so for religious reasons, most of these people were considered sacred because they bridged the two worlds of fe/male on their own.

Gynocentrism views the world through the importance of women, especially through an understanding of their experiences—from sexual aspects to crafting; in other words, as those with spiritual powers to reach across liminal spaces and as creators. Spiritually, gynocentrism is the belief that women are naturally more spiritually connected to all things than men because women's menstrual cycles are not only similar to the cycles of the moon, which is a supernatural entity, but also because menstruation causes women to bleed three to five days a month without dying, a magical skill many different androcentric cultures tried to imitate through actions like circumcision, penile lancing, cross dressing, and castration. A gynocentric culture credits women with more (if not all) of the responsibility for creating children, not just bearing them, since pregnancy and birth would have seemed magical to the ancients because women appeared to give birth to whole other humans seemingly on their own. For some cultures, women who gave birth were warriors who wrestled the spirit of another human being into this world from another realm. Distinguishing this early human belief from later androcentric beliefs in parthenogenesis, what is often called "virgin birth," occurs in the chapter on those ideas.

Creating human beings seemingly from their own bodies without aid from any visible outside forces was the ultimate outcome of Female Creative Magic. The act of parthenogenesis, which is the act of a female cloning herself, can actually occur among some animals, especially among birds.

Realize that ancient humans were not stupid just because they had not figured out that microscopic sperm cells exist in semen, nor that, once they discovered procreation (the idea that it takes two creatures to make a third or more), they first believed males were planting "seeds" in women's wombs. Instead, realize that some sort of scientific method, which began much earlier thus led to the creation of ceramic and metal production, was needed to prove that males contributed to procreation circa 2400 BCE. There are several indications that a particular branch of scientific discovery, which we now call biology, helped human beings discover the process (heterosexual intercourse) necessary for procreation, but it still took the ancients more time to discover that such coitus between closely related people caused malformed fetuses, or monsters, in ancient people parlance, to be born, so that **sanctions against incest do not occur until after** the knowledge of procreation becomes more common.

Most gynocentric cultures encourage women to contribute as fully as possible, and not just as mothers. We get terms like **homemaker** from cultures where women were/are, literally, the **people who constructed the homes**, whether they were teepees, wigwams, longhouses, rock lined, or wattle and daub frame houses. Women were still such esteemed builders that one Mesopotamian king required 285 enslaved women to build him a palace called Bûr-Sin (Gelb 81).

Gynocentric cultures tend to revere blood as a sacred creative element, some going so far as to burn their menstrual blood on moss or cloth as offerings to the sacred spirits or gods of their cultures. Most demonstrated their reverence for Female Creative Magic by sprinkling their dead with red ochre or cinnabar, imitating the birth blood to enable the deceased to be reborn into the next realm. They also buried their dead in mounds as symbols of female pudenda, the "mounds of creation" through which humanity passes as we are born into this realm. Many gynocentric cultures used many symbols for Female Creative Magic, including snakes, which not only shed their skins in such a way to appear as though they are giving birth to themselves, but also open their jaws to swallow other animals much larger than their mouths, just like vaginas do, and also resemble the umbilical cords that link the newborn to its mother. Similarly, any plant that bears fruit, even grains, could symbolize Female Creative Magic, as will any creature that is known to be a nurturing mother, like cattle, crocodiles, and monkeys.

Androcentrism is the belief that men are naturally more spiritual (often believed to be more spiritually pure, thus "ascendent" above women) than women, who gradually over time became considered "unclean" because of their menstrual cycles, so unworthy of direct contact with religious items or unworthy of preaching religious doctrine. Most androcentric cultures have had to purposely **reduce or negate** (make evil) the spiritual beliefs about women in order to assert the idea that women are inferior to men, just as the ancient Mycenaeans did when they replaced ancestor (mother) worship with the Cult of the Hero as they became what we now know as the ancient Greek culture, which I discuss in more depth in a chapter in Section 2. Most androcentric cultures *forbid discussion of women's issues*—making natural female processes like menstruation taboo subjects which were seen as "unclean" or "evil."

If you find yourself uncomfortable reading about sex or menstruation, realize that what you are reacting with is androcentric social programming, which has led you to believe that these things are nasty, unclean, or even evil.

Many androcentrists believe men are smarter and more capable than women, but androcentric cultures have had to rig the educational systems to make this logic arguable. Many androcentric cultures forbid women to interact in public and frown upon women working with or talking to men who are not blood relatives in order to forcefully preserve knowledge of who the fathers of the women's children are. Such restrictions are enacted because men do not trust their own gender and assume other men will have sex with any woman at any time. Instead of controlling men, however, androcentrists usually choose to limit women's rights, thereby impinging on the productivity of more than half of the human population.

Many androcentrists also believe that males are naturally stronger than women, so have the right to rule over women, but this belief flies in the face of what many ancient egalitarian cultures practiced. Archeological studies of Neolithic remains have demonstrated that women were as muscular as men because they did strenuous labor just like the men did. We also know that women were physically capable of hunting and fighting as warriors. Again, limiting modern women's abilities to strengthen their bodies is an attempt, of course, to keep them weaker than men, even though pound-for-pound women are stronger than men, with the strongest women able to lift four times their own weight, while men struggle to lift twice theirs.

In pre-literate cultures, various social methods to keep stronger individuals from bullying weaker individuals resulted in shaming the bully, but somewhere along the line men began to embrace the "might as right" point of view, which was held to be morally wrong by most egalitarian cultures. These men, seeking increasing amounts of power, prestige, and wealth, decided that power/strength was and should be the main "rule of law," so wrote laws accordingly, beginning a steady decline in women's rights during the Bronze Age (circa 3000 to 1500 BCE).

In fact, we have ample evidence that males, and females who support their skewed social structures, used written laws to egynovite[22] women's power by changing terms

---

[22] Egynovite (pronounced ee-guy-no-vite) means to removed feminine power, similar to how emasculate means to remove masculine power.

describing women's cosmic connections--their spiritual powers to transform blood into life, sex into healing, and to bring new life into this realm from another—into negative ideas in order to sully women and their feminine spiritual powers, making menstruation unclean, sex dirty, and allowing men to stone to death women who have sex with more than one man. Male spiritual transcendence depends upon making it illegal for women to have educations in order to shut us off from the knowledge that we were once the more spiritually powerful gender. Egynoviting our freedoms, they made us weaker, deemed us inferior, pronounced us unable to compete with them on every level. Without this active, often brutal, and "legal" oppression, men could never hope to be better than women at most things.

Note, however, that this demarcation does not mean that there were no female leaders who reacted violently to situations. Most likely, the original reason the Aztecs sacrificed human beings on their pyramids was an imitation of an original menstrual sacrifice meant to return the moon to full strength during a lunar eclipse (which is a blood red moon) or from its hidden phase of the month. I will discuss this idea in more depth in Section 2.

While any human being can exhibit violent behavior, it has been mostly patriarchal cultures that have routinely oppressed others, often choosing to do so violently and without shame or remorse.

ΔΔΔ

At this point, I would also like to address what happens when noted scholars avoid addressing the fact that women were much more influential in early, prepatriarchal cultures than most are currently willing to give them credit for. In particular, we need to focus on some desperate mansplaining that occurs in David Graeber and David Wengrow's new book, *The Dawn of Everything: A New History of Humanity*.

Graeber and Wengrow's primary weakness in viewing definitions and conditions of **equality** or egalitarianism are because they approach the ideas from mostly an economic and property-based point of view, further demonstrating that they have not personally dealt with issues of being unequal to most other people in the same way most women have most of our lives. They despairingly proclaim that searching for the "'origins of inequality'...is little more than chasing a phantasm" (427), despite the fact that a little biological research could have yielded volumes for them.

If we admit that human beings are, at our most basic level, herd animals, we have to admit that **human groups will always have some sort of hierarchy**. That hierarchy could be as simple as having a matron cow leading the herd into the barn every night, or it could be as ephemeral as the lead goose at the point of their V in a migratory flock who eventually tires so is replaced by another goose capable of leading—an action that makes flying a bit easier on the geese that follow. Graeber and Wengrow would probably agree that human beings are clever enough and flexible enough in most of our social units that we can make the most of either of these types of hierarchies.

What we have to recognize is that basic mammalian hierarchies do not have to "evolve" or devolve into non-egalitarian societies wherein "might makes right," so that we have to allow violence, not even the threat of violence, to dominate our social spheres. Typically, among herd mammals, the primary threat of violence that occurs comes from the outside, unless we count male battles during rutting season. **Is this sometimes violent competition over sexual rights among herd mammals a reason why some human males believe violence is something they can employ any time they like?**

While examining what another colorless man said about plains Indian tribes, Graeber and Wengrow assert that "authority" only comes from exhibiting violence, or at least the ability to do violence (make threats) to control other people's behavior (110), overlooking how well various social pressures work without threatening violence. Graeber and Wengrow repeatedly assume the use of violence is necessary before a society can be called a **state**, calling the controlled use of violence one of three "basic" principles of statehood (408), what they later admit is an "elementary form of domination" (413). They refuse to believe an egalitarian culture which prevents any one person, not even the person at the highest point in the hierarchy, from using violence to control individuals or whole populations is worthy of being called a state or operable governing force, despite the fact that they provide plenty of examples of societies that did just that.

What these males refuse to admit, refuse to see, is that cultures led by women, while they, too, can devolve into violent regimes, tend to follow **the real first principle of equality** (indeed, the real first principle of being a mammal)—**nurturing** the best out of others. Here is one example of where Graeber and Wengrow completely miss the message from our ancient ancestors. They willingly note that the ancient Sumerians' word for freedom was "*ama(r)-gi,* which literally means 'return to mother'" (426). Again, using Graeber's preferred economic bias, they interpret this word to signal a cancelling of most debts, "and in some cases allowing those held as debt peons in their creditors' households to return home to their kin" (426). What they refuse to admit is the fact that these "debt peons" were actual slaves, captured by force or sold into what was often lifelong **slavery** to pay debts. While these seemingly magnanimous acts of freeing the enslaved undoubtedly occurred from some Sumerian leaders' commands, Graeber and Wengrow miss two important points about this origin of the word freedom and what it has to do with the social ties to our mothers.

First, they are acknowledging that the "kin" the enslaved Sumerians—most enslaved because the father[23] of the family got too deeply into debt to a land owner, so had to sell his children, his wife, himself or all of the above into bonded labor in order to pay his debts, all actions which became legal after the development of written laws in Assyria (we have no evidence that it happened in ancient Sumer prior to males learning they have a biological role in procreation, mind you)—are going home to their ***mothers***, which clearly indicates **matrilineal** ties were still valued, even at this late date in Sumerian/Mesopotamian history.

Second, the leaders in question were allowing the "debt peons," as Graeber and Wengrow prefer to call them, clearly avoiding the term "slave," their physical and financial freedom as a gesture of ***nurturing*** the population.[24] Repeatedly, throughout the history of ancient Mesopotamia, leaders became popular and were highly supported by the populations because they rectified wrongs done to individuals from previous tyrants by either freeing them from slavery or by compensating for the loss of land, crops, and even human beings who were punished or killed for some perceived slight thereby harming the financial independence of a family or household. One of the most famous of these Mesopotamian releases came in the Iron Age (586-537 BCE) when Cyrus the Great freed the exiled Hebrews from Babylon after conquering that city, allowing them to return home to Judea 49 years after their cities were sacked and most of their royalty and priests rounded up and kept hostage by Nebuchanezzar II, king of Babylon.

---

[23] Laws allowing men to sell their wives and children in Sumeria arose after humanity learned of male contributions to procreation.

[24] Ironically, Graeber and Wengrow claim that the "hierarchy of property may derive from notions of the sacred, but the most brutal forms of exploitation have their origins in the most intimate of social relations: as **perversions of nurture, love, and caring**" (208), although they fail to provide examples to explain this claim that nurturing actions were perverted in the creation of property or hierarchies.

Most strangely, Graeber and Wengrow associate **egalitarianism** with being *forced* to conform to a uniform social identity, wherein everyone lives in similar housing, eats similar foods, and uses the same sort of patterns in their clothing and pottery. These ideas clearly grate against their ideas about the importance of individualism. They believe that systematic production of goods, which creates more uniformity in design, thus streamlining the production process, was done purposely to "prevent innovation"[25] (422) in order to prevent differentiation in status. They repeatedly lambast this "overt ideology of equality" (124 & 423) because they view such designs as a rapid means of identifying "us" from "them," instead of seeing them as some cultures view them: as beauty choices. While some people might view the same basic jacket[26] design used by men for well over 400 years now, for instance, as outdated, others might adopt that style to enhance their own social status or find ways to alter the design to demonstrate her/his/xis own individual tastes. Just because a community of potters decides to feature similar colors and designs on their wares, which, ironically, allows

---

[25] Ironically, this same form of actions, which occur quite often in order to produce more goods more quickly, occurred during the European Industrial Revolution, but we could argue that the actions objected to by the two Davids marked the beginning of the Bronze Age Industrial Revolution, which I will discuss in more depth later. My point is that the regularization of goods in order to produce more goods more quickly was an innovation, not the suppression of innovation. However, the two Davids appear to be speaking from personal experience, so they are possibly reacting to what Wengrow found at an archaeological dig with the personal indignity that comes from someone shutting down their creative ideas, instead of viewing the ideas as creditable. I have been there and know that feeling well.

[26] The basic lapel jacket most men wear as part of their business suits, for instance, are a basic design that has not altered for more than 400 years, except to widen the lapels, double the buttons, add or remove height to the collar, or to change the location of pockets. Yet men choose to wear these same garments every day as a signal that they "mean business," so many women have adopted that style in order to be taken more seriously, too.

many archeologists to immediately date and trace said pots' cultures of origin, does not mean anyone stood over them with a whip and forced them to create their collective cultural artistic attributes instead of developing their own styles.

Further, in a cooperative community that has planned their housing structures together, no one's individual house will be bigger than another's because they would not be able to justify asking the other people of the community to haul more rocks or tan more hides or produce more lime for the white washing of the interior or exterior. Simple humility—seeing oneself as no better than anyone else—accounts for such planned community homes, a concept that seems to escape Graeber and Wengrow.

Perhaps more importantly for my argument in this book, at one point, Graeber[27] and Wengrow clearly assume that the idea of "the patriarchy" began in either the Pre-dynastic or Proto-dynastic periods before the start of Egyptian culture proper, simply because they believe only patriarchal cultures herded cattle (404), ergo the predynastic people of northern Africa who herded cattle simply must be patriarchal, in their views. This assumption completely disregards the Egyptian reverence for the goddess Hathor, the cow goddess, including ignoring the increasing evidence that she was worshipped by these ancient North African cow herders long before ancient Egypt was conceived. While some scholars will respond to this point by arguing that patriarchal cultures can still worship female gods, I remind them that an androcentric cattle-based culture would most likely prefer a bull god, not a cow god, if their values were full on patriarchal (emphasizing

---

[27] David Graeber actually admits that the patriarchy did not rise until the Bronze Age in his book, *Debt: The First 5000 Years*. He actually pinpoints the start of the patriarchy in Mesopotamia (176) because women were "everywhere" in early Sumerian texts, named as "rulers...doctors, merchants, scribes, and public officials" (178), even going so far as to admit that women disappear from such records after 2500 BCE. However, while he seeks out potential reasons for why this change occurred, he never entertains the idea that humanity had discovered procreation at this time.

brute strength) rather than leaning toward matriarchal values that emphasize nurturing. Hathor, in her human-cow hybrid head, appears at the top of both sides of the Narmer Palette, often considered the oldest document in the world, so she was clearly very important to the early dynasties of Egypt, something I discuss in more depth in the chapter on Egypt. In fact, she is so important to the Egyptians that many goddesses will adopt her horns as their headdresses, often framing a blood red moon.[28]

Ironically, Graeber and Wengrow also admit that the concept of patriarchy came much later when they discuss how impersonal administrators created laws that created impersonal standards of equivalence that were, in the end, stacked heavily against the poorer inhabitants of the city of Uruk. They state, "the same laws, the same rights, the same responsibilities applied to all of them, whether as individuals or, ***in later*** and ***more patriarchal times***, as families under the aegis of some *paterfamilias*" (emphasis mine, Graeber and Wengrow 425). They attempt to argue against egalitarian ideals by asserting that treating everyone the same inevitably creates unjust hierarchies, but fail, in the end, to be convincing, largely because they go out of their way, seemingly, to avoid discussing laws that restrained only women's rights.

Perhaps, overall, it is Graeber and Wengrow's dual ignorance both of biology, particularly that concerning other mammalian species, and of gynocentric cultures, that makes them miss some of the most important interpretations in the evidence they examine. They begin, for instance, to object to humans being compared to other ape species on p. 92 because they arrogantly assume human beings are above such comparisons, and they struggle with understanding actual

---

[28] The lunar eclipse produces a blood red moon. Since the moon was already associated with women's menstrual cycles, the blood red moon would have held particular spiritual significance.

herding of animals and what that requires, especially why such herding lends itself well to a foraging lifestyle on p. 105.

Their reactions against comparing humans to other animals possibly comes because of scholarship, like Nancy Makepeace Tanner's *On Becoming Human*, which demonstrates that chimpanzees were a lot less brutal, and a lot less male dominant than many androcentric scholars would like to believe. Tanner notes that both "male and female chimpanzees ordinarily copulate with many individuals, and mating may be initiated by either sex. Females approach and present to a male for copulation," describing in detail how males will indicate to females that they are interested in coitus, noting that "females may ignore willing males, and males may ignore female approaches" (95). Tanner even relates how females deal with persistent males who do not want to take "no" for an answer, with the simplest way to simply ignore them, noting that the most successful males spend ample amounts of time "grooming females" and sharing food with them (96). Are most chimpanzee males more gentlemanly than human males?

Perhaps if they spent more time studying animals' social behaviors, including how matrons work in creating cooperation among most herd mammals, which I will discuss in more depth in the animal domestication chapter, they would better understand human ones.

The first thing we all need to do, though, is to take off our androcentric glasses to see the world as the ancients would have seen it—full of Female Magic.

These ideas and more will be discussed more thoroughly throughout the book regarding each culture I examine.

Chapter Three
An Important Myth About Gynocentric Communities to Debunk

While most of the earliest human societies--such as those in early Sumeria, or at the Orkney Islands' Skara Brae, or among communities in the Totonac area of Mexico--were more egalitarian with the people having a more equal say (aka more democratic) in the structure and nature of their communities, so that most lived in similar homes in similar conditions, consuming similar foods and producing similar goods to sustain them, one myth we must confront about these more gynocentric cultures is that they lacked hierarchies. Possibly because so many androcentric archeologists confer the term "civilization" on communities that demonstrate hierarchies, many feminist scholars have desperately attempted to prove that gynocentric cultures lacked hierarchies, but this belief is not only impossible to prove, but also improbable biologically.

As a farm girl, I can report that even herd animals have hierarchies, which might not be evident to the unschooled observer. Every matron of a herd is looked to for guidance, especially in times of crisis, but she is also usually the first to put herself between her herd and danger. Should she not be apprised of an impending crisis, the matron has deputies among the next oldest in her herd who will insinuate themselves between the herd and danger, if necessary. These are the females who will carry on after the matron passes. This generalization holds whether we are talking elephants, bison, horses, domestic cattle, or even lions.

Yet humanity has known about these animal pecking orders for several millennia, probably since the beginning of animal domestication, if not earlier. However, human beings have been projecting their own androcentrism on the animal world for at least the last two hundred years, since we still have many biologists who insist that male lions kill offspring in order to procreate as though they consciously desire to pass on their genes (lions understand genetics, really?), when, in fact, the male lion simply remembers with whom he has had sex, so will kill offspring in order to bring that reluctant female lion into heat. Yet the androcentrists have people believing lions understand genetics and how offspring are created, concepts human beings still struggle with.

Ask any dairy farmer about her herd, and she can tell you which cows get milked first every day, and which ones might have worked their way up that hierarchy recently.

We call this animal self-ordering system a "pecking order" because our tamed foul do it, too—chickens, turkeys, ducks, geese, and pigeons.

Therefore, hierarchies are not unnatural, and they exist among both wild and domestic animals in varying degrees of complexity. Expecting humans to act differently is unrealistic.

Some Neolithic and late Paleolithic scholars, for instance, argue that some 30,000 years ago, a major change in human longevity occurred for reasons yet to be agreed upon. Chances are great that this increase in human longevity resulted from an increase in eating meat along with a variety of other fruits and vegetables. This longevity created, for the first time in human evolution a three-generational family, so that grandmothers stuck around long enough to do babysitting, freeing up other younger women to do more work. This phenomenon is often "called the Grandmother Hypothesis" (Adovasio, Soffer & Page 164), and demonstrates for humans what we know already happens for other animals: the "older people's long memories serv[ed] as living repositories of useful information" (Adovasio, Soffer & Page 164). Many scholars attribute the population increase that occurs in the next 20,000 years to the fact that more experienced people lived longer, passing along their wisdom to new generations, which further enhanced longevity and reduced infant mortality, leading to gradual increases in population in many parts of the world as well as freeing up more people to create more tools and artworks as they are able to dedicate more time to experimenting with various mediums, such as weaving and ceramics (Adovasio, Soffer & Page 165).

The next big growth in human populations come between the Neolithic and just prior to the Bronze Age, after a greater variety of animals, especially goats and sheep, have been domesticated. Sheep and goat milk fed to human infants by doting grandparents would have helped more infants survive into adulthood. In this cooperative way, many ancient humans led slightly hierarchical, but more egalitarian lives by caring for each other in mutually beneficial ways. Margaret Mead famously observed that civilization began when some humans helped an injured human who had broken his leg heal from that fracture—a healing that would have not only required physical assistance, but also sympathy and compassion.

We all know that using one's hierarchical power to abuse others, either through overt violence or via threatened violence, simply for the sake of increasing that power, is inhumane. It is this abuse of hierarchical power that feminists want to erase, but this kind of might-makes-right abuse tends to come from androcentric and outright patriarchal cultures that seek to dominate their "lessers," so it is important to note that natural female hierarchies are not usually designed to put others down, but to elevate those with experience and more knowledge by naturally having the younger animals look to the older ones for guidance in difficult situations, as well as to nurture the young and to protect the injured. In other words, human beings have been teaching the importance of respecting elders, nurturing the young, and caring for the ill for some time, just as many animals do. An elephant matron, for instance, can lead her herd hundreds of miles to find water or food, and that knowledge is vital to the herd's survival. The matron and her deputies will lead the way all while taking care of the young, assisting the injured, and even remembering the dead as they pass their bones.

Human beings are similarly gregarious by nature, meaning we work well together when everyone is of a mind to work together, rather than at odds with each other. Brian Hare and Vanessa Woods assert, "The most sophisticated social understanding, memory, or strategy will not facilitate innovation unless it is paired with the ability to communicate cooperatively with others. This friendliness evolved through self-domestication" (xxv). According to Hare and Woods, it was not competition with others that gave rise to civilization, but cooperation and communication through group action, or what they call "self-domestication." In essence, they argue, we "domesticated" or taught ourselves to be friendly and cooperative in order

> to make the shift from living in small bands of ten to fifteen individuals...to living in larger groups of a hundred or more.... Our sensitivity to others allowed us to cooperate and communicate in increasingly complex ways that put our cultural abilities on a new trajectory. We could innovate and share those innovations more rapidly than anyone else[, so that] other human species did not stand a chance (Hare & Woods xxvi).

While many androcentric archeologists and historians love to argue that a clear hierarchy is necessary to achieve civilization and to develop culture, few explore what they mean, often insinuating that gynocentric cultures were not sophisticated enough to create much of anything, despite the evidence found from around the globe. Scholarly attitudes about hierarchy is an issue I will address more specifically soon.

What we must acknowledge is that most human beings enjoy being around other people, as long as we are not feeling pressured to do something we do not want to do or are not feeling the need to be elsewhere doing something else. However, human beings, like domesticated cats, tend to be relatively self-centered. We want to be comfortable, and, if we are not comfortable for whatever reason, we tend to take our discomfort out on others, creating conflict.

Various ancient cultures developed mechanisms to keep such self-centered behaviors in control. As Hans van Wees points out, shame was a useful tool to keep people in line when they exhibited socially unaccepted behavior (21). Of course, van Wees was referring to how the warriors in the *Iliad* create a unifying force using shame: "solidarity is reinforced by fear of the criticism and ridicule" for failing to fight and to support soldiers on their side, which could result in "shame at losing reputation" (21). Shame, however, has been a useful tool for most societies, especially as a form of peer pressure.

The most common method of garnering support and of issuing approbation was through peer pressure, which humans still feel in modern times. I recommend you research the Asch Conformity Experiment if you believe you are above such influences. Psychologically, we humans feel the need to conform to a collective pressure, and most of us do so willingly, even unconsciously, even if we know that the collective idea is wrong. Do you have a bitter person in your group who keeps bringing the energy of the work down, making others miserable? Ancient humans would hold a type of intervention, with some cultures, like the Iroquois, so knowledgeable about human psychology even before it was officially a science that they could diagnose and treat such emotional issues through various methods, including prescribing more sexual intercourse. Sigmund Freud was really behind the times, but he was a European male, so he was listened to, unlike the Iroquois.

However, increasing numbers of people seem willing to risk ostracism in order to do what they want to do. That ability to stubbornly follow one's own counsel speaks both to defying social pressures and to human individuality. While we would never have any social changes if individuals did not stick up for what they believed was fair, we have to admit that there are many individuals and small groups who only seek power because they want what they want, not caring what others need. As Hare and Woods warn, "when we feel that the group we love is threatened by a different social group, we are capable of unplugging the threatening group from our mental network—which allows us to dehumanize them" (xxvi). Essentially, human beings are capable of ignoring our natural senses of empathy and compassion, allowing us "to become capable of the worst forms of cruelty," so that we are, essentially, "both the most tolerant and the most merciless species on the planet" (Hare and Woods xxvii).

Because of this ability to turn off our senses of empathy and compassion, if the subtleties of peer pressure did not work for the ancients, there were more effective methods of punishing individuals for their aberrant behavior, such as literally turning the community's backs on the person, acting as though she or he was dead, no longer existed. When human beings were still foragers, such a shunning often meant actual death from exposure, starvation, or animal predation for that individual, so most people would do whatever they needed to do to avoid such a punishment.

At some point, maiming or dismemberment became forms of justice against people who went against the community's norms. As

I explore in another chapter, physical castration, which probably meant the complete removal of penis and testicles for human males, was probably used frequently to subdue men who sought to dominate or to disrupt a community's social mores or who might challenge a person in power's political agenda.

It is also quite possible some of the first sacrificial victims—whether we are speaking of the thrice-killed found in British bogs or among the Toltecs in ancient Mexico—were people singled out for their aberrant behavior with some sly shaman insisting that the person was chosen by one divinity or another for the honor of dying for her/his/xir village in order to keep the invaders at bay or to help the moon return to fullness. We know that the Inca leaders often required the sacrifice of children from outlying communities before those communities could receive any form of community assistance, a terrible way for a small community to prove to its overlord that they deserved his assistance in their time of need.

The threat of physical punishment or even death could be, and undoubtedly was, a great deterrent of negative individual behaviors, unless the individual had the support of her/his/xir peers.

So how, then, did some people rise to power, influencing their communities to follow them down such darkly oppressive human avenues?

The threat of death or dismemberment, of course, became the chosen tools of oppressors. Fear mongering is what dictators do to keep their populaces in line with their personal agendas, especially with creating false lines of difference, i.e. those people are not us, so do not deserve to have the same comfortable lifestyle we do. As Hare and Woods point out, creating such us-versus-them dichotomies allows us to dehumanize "others."

Which brings us to what many historians and archeologists call "complex" hierarchical societies.

Jan Driessen, a Belgian archeology professor who has focused mostly on the so-called Minoan[29] societies of ancient Crete,

---

[29] British archaeologist Arthur Evans called the ancient people of Crete the Minoans because he assumed the mythological King Minos had to be the person responsible for the splendor of the archaeological remains he found there. Since increasing archaeological evidence is making it clear Crete was more matriarchal

facilely avoids admitting that the Cretans were matrifocal and likely matrilineal, clearly gynocentric, in most of his works, especially in his article on hierarchies, "History and Hierarchy: Preliminary Observations on the Settlement Pattern of Minoan Crete," published in 2001. While he admits many caveats in this article, including that the method he uses to assess the hierarchical complexity of Minoan cultures through its many phases of development is problematic, he refrains from admitting anywhere in the article that women might be the ones organizing the culture when he argues that "the existence and size of 'public' structures at some sites evidently manifest the presence of some kind of 'power' at these settlements," citing his own argument from 1999 on this point. He goes on to claim that "the size of these structures requires both a considerable population and a social cohesion and their mere existence implies an energy input and hence the presence of administrative personnel" (Driessen 64). What, exactly, he imagines the "administrative personnel" to be like or to be doing is anybody's guess, but he clearly conflates cooperation and coordination, the only two factors necessary to build anything, with administration, a heftier androcentric word that implies a kind of social power that goes beyond cooperation and coordination into the realm of actually controlling goods and people. The fact that he repeatedly fails to admit that most of the coordination and organization of the culture from early on would have been completed by women looms large.

Driessen, to support his shaky foundation about the increasing levels of complexity among Cretan settlements argues that "chiefdoms are assumed to have two or a maximum of three levels [of complexity; while] states [have] at least four: cities, towns, large villages, and small villages" (57). What is a "chief"[30] in his view? He never says. What, exactly, manifests a state is also lacking clarification. He does provide a chart to help others determine levels of complexity merely by counting hectares and possible population size with a "small family farm" being the smallest unit of less than

---

than patriarchal, we should find a better term to describe the culture, so I will call them Cretans.

[30] See my discussion of the origins of this word in other chapters.

50 people. That's some family, and I grew up in a family with 39 people in the immediate (grandparents, parents, children) family!

He then uses European concepts to break down the rest of the community sizes into categories: a Hamlet has 70 plus people; a Small Village has more than 250; a Village has more than 600; a Town has more than 875, a Large Town has at least 1750, and a Capital Town has at least 6250 (64). While he cites sources for these numbers, most are being assigned rather arbitrarily in order to address questions often asked, but without actual examples as evidence.

Having grown up in a town of less than 7000 people, I am well aware that that particular small town (by today's standards) is the "capital" of its county, but I am also steeped in that county's history, knowing that other towns vied vigorously for the title of county seat. The other towns lost, in the end, and their populations did plummet, but not just because their town did not become the county seat. Their populations plummeted because the railroads and highways bypassed those towns. So large entities outside my hometown, outside the county and state even, have directly affected the county in which this small county seat is located.

Similar possible external influences existed for Crete, as well, despite it being an island in a very busy sea, but such factoring of outside influences did not enter into Driessen's accounting.

Driessen admits, toward the end of his conclusions, "that Crete, because of its insularity [in the midst of the Mediterranean Sea], had developed a different type of territorial organization, perhaps largely ritually motivated, which reduced intra-insular tensions" (70). This claim merely speculates that the Minoans had some sort of social method for mitigating conflicts between communities on the island. What is he omitting? The fact that a gynocentric culture, organized by women whose cooperative instincts are often stronger than many modern day men's are, could have coordinated the communities of their island much more efficiently than the androcentric cultures managed their own communities in other parts of the Mediterranean. I would absolutely love to see an androcentrist like Driessen admit such a possibility.

What we need to take from this discussion is the fact that women, too, recognize hierarchies among ourselves. In our modern cultures, these "pecking orders"[31] tend to begin to organize in our schoolyards, where young girls learn to play together while navigating issues of social acceptance.

Some scholars, such as Lull, Rihuete-Herrada, Risch, Bonora, Celdrán-Beltrán, Fegeiro. Molero, Moreno, Olieart, and Velasco-Felipe, exhibit frustration and confusion when confronted by obvious signs that women were rulers—that they wielded power in hierarchical cultures long before their cultures became full on patriarchies. Could, these scholars ask, "a class-based society...be ruled by women?"

To be fair, many feminists' readings of ancient cultures push the idea that earlier societies, because smaller and less settled, were organized in more egalitarian social structures, which is verifiable by many means of measurement. However, to assume that all ancient complex societies were either egalitarian or patriarchally hierarchical misses most other forms of social organization, social organizations that were led by women that also developed hierarchical social structures. Surely it would be as "natural" for a gynocentric culture to view certain women as more divine or more powerful than other women, including women who menstruate longer than most women do. See the later chapter on the Venus figures regarding that phenomenon and what it undoubtedly did for some human communities.

Because of patriarchal pressures to be the dominant person socially, though, this "natural" mammalian hierarchical tendency, which seems to come from a need by a few to control others or to always be first, can morph into sadistic cruelty by puberty, a time which also often means girls are limited by their parents, their peers, and their teachers regarding how they view other females, and what they as females in a male dominated society can become as professionals in adulthood.

However, just because a group establishes a hierarchy does not mean that hierarchy is unmalleable or so deeply entrenched that

---

[31] Such a concept as "pecking," which implies birds, is sexist when used to describe women only, but I am using it here to connect the concept with my earlier arguments about animal hierarchies.

force must be used to create cooperation. By learning to give and to take, to opportunely challenge or to side with one another, an egalitarian culture guided by women is just as productive, if not more so, than one guided by males who believe women are subordinate and must be kept subordinate to men, so uses fear and obstruction as its handiest tools. Such an egalitarian gynocentric society would also be more flexible, changing when necessary to in order to meet its people's needs.

While no woman is perfect, women have proven, time and again, that we can effectively persuade others to coordinate and to cooperate with us in order to achieve something wonderful for our families and our communities, without having to resort to violence or the threat of violence. Recently, it has been noted that egalitarian countries (Ostergard) led by women survived the COVID pandemic better than those led by men (Zenger & Folkman). Women often manage coordinating groups by drawing on our natural communal instincts to nurture, to draw out the best of others through care and consideration, instead of relying on competition or threats.

Women, however, can be just as competitive as men, but, as our prison systems attest, fewer women take competition to the deadly conclusions that some men do.

Think twice, however, about assuming a society is at a "higher" level of sophistication if it must rely on violence to maintain control.

Chapter Four
Imagining What Ancient Humans Believed About Sex in Order to Understand When They Discovered Procreation

If you are like I used to be, you probably assume humanity has "always" known about the connection between heterosexual intercourse (aka coitus) and procreation, but there are many indications that this process, which seems obvious to most of us in the 21st century, was not obvious to our ancestors prior to 2400 BCE, not to mention not obvious to many people even after the Egyptians, probably, discovered the connection around that time.

Being able to imagine that human beings had yet to discover procreation, which is the act of two living beings being necessary to create offspring as opposed to parthenogenesis wherein the female creates offspring completely on her own, is vital to my arguments. Otherwise, readers might assume, as I once did, that humanity has always had it all figured out.

However, there are two important, among many, indicators that human beings had not yet determined that heterosexual coitus creates offspring, and both involve children: incest and pedophilia. While it is accurate to argue that both still occur today, almost every culture on every continent today has prohibitions against both. Such prohibitions demonstrate that not only do we recognize that hetero coitus can create pregnancy, even in premenarchal girls, but also that we recognize that the act of adults enticing children into sexual relationships is morally wrong. Even though royalty among some cultures continued to practice incest, such as the marriage between a father and his daughter or between siblings, for centuries After the Common Era (ACE), the earliest known written laws prohibiting incest occur in Hammurabi's Code of Laws, circa 1750 BCE. Since human beings had been writing since approximately 3000 BCE, we have to assume that something happened in the intervening 1250 years to clue our ancestors in to the fact that hetero coitus causes pregnancy, and that such coitus with a close relative can cause birth defects and/or miscarriage.

Unfortunately, there are still many countries, including some states in the United States, that still allow adults over 17 to have sex with children younger than 17.

The many scholars who simply assume human beings have "always" known about the connections between heterosexual coitus and procreation are actively ignoring millennia of scientific evidence that belie this assumption.

For instance, Edwin Sidney Hartland elucidated the many reasons why ancient human beings could not possibly have made this cause-effect connection between heterosexual coitus and the creation of babies way back in 1910 in his second volume of *Primitive Paternity: The Myth of Supernatural Birth in Relation to the History of the Family*. Hartland attempts to remind his readers of how the ancients viewed even the most natural of events with several examples, but one of his best examples describes how a hunter might assume that "his magical or religious ceremonies draw from unknown distances the herd of bison which he desires to hunt, or [that certain rituals] gather the clouds and bring down the rain upon the parching and aching land" (250). For the ancients, everything in their lives revolved around appeasing natural spiritual forces, often regarded by scholars as outright deities, in order to achieve success in their many endeavors—from firing pots successfully to giving birth successfully.

Among the many reasons Hartland lists for elements that would get in the way of early humans understanding the connections between coitus and pregnancy was a rather detailed explanation about how different cultures around the world initiate sex with children, noting that "premature intercourse....practised [sic] by immature girls with adult men...often results in such injury to the sexual organs as may seriously affect the reproductive powers after maturity is reached" (254). While we could fret about the way this man from 1910 seemed to make the girls the initiators of the heterosexual intercourse, his point still stands—many young girls, molested, raped, or even willing participants who experience sex prior to the natural maturing of their bodies make them ready,[32]

---

[32] Some of the indigenous nations of Australia supposedly still practice adult male penetration of young girls in order, they believe, to bring on menses (Merlan 478), we have to wonder how many women were led to believe that their inability to

often become unable to carry pregnancies to term for many reasons—from lesions in the vagina and uterus to venereal diseases. If young girls, males of any age, women who had suffered vaginal penetration at too young of an age, and old women were not getting pregnant from coitus, why should the ancients have assumed coitus was impregnating women who were "of child-bearing years"?

Scholars also ignore the sequence of scientific discoveries that informed human beings about how offspring are born. In fact, until Thomas Henry Huxley refuted the belief in spontaneous generation, called abiogenesis in 1870 ACE, men who called themselves scientists fervently believed in the phenomenon. The belief was that some animals could be created through mysterious spontaneous processes, such as rain, wind, and lightning, with the sometimes occurring rains of fish and frogs. As Eugene S. McCartney points out, it was not until "Huxley gave a summary of the investigations by which [abiogenesis] was refuted" as president of the British Association for the Advancement of Science in 1870, was serious doubt thrown on the idea. In fact, even as late as 1912 ACE, nearly 250 years **after** the invention of the compound microscope, Sir Henry H. Cunynghame cautioned scientists attending his presidential lecture of the British Association for the Advancement of Science that "we can by no means be sure that the evolution of non-living substance into living may not be happening still" (McCartney).

Therefore, the following is a way to get into the pre-procreation mind of our ancient ancestors to see the evidence I produce in this book from such a mindset. A warning for the Puritans among us: there are many descriptions of sex.

ΔΔΔΔ

As an ancient human being, you would have grown up with a close relationship with your mother and her relatives, fairly secure in the idea that they are there to protect you and to educate you about how to live in the world, which can be unpredictable at times. As an infant, you would have probably been strapped to your

---

carry pregnancies to term meant they were "spiritually" tainted, when, in fact, the fault would lie with this form of pedophilia.

53

mother's chest for easy nursing, even while she foraged for food or worked on hides, pottery, or shaping rocks, wood, or bones. If she needed to concentrate on her work, she might have strapped you to a cradle board, which she could carry on her back, hook to a tree, or stand upright next to where she was working. Your diapers while strapped to her or to a cradle board would have been dried mosses, easily changed when soiled. To teach you "potty" training earlier than most children are taught today, you would have been allowed to run around the community devoid of clothing, allowing you to squat and defecate or urinate whenever you needed to, but many people would have been around to guide you to avoid heavily trafficked areas for such evacuations, just to keep down on the smell. Remember, ancient humans had not yet learned about the bacteria and parasites that are transmitted through urine and feces, or from drinking water that is contaminated with urine and feces, so few ancient people made concerted efforts prior to the Bronze Age (circa 3000 BCE) to create collective sewage systems.

If you survive long enough, your mother would have bestowed on you a permanent name to replace the childish nickname she called you by. This kind of naming ceremony tended to happen between the ages of five and ten because the majority of children die before they reached these ages because of the challenging circumstances of life.

By the time you are about ten years old, your awareness of what is happening around you will have become more certain, although everything is interpreted via its magical or spiritual elements. For instance, the stream is fun to play in because it is too shallow to swallow you whole, like the spirits of the faster and deeper moving river might. The fact that you exist would be explained as an act of Female Creative Magic, and you would witness your mother, her sisters, her female cousins, all secreting themselves away, possibly in a building built just for them during their "moon time," so they can prevent their stronger magic from interfering with the natural magic everyone else imbues. Having gone with your mother into such a building as an infant, you might even have foggy memories of what this secluded time was like. Whether or not you remember that seclusion with her, if you are female you will be both anxious and excited to wait for your time to become a woman, which we now call menarche,[33] when you begin menstruating. You will be anxious about that time because you know it means an ongoing, monthly battle with divine forces that are so fierce you will experience pain and bleeding. You know, as a female, though, that you will not die from this monthly battle, and you might be aware that males in the community envy this ability to battle, bleed, and not die, so that your community might have developed a sympathetic magic for men in the form of circumcision[34] or penile piercing—ceremonies done in secret with just men in attendance to avoid having women's superior magical abilities interfere with men's magic.

You will be excited by the prospect of becoming a woman because it will mean you will be trusted with more responsibilities, after you pass through your initial rituals welcoming and testing your abilities to win the constant battles with the divine forces, so that, one day, you, too, will be able to pull a child from the Other Realm into this world. The risks in this divine battle to create new humans are great, since you could die, or since the child might die

---

[33] Due to less steady diets, many girls prior to and during the Bronze Age probably did not begin menstruating until the mid to late teens, and celebrations like the Latino Quinceañeras would have announced this stage in your growth. Many people around the world mark this change with changes in hairstyles, tattooing, and different methods of dressing, but some, like the people of Crete, marked the change with a girl's first extrasensory experience induced by drug laced beer or wine in order to draw the developing woman's attention to the more spiritual or magical elements of life.

[34] David Friedman notes that in "*The Golden Legend*, a popular religious book of the Renaissance era, declared that the day of Jesus' circumcision, when 'he began to shed his blood for us...was the beginning of our redemption,' a view shared by Aquinas" (52), demonstrating that even after the patriarchy's rise, some scholars recognized the symbolic blood loss engendered on men through circumcision as sacred. Whether or not they admitted that it was an imitation of menstruation is unknown, but he does note a common belief by early Christians "that Jews menstruated through their penises" (159), but their disgust of Jews was compounded by their disgust of circumcision, a ritual which they believe "proved the Jew's perversity" (159). At what point, then, did American Christians embrace circumcision as a necessary ritual? Apparently, after the publication in 1896 of the book *All About Baby*, which warned that, if the foreskin was not removed, boys would be plagued by "'priapism, masturbation,...and most of the functional nervous diseases of childhood'" (Friedman 92-3). Friedman also claims that circumcision is not common outside Jewish communities in Europe (93).

in the process of your retrieving it for life in this world, thus sending it straight back to the Other Realm until someone stronger can bring it forth.

Depending on which people are raising you, you might call at least one man "father." He would be a man who is very fond of your mother, who might bring her stone tools, precious clays to make pots, meat to cook in those pots, or soft hides for her to make into beautiful clothes for you.

More than likely, though, you call all the men your mother's age in the community either "father" or "uncle" as a sign of respect for their being older than you. Men who are even older than your mother's age would be "uncles" or "grandfathers," much the same way many people around the world today use those terms as titles of respect for close friends of our families or for the elderly in their communities to show respect and connection.

Similarly, your mother's sisters might be called "mother," too, with her cousins called "aunt," and your mother's mother and her sisters or cousins would be called "grandmothers" or "aunt," depending on how your mother regards them in terms of their importance to you.

In this way, you learn that everyone in the community is related, even if not directly with blood ties because, as far as you know, you are only related to your mother and any other children she gave birth to, and by matrilineal succession also to your mother's mother.

If you are male, you would have been taken care of by the many females in your community, your "aunties" or other "mothers," every month when your birth mother secluded herself for three to five days. You would understand that she left you in their care in order to protect you from the magical, spiritual battle she would be waging during those days. You would understand that the blood from female spiritual battles is powerful magically, so you would be cautioned from ever touching it.

By the age of ten or so, your mother's brother will begin to take you in hand to teach you more "manly" things, such as why your penis might appear to leak at night or get hard when you do not want it to.

Whatever gender you are, your behavior will largely be indulged, and you will probably have witnessed many sexual acts, both heterosexual and homosexual, so that by the time you enter puberty, you will understand how both women and men give pleasure to others. Sex is viewed as a form of social pleasure, not a private act.

For your community, there will probably be annual spring ceremonies wherein the human beings of all ages around you imitate the animals of the field, particularly the birds, coupling for coitus, cunnilingus, and fellatio in ecstatic public or sacred rituals that make you want to experiment with sex, too, because sex is pleasurable, social, and even spiritual. Your people might even prepare for these festivals over the winter by erecting huge pillars in circles, carving them with animals and their offspring, which circle around two human beings touching their own genitals, like the sacred sex circles of Göbekli Tepe or Stonehenge. These spring rituals, and autumnal ones[35] as well, will probably be accompanied

---

[35] An interesting hypothesis, proposed by an independent scholar, Bennett Bacon, to paleolithic scholars who published their proposed ideas in early 2023, is that ancient humans noted time periods—moons associated with specific activities, which is how most ancient people tracked time—when animals depicted on cave walls were grouped together for both sexual and birthing purposes, with a rational explanation that the humans would have noted when these groupings occur to improve their hunts (Killgrove). The most controversy about this theory, so far, is a contention that the Y markings, in addition to dots and straight lines, could indicate the times when the animals gather to give birth because the authors believe the Y symbolizes "two parted legs," with some scholars arguing that the Y could mean other things besides birth (Bacon, Khatiri, Palmer, Freeth, Pettitt, & Kentridge). However, as you will see, this basic shape **Y** is also that of the skull and horns of many herd mammals, such as bison and sheep/goats, which also become symbols of the creative powers demonstrated by females giving birth because that shape mirrors the shape of every set of mammalian female reproductive organs, with the horns being the fallopian tubes and the skull the vagina and womb. Ancient people would have undoubtedly viewed these spring/autumnal moments when large herd animals they hunt get together for sex and safe birthing as liminal moments of time—transitional sacred moments that they, too, would mark with feasts and celebrations that include sex, and, quite possibly, celebrations of their own successful births. This shape also explains why so many gods, especially Egyptian goddesses who also have blood moons between the uprights, wear horns.

by singing, dancing, bonfires, and lots of food, so that everyone looks forward to these seasons and these festivals.

You might learn that people who are more shy might make ritual emblems or small tokens to give to a person they desire to have sex with; these tokens could be inscribed with images of what they desire most. The narrator of the Ancient Architects YouTube channel. Matt Sibson, is headed in the right direction when he outlines why he believes a small carved bone spatula found at Göbekli Tepe might depict a vulture, which is seen as a motherly figure in many ancient Turkish archeological sites, although he is rather optimistic when he believes that these ancient human beings understood coitus' connections to procreation. Instead, the spatula most likely depicts a vulture in the process of feeding its offspring, so that the "phallus" Sibson sees is actually the beak of a baby bird. This kind of token would have easily been given to another person during such a spring rite at Göbekli Tepe in order to indicate to the spirits of the place and to the person s/he gave it to that motherhood was the desire goal of the spring ritual. Remember, since our ancestors did not connect any form of coitus with pregnancy, while sex was part of the spring ritual, all that ejaculated semen and female lubricant were helping to generate spring rains and flooding rivers to water the plants and animals the people rely on. The sexual rituals, though, would be seen as just one numinous (spiritual) activity that could heighten a woman's ability to pull another soul from the other realm into this one, among other magical abilities.

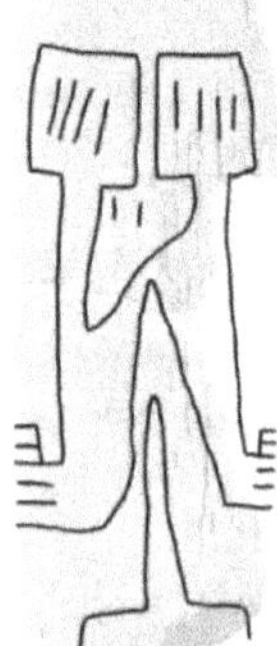

"The 11,000-Year-Old Göbekli Tepe Bone Plaque"
Ancient Architects, Matt Sibson

Women's spiritual battles to produce babies are enhanced, your people believe, because of sacred rites in sacred places like Göbekli Tepe. Women's monthly spiritual battles would mean more to your community than any of the actual acts of sex that take place because the blood is proof of these battles. Most will dismiss any idea that sex of any kind causes pregnancy because, if men are "planting" seeds inside the people they have intercourse with, other males, young girls and boys, and older women they have sex with would also become pregnant. Therefore, it is likely that many ancient people, like yours, viewed semen either as milk meant to feed babies while a woman's body transports them between worlds, or as water, like the spring rains or river floods, which make plants grow. The idea that semen was seen as milk (thousands of years before humans knew that sperm resided in semen because we could see them through compound microscopes) is reflected in the many sex poems dedicated to the goddess Inanna of ancient Sumeria, as well as in the milky lettuce associated with the god Min in Egypt—a god always depicted with a constant erection. Even scholar Thorkild Jacobsen admitted that the flood waters of the Tigris River were believed to be the semen of the god Ningirsu (qtd in Kriwaczek, 28).

However, sex as a spiritual activity would have been one action among many others that would have been believed responsible for the coagulating of women's menstrual blood into new humans. As Erich Neumann noted in his 1955 book *The Great Mother*, ancient people believed "the embryo is built up from the blood, which, as the cessation of menstruation indicates, does not flow outward in the period of pregnancy" (31). Even Lakota writer, Charles A. Eastman relates the story of "Blood-Clot Boy," who is formed from a mere drop of blood (*Wigwam Evenings* 68), demonstrating an old idea about how humans were formed in the womb: magic plus blood becomes new beings.

Spring festivals would have been rife with sex because of the associations of men's flowing semen with rain and river floods, allowing you to discover just how spiritual sex can be during your first experimentations with it, marveling at how your body can create transformative orgasms simply by joining with another in some way. Or even on your own.

Your people will have stories about how beasts have sprung fully formed from the earth when rain, lightning, or sacred libations were poured, ideas that will last well into the Victorian era. They will have stories about how beasts have been transformed into human beings, like Enkidu in the *Epic of Gilgamesh*, through a process of sacred sexual interactions, the *hashadu*, that can be healing or transformational.

Your people will have stories about why they live where they live and why that particular ground is sacred. Associating yourselves with that sacred space would help your people develop a cultural identity. As Paul Kriwaczek describes the process of discovering and transforming sacred spaces:

> It is easy to guess that those who came to the sacred Apsu [springs] would have joined together in ritually feasting on the rich harvest of the marsh; great shoals of freshwater mussel shells have been found among the earliest layers of the site. To our forebears, food never lost its ritual significance (as it still hasn't to the religious-minded of our own day). Here at Eridu, with its numinous associations, the sacred meal would have been a serious, although not necessarily solemn, occasion. And from this regular event, perhaps yearly, perhaps monthly, by the holy marsh at the edge of (25) the sea, would slowly have grown an entirely new group identity: 'those who come to the Apsu.' Drawn from the pioneer settlers in southernmost Mesopotamia, their very presence and survival demonstrated a commitment to changing the face of the land, and to securing a different and better future. The religious rites they performed at the water's edge would forever associate the divine spirit of the Apsu with that belief (26).

Various sacred celebrations, including feasting, dancing, music, and other forms of revelry, would have happened at various times throughout the year, and you would have looked forward to each of them for different reasons, such as the great foods that would be offered at different times of the year or your chance to meet someone you fancy from another, more distance community.

Your community might even use coitus for healing purposes, since desiring someone who has paired with another might prove to be too painful for the rejected person to bear. Compensation for coitus with the desired person might be offered by the entire community, just to heal the person of this social sickness.

ΔΔΔΔ

In short, as modern humans, we have to discard all Puritanical and patriarchal thoughts about sex in order to understand how our ancient ancestors viewed its magical and spiritual qualities that allow human beings and animals to transcend the plane of existence on which we dwell, putting both women and men in touch with other spiritual realms, while, at the same time, demonstrating our integral, immanent place on this earth.

As Paul Kriwaczek[36] admits, human "belief in the transcendence rather than immanence of the divine had important consequences. Nature came to be desacralized, deconsecrated .... Rather than an integral part of the natural earth, the human race was now her superior and ruler" (229-30). He also admits that the belief in a transcendent god

> is all very well for men, [who are] explicitly singled out in [Genesis 1:26]. But for women it poses an insurmountable difficulty. While males can delude themselves and each other that they are outside, above, and superior to nature, women cannot so distance themselves, for their physiology makes them clearly and obviously part of the natural world. They bring forth children from out of their wombs and produce

---

[36] Kriwaczek's Abrahamic religious bias escapes repeatedly in his book because he clearly believed that these religions were a more perfect form of religion than what had come earlier. If there was such as thing as religious evolution, Christianity seems to be the pinnacle for Kriwaczek, who states, "The female sex would not begin to regain a measure of religious respect until Christians came to believe in a God who was born naturally, as a human being, into the physical world out of the womb of an earthly woman" (230). How, exactly, he saw women benefiting from Christianity's ideas about the Virgin Mary is unclear, but he clearly ignores the history of the Catholic and Puritan churches, both of whom were responsible for countless women's deaths as witches—women wielding magic who threatened the patriarchal order.

61

food for their babies from their breasts. Their menstrual cycles link them to the moon (230).

While Kriwaczek admits these facts, he fails to admit directly that women were, once upon a time, superior in spirituality because of their links to nature and the cosmos were stronger, more obvious than the ones men experienced, so that the discovery of procreation circa 2400 BCE was the actual tipping point when men began to "transcend" women and all other elements of the natural cosmos to make themselves more god-like than women, thus superior to women. I explore exactly what had to happen to prove to ancient humans that hetero coitus caused pregnancy in the next chapter.

To Kriwaczek's credit, he does admit that by "Assyrian times, [male transcendence to a god-like status while women remained linked to natural magic] was a self-evident fact that debarred [women] from full humanity" (230). He acknowledges that men elevated their status above women on purpose, tacitly acknowledging that humans had probably discovered procreation circa 2400 BCE, the only thing that allowed men to transcend women in importance since they came to believe that men were "planting seeds" in women, who were merely fertile fields to be "plowed." Men, then, became the Actors and Creators, while women were relegated to the role of Controlled Vessels, despite women's many contributions to the cultures that men then overtook and commanded as they elevated themselves to their mythical status as gods-on-earth.

Chapter Five
When Magic Became Biology:
Human Understanding of Procreation

I grew up on a farm, so I witnessed, firsthand, sexual intercourse for a number of different kinds of farm animals: horses, cattle, pigs, rabbits, ducks, chickens, dogs, and cats. After I started 4-H when I was eight, my father put me in charge of breeding the rabbits.

To my embarrassment, breeding the rabbits meant learning to tell when a buck successfully ejaculated into the doe. Unlike most other species, when a rabbit climaxes, the buck reflexively curls up, literally falling off of the female in a stiffened ball. So, when the male, overcome by his climax, falls off, you know he ejaculated successfully. Domesticated rabbits have a short gestation period of approximately 30 days, so the easiest way to tell if the breeding "took" is to put the doe back in the buck's pen a week later to see if she is receptive to his sexual needs. If she resists, she's pregnant.

I became so intrigued with different animals' behaviors, especially when it came to their interpersonal relationships, particularly that with humans, that I entered college, originally, to become a veterinarian. I ended up, however, following my real passion: literary analysis. In the last few years, however, I have found myself exploring biology once again as I unraveled a mystery few people other than scholars ever talk about: **when did human beings finally learn males have a biological role in procreation?** In other words, when did we prove beyond reasonable doubt that heterosexual intercourse, of some sort, is necessary to create offspring?

This question is crucial to address because, without the knowledge of procreation, there would have been no patriarchy.[37] And, without the oppressiveness of the patriarchy, humanity would

---

[37] Scholars like Daniel C. Snell, however, admit that "the patriarchal period cannot be dated with any precision, and though some figures like the patriarchs [mention in the bible] may have wandered into the land [circa 1900 BCE; his date], others must have come later" (46).

not be in the particular "might-makes-right" mess we are currently in.

To be honest, I had grown up around animals and had assumed humans must have learned about this important connection between heterosexual intercourse and procreation ages ago, but then I read an article that talked about how humans became more aware that males might have a role in procreation once our ancestors began domesticating plants and animals approximately 10,000 years ago (Stiebing 28)). This idea did not make much sense to me, largely because I had studied many pre-literate and neo-literate cultures in the intervening years between being a farm girl and becoming a doctorate holder, so I knew ancient hunters would have taken the time to observe animals in the wild just as much and as often as the people who finally domesticated them.

So why had humans not figured out that heterosexual intercourse can lead to pregnancy earlier than 10,000 years ago? After all, domesticated animals exhibit much the same seasonal/cyclical needs for sex and birth that wild animals do.

I found several other articles that made similar claims, but the article that truly made me determined to pinpoint a more exact time when humanity first made this biological discovery was written by a famous colorless[38] male supremacist, aka rabid patriarchy supporter, who freely admitted that humans probably did not make the connection until after animal domestication. If he was so willing, I thought, to admit that humans have only known about this vital biological connection for approximately 10,000 years, a connection upon which the whole idea of the patriarchy is built, the truth had to be much more recent than that, so I began exploring the earliest literature written in Sumeria and Egypt for clues about what those early civilizations thought about sex and procreation.

What I discovered shocked me, and it might shock you, too.

ΔΔΔΔ

---

[38] I hope I do not offend anyone, but I do not use racial terms like "white" and "black" to describe people, simply because in the Abrahamic religions such colors symbolize purity and evil or goodness and corruption. I refuse to perpetuate such stereotypes.

Before we delve into the literature, however, we must examine why humans took so long to figure out the sex-procreation connection.

The first problem, of course, is that human beings have long had sex with whomever they have wanted whenever they have wanted. From heterosexual and homosexual interactions have come acts like fellatio and cunnilingus. In fact, we can observe sexual behaviors among our primate cousins and know, without a doubt, that human beings would have been just as sexual as our cousins. Having that much sex and that many different kinds of sex means that human beings had no reason to associate sex, let alone just one type of sexual activity, heterosexual intercourse (aka coitus), with pregnancy.

Complicating those important variables is the fact that most herd animals have sex in cycles, usually seasonal cycles, with varying lengths of pregnancy or gestation, and we can truly begin to see why isolating sex as a factor, let alone a single kind of sexual act as a cause for pregnancy would have eluded humanity for a very long time.

Some of the first animals domesticated in the Eastern hemisphere were sheep and goats, and the gestation period for both is approximately 150 days or roughly five months. While that gestation period is not as long as it is for cattle, 283 days, or horses, 362 days (nearly a whole year!), it still puts a considerable amount of time between the act(s) of sexual intercourse and the birth of babies, so making a causal connection between heterosexual intercourse and pregnancy and to birth would not have been a natural cause-effect rationalization.

Factor in the fact that females of each of these four species—horses, cattle, goats, and sheep—will mate with as many males as they can, so even if a herd of cattle or horses or a group of goats or a flock of sheep has one dominant male[39] who tries to keep other

---

[39] Apparently, the information that only a few males are necessary to fertilize an entire flock of sheep is still unknown or ignored among some sheep herders today because H. Epstein reports that, among Awassi sheep raised in the eastern Mediterranean, the ratio of males to females is almost always equal, even though male lambs die more often than female ones do, and male sheep are often

males from mating with the females, the females often find a way to mate with males on the periphery of their territories because when a female is in estrous, often called being in heat, her body is driving her to really want to have sex, and estrous lasts for 24 to 48 hours for most of these species, although mares can remain in heat for more than a week. Biologically speaking, their bodies are driving them to have lots of sex with any male during this time period. Often, the result of this sexual frenzy is pregnancy, but getting pregnant is not always as easy as it sounds for some species, so one other problem with the cause-effect logic becomes the irrefutable fact that not every female gets pregnant after heterosexual intercourse.

Note another important logical point here: the animals are driven by Mother Nature to want to have sex in order to propagate their species; the animals themselves are not thinking, as they seek out sex, that they have to make babies; they are merely following up on their natural physical urges. Therefore, it is important to remember that sex only feels good for most mammal females when they are in estrous, so they will not allow males to copulate with them otherwise. Horns and hooves make great rape deterrents for most herd animals.

We must recognize, then, that humans would have viewed the seasonal flurry of sexual activity, for some species in the spring, for many species in the autumn, as a sign of that particular season, rather than as a drive to procreate. And they would not have been able to tell in such limited animal herds which genetic traits were being passed from which sheep, goat, cow, or horse, since most of the animals in the herd, group, or flock would have looked very similar to each other, so the argument that we can tell who a father is by the physical features of the offspring is also questionable.

Some ancient societies that had an inkling that men somehow contributed to offspring often believed (some still believe) that all males of the community contributed to procreation spiritually, which would explain any unique physical similarities between one man in the community and any children born into the

---

slaughtered before they reach a year old for food. Such a practice is actually more genetically sound than limiting the number of males who can procreate, so keeping male-to-female ratios higher would result in fewer birth defects.

community. Some scholars attempt to argue that neolithic groups in western Eurasia already knew males contributed to procreation because DNA studies have shown that, for some of these groups, "females moved from the birthplaces during adolescence or later," as far as a couple hundred miles, anyway, less than most trade route distances, with some scholars noting that "most males' descendants shared their ancestry with a single female," which they believe indicates not only patrilineal practices, but also monogamous marriage (Lalueza-Fox). These beliefs could be accurate, but there are several other possible explanations. For instance, we know now that female chimpanzees often leave their troops to join other troops (Dunphy-Lell). No one runs them out, and most leave just prior to starting menstruation. This kind of female-departure could have been common among ancient humans as well, with many young women beginning their "adult liv[es] knowing no one at all" (Dunphy-Lell).

The findings could also indicate that women were sought, and valued, for their perceived spiritual strengths, leadership abilities, or technical skills. The fact that children were related by DNA to a woman outside the larger social group could also reveal a cyclical form of exchange where their women, who were dying in childbirth or from diseases, needed to be replaced, so the groups either wooed, negotiated, or kidnapped the women they needed from other towns to replace the ones they themselves lost to death or kidnapping. While DNA tracing is helpful, there are many variables involved that must be considered.

Factor in the small populations of most ancient communities, and we have to admit that believing any individual man was a biological father of a child, when nearly everyone in that small community looked similar, would have been hard to accept. Even modern human males cannot recognize their own children just by features alone, and many children, we have learned with the recent popularity of DNA testing, are not the biological children of the men they long thought were their fathers, despite similar behaviors and appearances.[40]

---

[40] Think about this problem carefully. For instance, I have a friend who is a serial monogamist, who has been married three times now, but all of the men she has married have looked very similar to one another, so much so, I can easily identify

We also have to admit that, prior to the discovery that men play a biological role in pregnancy, incest was very common. Even at Newgrange, which dates to the early Bronze Age, one of the interred individuals was a male who was a product of an incestuous relationship, since "a quarter of his genome had no genetic variation" (Lalueza-Fox). Had these ancient humans been observant enough to determine heterosexual intercourse created babies, they would have also been able to recognize that such incestuous relationships increase the chances of miscarriage and birth defects.

In order for human beings to verify, beyond a shadow of doubt, that males actually have a biological role in procreation, they would have needed to develop a scientific method[41] for sorting out all the potential factors. The only way that scientific method could work was to domesticate a species 1) that is monogamous and 2) that engages in sexual intercourse which is followed almost immediately by the presentation of offspring. I use the word "presentation" here on purpose, since "birth" does not have to be an immediate process, especially when the species lays eggs.

The animal must be monogamous in order to rule out other sexual partners, female or male, so that the experimenters could witness first hand the heterosexual intercourse that happened just between those two paired animals.

The presentation of offspring in a rapid way is vital because the intervening activity the animal experiences between coitus and birth of offspring could have been viewed ascontributing factors to the pregnancy, as well. Our ancestors believed in magic, and many tales confirm that some magic elements played a role in pregnancy, including how the baby formed, such as tales of rabbits causing women to give birth to children with harelips.

While chickens might have been domesticated in China circa 10,000 BCE, possibly even before most other species except,

---

which men she finds attractive. So, even if women are having lots of sex with different men, chances are great that most of those men bear significant physical similarities.

[41] While many will scoff at the idea that ancient humans used any form of scientific experimentation, such scoffers have failed to realize that most of what we know about the world has long been from such experiments.

perhaps, dogs, they were most likely bred for fighting, with some speculating that such fighting was a sacred ritual, although it could have been for entertainment and for gambling[42] purposes (Lawler 12). The chickens themselves could have served as convenient danger alarm systems; however, few believe they were originally used for food (Lawler 45), although it is hard to believe that foraging humans did not gravitate toward birds' nests for a tasty and protein-filled addition to their diets from eggs. While chickens have a short gestation period, 21 days, they are not monogamous, and hens will mate with as many roosters as possible, and, since chickens will continue to lay eggs all year round (many of which remain unfertilized, so never hatch, creating another disconnect between coitus and procreation), they will mate hundreds of times a year, so humans would not have made the connection between heterosexual intercourse and procreation from watching chickens, either. There was simply too much coitus, including homosexual intercourse, which is natural among animals, including poultry, and too much egg laying for humans to determine that one action caused another. After all, if heterosexual intercourse causes pregnancy with males simply laying "seeds" inside another person, then why does homosexual intercourse not cause pregnancy? Remember, the ancients knew nothing about where eggs come from[43] and had no idea that semen contained sperm.[44] These individual aspects of procreation would not be revealed until well After the Common Era (ACE) and the invention of microscopes.

---

[42] We think of gambling as an addictive habit, but our ancestors would have seen "luck" as both magical and spiritual, with individuals who experience the best luck being "blessed" by one deity or another, just like the Prosperity Gospel believed by so many Christians today.

[43] In fact, when the ancients embraced skulls of horned herd animals as symbols, they probably did so because the skulls with horns resemble all mammalian female reproductive organs except the ovaries. So, even though butchering herded mammals would have demonstrated that the female reproductive organs in those animals were similar, nearly identical, in structure to human female reproductive organs, which they would have seen less frequently, but possibly during war, the ovaries remained mysterious, since they are only tenuously attached to the rest of the reproductive organs.

[44] Later, the Akkadians come to believe that males plant a "seed" inside women, who begin to be seen as merely fertile fields to be plowed and seeded, so that "the male semen is received by the womb, which gives birth" (Lieck 28).

The only monogamous animal ever domesticated by human beings is the swan. Most swans choose a mate for life, which is why they are often used as a symbol of monogamous relationships. Usually, the pen (female) will lay eggs in late spring, with an average of seven eggs per pen. The pen must copulate with the cob (male) in order to fertilize each egg she lays each day, and the cob and pen will work together to incubate the eggs, which hatch between 35 and 41 days after being laid. Humans working with the domesticated swans would have been able to watch the couple copulate, then lay and incubate the eggs in a short enough time that they would have, finally, recognized that heterosexual intercourse must happen in order to create offspring.

It is also distinctly possible that this observation of birds who create pair bonds in order to produce offspring and raise them to be old enough to take care of themselves was what convinced human beings to create "marriages" where two people paired off (becoming bond-mates) in order to care for offspring, so that we could owe our modern human concepts of monogamous marriage[45] to swans, as well. Therefore, while humans had once thought women magical because we seemed to be able to produce offspring all on our own, humanity learned males played a physical or biological role in procreation circa 2400 BCE, probably in ancient Egypt, from observing swans under these controlled conditions.

That date, 2400 BCE, is the earliest we can identify the domestication of swans. We know from text written in Egyptian Vizier Ptah-hotep's tomb that he owned 1225 swans in Saqqara, Egypt circa 2400 BCE[46], so we know that humans finally had

---

[45] Another plausible reason why powerful men like Sargon transformed the sacred *qursu* ritual into marital vows for the common people to emulate is because creating a social institution based on religious rites, aka marriage, would not only limit people's sex lives, restricting some spontaneous moments of joy and different avenues for spiritual transformation, but also emphasize the importance of the *qursu* ritual itself—stressing the fact that only one man (and only a male) can become a divinely sanctioned king. I would not be surprised to learn that the idea that "a house is a man's castle" idea originated at about the same time.

[46] While some waterfowl, especially geese, ducks, and swans, were easily domesticated, studies have been conducted to determine if birds, like ibises and birds of prey, were farmed (aka domesticated) in order to have enough birds for

domesticated the perfect biological observation species by that time. How long between the domestication of that species, the analysis of their findings, and the spread of the information to other cultures still remains unverified, but we now have a benchmark time to work with to determine when humanity finally discovered that pregnancy for humans is a dual act, thus called procreation, not a solo one.

Realize, however, that ancient human beings had no idea what was happening inside the female animal's body, exactly. They would have found fetuses inside both wild and domestic female mammals they butchered for consumption, but how did they get there?

Since all mammals have similarly shaped reproductive organs, ancient humans would have associated the vagina, uterus, and fallopian tubes with Female Creative Magic, which is one reason why skulls of cattle and goats should be seen as symbols of such magic, since their shape is similar to that of these creative organs (Cameron, Kvilhaug). A huge problem, however, is that, even if they had discovered mammalian ovaries, which are not clearly attached to the fallopian tubes, they would not have been able to see the individual eggs except as follicles about to burst from the ovaries when a female was near ovulation. I will discuss these symbols in more depth in later chapters.

While they discovered caches of fish eggs inside many fish they caught, how were they to witness what happened in the depths of streams when fish spawn? While bird and reptile eggs are easy to see, for most of those species only the female makes the nest and sticks around to raise the young.[47] Realize that ancient humans witnessing these ancient maternal acts by birds and reptiles would mean they would use these animals as symbols of Female Creative Magic.

Ergo, it is not a leap of logic to assume that most ancient cultures viewed pregnancy and childbirth as magical or spiritual

------

mummification, which became a popular trend in later Egypt. According to Linglin et al, ibises and birds of prey were all still hunted from the wild, so were not domesticated.

[47] One important exception are quetzal birds, which I will discuss in the chapter on Aztecs.

events, with most believing that women were solely responsible for creating these mini-humans, which would have created clear associations of Female Creative Magic with godlike powers. In fact, the ancient Aztecs believed women bled monthly as part of their ongoing spiritual battle to pull a human soul from another realm into this one. They even honored women who died in childbirth as warriors.

When we understand how ancient people's viewed the world through a magical or spiritual lens, associating Female Creative Magic with various divinities or spirits found in nature, such as the moon, we can understand how some cultures would have undoubtedly believed pregnancy occurred with a little divine assistance.[48]

Just as birth is a transition from one state to another, heterosexual intercourse was also viewed as a liminal (crossing boundaries) act. Sex was sacred.

Many cultures demonstrate repeatedly that sexual acts—from fellatio and cunnilingus to coitus and anal sex—represented moments of liminal connections to other worlds, other spiritual realms.

Take, for instance, the Moche sex pots. Since the Spanish conquistadors first stumbled upon these mostly stirrup pots used for pouring liquids with their vivid acts of sexual congress depicted as part of the ceramic vessels' shape, archeologists have struggled with whether or not the pots were simple pornography or symbolized something more important (atlasobscura).

As quoted by the archeologists in the blog post, "The "Moche' Erotic Pottery: Ancient Pornography or Something Else?" speculation has ranged from the pots as being vessels meant to titillate, to educate, or to being symbols of hierarchical oppression.

The quoted archeologists seem to assume that the Moche knew that coitus caused pregnancy because some even speculate that the pots are ancient teaching tools to demonstrate how to prevent pregnancy, despite the fact that so few of the vessels actually show heterosexual intercourse (The Archeologist).

---

[48] See the chapter on the origins of divine birth.

However, it is highly unlikely that the Moche, who might have domesticated llamas or other camelids, but few other animals, would have determined that coitus causes pregnancy, in and of itself, by this time (circa 100 CE to 800 CE). While some modern humans practice heterosexual anal sex as a form of "birth control," the ancients would have viewed pregnancy from the other direction—as a stoppage of menstruation, which is not necessarily a pregnancy indicator because menstrual cessation can be caused by a number of different factors. Since 1) they more than likely did not associate heterosexual coitus with pregnancy, and 2) thought of sex as a social, bonding activity, rather than knew it could cause procreation, they would **not** have tried other forms of sex to **avoid** pregnancy,[49] but would have tried them simply for the pleasure of trying them.

Stephen E. Nash, in his column "Skeleton Sex Pots," believes that the pots depicting skeletons having sex are "about establishing and maintaining continuity between the living and the dead," believing, without much explanation, that the fact that at least one of the skeleton[50] figures has a penis, "that organ's presence on a skeleton alludes to the fertility, and therefore to the power, of the ancestors." Nash indicates that the sex served a social function for

---

[49] We also have to address a biased myth that many archaeologists still try to pass off as fact—the idea that women breast fed their children longer in order to prevent another pregnancy. This myth is dangerous and has been proven repeatedly to not be accurate for most women. Ancient women used many plants to return their menstrual cycles, which would have also ended any pregnancy. Perhaps there was more power to having regular monthly blood flow than there was to being pregnant? While many male scholars struggle to explain how the ancients kept their populations so low, few take into consideration that it is quite natural for one out of every four conceptions to end in spontaneous abortion (aka miscarriage), that physical stressors, such as malnutrition, can cause women to miscarry, and that all that coitus probably caused a number of sexually transmitted diseases, which would have also limited women's ability to conceive.

[50] While Starr Goode notes that ancients would have understood the symbolism of "the tomb of the winter death goddess being regenerated to become the womb that brings new life in the spring" (51), she struggles a bit with understanding *sheila na gigs* who look like "a monstrous hag with a skull of death," overlooking the regenerative ability of these figures, who graphically symbolize being "reborn" in their positions over church doors. One has to pass through a form of death to be reborn, after all.

both the living and the dead (presumably in the Other Realm) and between the living and the dead, but he does not explain why the dead ancestors would be interested in human fertility. While he is leaning in an interesting direction, he does not pursue the idea far enough, since many tribal cultures routinely appealed to their ancestors for assistance in conceiving children, with some actually believing that children were reincarnated ancestors, thus naming their children after the ancestor she/he/xi most likely resembled.

While Nash is close to a real breakthrough with his theory, the Moche sex pots, which were mostly found in "spiritual temples and royal tombs" (Babe), were likely used to hold psychedelic or psychotropic liquids and have been depicted being used to rinse sinus cavities, which would be the fastest way to imbibe without swallowing (which could induce vomiting) these drugs. The sex pots' connections with the spiritual realms associated with (and through) both temples and burial grounds means that they most likely symbolize the sexual acts used in conjunction with the drugs to put their participants more deeply into a liminal, more spiritual, more religious space in order to enhance their experience in "another realm." As Babe notes, "the sex [depicted on the vessels] appears pleasurable and unabashed."

Like the ancient Sumerians, the Moche viewed sex as not only a great pleasure to participate in, but also a liminal activity which helped its participants access other worlds beyond the physical one we all inhabit. Sex was both sacred and secular.

Because the ancients did not have access to microscopes or pregnancy tests, it took human beings a long time to determine that the male of the species contributes biologically to pregnancy. Even after domesticating swans[51] and making that connection between heterosexual intercourse and pregnancy, for a time even **after** they learned males play a biological role in procreation, humans believed several different theories about semen:

1. that semen coagulated menstrual fluids (for humans, menstrual blood) in order to produce a "seed" (something

---

[51] Most likely, even the most observant Aztecs would not have surmised that coitus was necessary for creating offspring until after quetzal birds, which are monogamous like swans, were caught and brought into the zoo kept at the leader's palace.

they could see outside the body that created copies of the parent plant, which they knew needed water to sprout; for swans, that "seed" was the egg) that would grow to become offspring; This theory, which combined menstrual fluids and semen, would explain for them why homosexual intercourse did not produce babies because menstrual blood was not present;

2. that semen, which some believed contained invisible seeds or was like water that activates seeds, activated during the heat of intercourse to release a mini-version of the animal (homunculi[52] for humans) or an egg for birds that still must be heated by the parents' bodies in order to eventually create a baby of that species, so that the females contribute nothing to birth, but began to be viewed as mere carrying vessels or fertile fields to be plowed and planted; this androcentric belief is still embraced in parts of the Western world today in order to argue male superiority;

3. that semen was milk meant to feed the magically forming babies within the womb, which is possibly a hold over theory prior to understanding the physics of procreation; this belief can still be heard in parts of the world today, explaining why couples needed to have coitus every day

---

[52] As Bryan Sykes, DNA researcher, points out in his book, *The Seven Daughters of Eve*,

> "If children are born with their father's design, how was it that men had daughters? Aristotle was challenged on this point during his lifetime, and his answer was that all babies would be the same as their fathers in every respect, including being male, unless they were somehow 'interfered with' in the womb. This 'interference' could be relatively minor, leading to such trivial variations as a child having red hair instead of black like his father; or it could be more substantial—leading to major ones such as being deformed or female. This attitude has had serious consequences for many women throughout history who have found themselves discarded and replaced because they failed to produce sons. This ancient theory developed into the notion of the *homunculus*, a tiny, preformed being that was inoculated into the woman during sexual intercourse. Even as late as the beginning of the eighteenth century the pioneer of microscopy, Anthony van Leewenhoek, imagined he could see tiny homunculi curled up in the heads of sperm" (23).

to feed the growing fetus. Breaking fertilized eggs would have allowed humans to see the yolk or vitellus, which is what actually feeds the growing embryo, but probably associated the egg white or albumen with the semen the male inserts into the female through intercourse. How exactly early humanity explained what happens inside birds' eggs is still a mystery.

However, what we do know is that several cultures, including Hindus, one of the world's oldest, still existing religions, adopted the egg as their favored symbol for the universe, so that each egg is a microcosm in and of itself. Egyptians also came to believe the egg symbolized primeval life from which a lotus was born, which in turn gave birth to the sun god Ra, which is why Western Easter celebrations include eggs and lilies as part of those Christian rituals. For some cultures, then, the discovery of how swan eggs are created through heterosexual intercourse revolutionized their thinking of how the cosmos formed. The Greeks and Romans even began having some goddesses and gods born from eggs, such as Helen, who causes so much trouble for the city of Troy, who was reportedly fathered by Zeus in the form of a swan[53] when he raped her mother Leda, causing Helen and her twin brothers to be born from eggs.

Remember, sperm, those single celled organisms that live in semen that are required to fertilize the eggs carried inside females, were not **seen** by human beings until after the compound microscope was invented in the 17th century ACE.[54] While it was long speculated that human women stored eggs in our bodies, ovulation—the act of an egg leaving an ovary before it moves through one of the fallopian tubes into the uterus—was not photographed until late in the 20th century ACE.

Prior to this monumental shift in understanding about how procreation occurs after heterosexual intercourse sometime around 2400 BCE, both Egyptians and Sumerians associated the creation of life with the Mound of Creation. This mound was not a specific

---

[53] The symbolism of Zeus, the least monogamous god in history, as a swan has to be hyperbole on the part of the original storyteller.

[54] As Bryan Sykes notes, "Observance of fertilization was not visible until the 19th century when new chemical dyes were invented and when microscope lenses were ground" (24).

hill they could point to, although many tried,[55] but was actually a symbol of the human female pudenda, which we still euphemistically call the Mons Venus or the Mound of Venus. The pyramids[56] of Egypt, which started as mastabas meant to symbolize the Mound of Creation in order to usher the dead into new life in the Hidden Place,[57] as well as the ziggurats of Sumeria symbolically represented this Mound of Creation. The ziggurat, in particular, is structured just like women's reproductive organs, with a ramp (aka vagina) leading up to the Holy of Holies (aka the uterus), where life itself is created and where the sacred *me*, the knowledge and powers of the gods, are passed from the gods to humanity through the ancient *qursu* ritual, which was a sexual ritual that had nothing to do with "marriage,"[58] originally, and everything to do with magic.[59]

---

[55] The Egyptian city of Khmunu, meaning "The City of Eight," was believed to be the original site of the Mound of Creation, also called the Isle of Flame. The Greeks renamed Khmunu Hermopolis because they associated it with their god Hermes, but the original primeval mound of creation rose from the waters of the Nile after the original eight gods known later as the Ogdoads—four heterosexual couples who represented the unknowable (formlessness or the eternal divine, what the Lakota called the Great Mystery), darkness, unseen forces of air, and primal waters—interacted, possibly with a chemical or explosive reaction (some descriptions sound volcanic). See Ancient Egypt Online for descriptions of the four central creation myths of Egypt. Various Egyptian creation myths include references to a cosmic egg, as well (Pinch 60). My favorite myth involves the goddess Mehet-Weret who, like Hathor, "was usually shown as a cow and was considered the mother of all the primeval beings" (Pinch 59). Their horns were symbols of Female Spiritual Power.

[56] Later, in the Middle and New Kingdoms in Egypt, female paddle dolls will be created and left in women's tombs. Notably, many of the doll's pudenda appear to be upside down pyramids marked in much the same way the Akkadian cuneiform for woman is marked with a slit at the pointed "top" of the inverted pyramid. See Megan Clark's lecture at The Egypt Center, "Paddle Dolls in Ancient Egypt: Gaudy or Godly?"

[57] The Afterlife was believed to be an invisible realm into which the dead step, starting a new journey. Usually associated with the Duat, which is the "invisible night sky" Ra travels through at night in order to be reborn each morning (Guilhou 3).

[58] Scholars mislabeled the *qursu* ritual as a form of *heiros gamos*, which is a Greek idea that formed several millennia later.

[59] William H. Steibing, Jr. admits, that the *qursu* ritual was probably "pre-dynastic" involving a citizen representative and the city's god or goddess (52) as an annual spring ritual, probably to invite spring rains.

This magical structure, wherein worshippers had to approach the sacred space of "the holy of holies" through a narrow hallway, is repeated in almost every Mesopotamian and Canaanite culture's temples, so the symbolism of Female Creative Magic was widespread. There are even churches in Europe where *sheila na gigs*[60] symbolic vaginal lips are spread open above doorways, so that worshippers enter the womb of the mother church, which is safe and secure, and precludes spiritual rebirth.

And, of course, the basic triangle shape[61] of the Egyptian pyramids is also a common symbol for females in many cultures because it mimics the shape of a woman's pudenda. The Akkadian symbol for woman was a triangle pointing down, with a small line indicating the vulva. The triangle, called delta, is also the symbol for change or transformation in mathematics and chemistry, and that idea is probably an ancient one, since women's pudenda were believed to have transformative powers.

So, if my theory is correct, and pyramids of all types symbolically represent Mother Earth's pudenda that can connect humanity with other realms beyond earth itself, why did the Aztecs spill human blood down the face of some of their pyramids? I explore that fascinating answer in another chapter.

Meanwhile, penis imagery abounds in most ancient cultures, encouraging androcentric scholars to assume an erect penis was viewed as a weapon or as a symbol of male power or fertility. They are only partially correct, since both men and women would have viewed penises as an important part of sex, not only in heterosexual

---

[60] Starr Goode had produced a marvelous collection of images and scholarship revolving around Sheila na gigs, but she misses one obvious linguistic connection between the word used for these figures and the term's literal meaning. Her book dedicates several pages (9-14) to discuss the possible interpretation of the name, but Goode misses the fact that the Australian term, "sheila," which means female, derives from the Irish immigrants to that continent. She also misses the nautical term from Gaelic, "ghig," which is a small boat used to assist large craft, or a tender in current nautical terms. Like the Sumerians, the ancient Gaelic people seemed to believe that one had to ride the Boat of Heaven, a woman's vaginal lips, to be reborn.

[61] See Elinor W. Gadon's examination of symbols indicating female pudenda or vulvas on pp. 14-20 of her book, *The Once & Future Goddess*. While Gadon understands this symbolism, I was startled to see that she never connected the triangle with the basic shape of the pyramids that proliferate around the world.

and homosexual sex, but also during solo sex or masturbation. However, we must remember that they did not yet associate heterosexual coitus with pregnancy. There are enough penis shaped and human-sized carvings discovered around the world to support the idea that they were used as dildos, but, like the common lingam and yoni symbols found in southeastern Eurasia, these monuments symbolize the liminal magic of sex, aka the "little death," which what allowed human beings to be like gods, so was a spiritual act. Penises are most decidedly not human fertility symbols. I explore Male Magic, specifically phallic magic, in more depth in another chapter.

Before humanity made the connection between heterosexual intercourse and procreation, they believed that sexual intercourse of any kind was magical. As A.A. Shokeir and M.I. Hussein put it, "sex is a basic human need, common to all people at all times" (Abstract). Sex was a social activity that created bonds between people,[62] but it was, on a higher plane, a liminal activity, which means it put the participants in a sacred, between worlds space, so sex in its many forms could be used not only to heal, but also to transform. In Sumerian literature, the sacred sex called *hashadu* was a healing and transformation magic performed by trained priestesses in Inanna's temple who transformed beasts into men (e.g. Enkidu), and the sacred sex of the *qursu* ritual transformed mere men into kings. I explore these concepts in more depth in the chapter on Inanna as well as the chapter on Egypt, since I compare Egypt's gods, Hathor and Min, to Inanna and Dumuzi.

Prior to the scientifically verifiable discovery that males have a physical, biological role to play in procreation during heterosexual intercourse, sex in all its manifestations conjured magic.

More importantly, prior to human beings verifying that heterosexual coitus causes pregnancy, something many cultures seemed reluctant to believe for millennia even after the discovery made in the eastern Mediterranean circa 2400 BCE, the term "father" was an honorary one—a form of respect to the males who bonded themselves to the women's households or became "husbands" (literally house bonded, men who have pledged to

---

[62] Bonobos, one of our closest mammalian relatives, still use sex in its many forms as a form of social bonding.

support a particular woman's house). Think about the number of males still called "father" who are not, supposedly, biological fathers at all today. That form of respectful title is what is usually meant by ancient people who wrote prior to the discovery of procreation. After the discovery of or verification of procreation requiring both male and female participation to create pregnancy, however, the terms "husband" and "father" both morph into different meanings in order to reflect men's increased spiritual and social status above women as androcentrism gave rise to patriarchies.

Through observation of monogamous swans' mating, which ruled out any problems with having multiple sex partners in the equation, and reduced observable gestational lengths because of how quickly these birds produced eggs, most of which successfully hatched, our ancestors moved from initial speculation and postulation that there must be a connection between sexual intercourse and pregnancy, through to verification that males play a biological role in the process of creating new beings. Because of a lack of writing that demonstrates any human experimentation, this discovery was proven, possibly for the first time anywhere, in Egypt circa 2400 BCE in what was probably one[63] of the first biology experiments in the world.

Once males realized they play a biological role in procreation, they ran with the idea, inflating their own importance in many cultures, both in regards to the process of child creation and in regards to spiritual power. They learned quickly that, in order to control the masses of people now filling their communities, they had to control the spiritual, magical contexts—the mythos—in order to convince people to do things they might not otherwise want to do. Confirmable knowledge of procreation led to a rise in androcentrism, so that, as time progressed, males found ways to make their roles in magic, in mystery, in religion more important, excluding women by relegating them to mere vessels carrying the

---

[63] Other biological experiments were probably accidental, such as learning how castration affects boys and adult men, not to mention cattle and horses, which were often castrated on purpose, something I explore further in the animal husbandry chapter. Some biology experiments were purposeful, however, such as using specifically shaped cradleboards to give babies a sloped forehead.

magical male's homunculi or to mere fields to be plowed[64] in order for men to plant their seeds, and declaring them "unclean" because of their menstrual cycles, cycles that had, once upon a time, clearly demonstrated women's superior spiritual connections to the moon, and to the cosmos overall. Most patriarchal cultures enacted laws, using writing as their weapon of choice, to forbid women to "marry" multiple men in order to better determine which man fathered which children, switching biological inheritance from being matrilineal to patrilineal. We see evidence of this switch in the Old Testament of the Christian bible, where women are convinced in the Book of Ruth to approve a new method of inheritance through the male line, instead of through the female line.

Not every culture followed the same trajectory, however, largely because many cultures were more reluctant to swap Female Creative Magic for more questionable Male Magical power, such as the Cretans and the Hittites, two cultures I explore more fully in Section Two.

It is even possible that a rebellion against the growing androcentrism that sought to submerge women and goddesses beneath all things male caused what is now called the Greek Dark Age. What becomes clear is that many cultures, time and again, rejected androcentrism and outright patriarchy for something more egalitarian, possibly even more gynocentric.

---

[64] This latter concept was probably just a step up from a much older view that women's vulvas were meant to be "hoed" or "plowed" by male penises, which were shaped very similarly to horn hoes and plows used in the fields.

Chapter Six
Mounds of Creation: Pyramids, Ziggurats, Boat Burials, and Stone Circles

One problem that arises when scholars do not do comparative studies of different ancient cultures is that they get tunnel vision, viewing discoveries, old and new, made in their specific fields only from those specific ancient cultural points of view, failing to recognize that they and their predecessor scholars have built their ideas upon a false narrative. This problem occurs, as an example, when archeologists studying Central or South American cultures do not compare those cultures with those of North America. However, the problem becomes especially troublesome when scholars interpret symbols from only an androcentric perspective.

Some scholars, like Judy Grahn, Gerda Lerner, Denise Carmody, Dorothy Cameron, Joseph Campbell, David Graeber, and David Wengrow, attempt to see patterns across the world to develop theories that are somewhat universal, but most are missing some important physical symbols in their interpretations, such as the importance of mountains, pyramids, ziggurats, and other "mounds of creation," as the Egyptians called their place of origin, not to mention stone circles and boat-shaped burials.

While many Egyptian scholars will admit that the early mastabas built to house the mortal remains of early leaders were imitations of the "mound of creation" from which Egyptians believed humanity emerged in the beginning, few carry that metaphor to its full extent, despite the fact that many scholars admit that the pyramids are considered vehicles that transmit the souls of pharaohs to the afterworld.

From what "mound" do human beings emerge at birth? The "mons venus," as it is still called, is the female pudendum—a body part that serves as a symbol in many forms for the power of Female Magic in most ancient cultures. The word pudendum itself is related to the Latin word *repudium,* which means to separate from (i.e. to divorce or to discharge), possibly even to give birth, but the root "pud," means "shame," demonstrating that even in ancient Rome the female pudenda, because of rising androcentrism, was becoming a shameful part of the body because the men (and women who believed them) were pushing a male-as-superior agenda, so that everything female was becoming negative, shameful, and even evil.

The Sumerian ziggurat was probably one of the most symbolic architectural monuments in awe of the female reproductive organs contained within the pudendum, since the stairs are like the vaginal canal which lead(s) to the holy of holies temple atop the structure, which symbolizes the uterus, where creation is formed, maintained, and kept in balance through spiritual connections with the gods.[65] The "holy of holies" originally stood as a symbol for the womb, the place of divine creation, the place of divine connection, through which women literally brought human souls forth into this world from whatever spiritual Realm they inhabited before. Note that the majority of temples in the Eastern Mediterranean followed the Sumerian example with interior "holy of holy" places wherein the gods they worshipped dwelled, viewing those sanctums as the earthly connection to Other Realms.

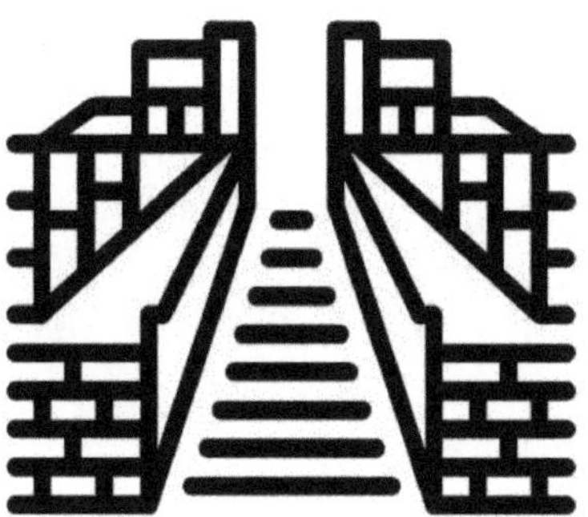

Basic Drawing of a Sumerian Ziggurat
Every journey up the stairs symbolized
the soul's spiritual journey back
through the womb to Another Realm.
Author

For the Egyptian and Chinese pyramids, indeed, for many mound building societies around the world, **the pudendum-**

---

[65] I will discuss both the concepts of virginity and of divine birth in other chapters.

burial mound is thus the vehicle through which a human being must travel to be birthed into the next world. This fact is often admitted to by archeologists, although few take their conclusions to their natural extent—that this vehicle imitates or symbolizes female reproductive organs. That this next world is often called the "nether" world and human genitalia is often called our "nether" regions should not escape you.

Note, also, that many mountains around the world are identified either as a primordial goddesses manifest in physical form, or as the connection between the Sky Realm (usually male) and Mother Earth, with the triangular or mound shape of the mountain associated with the Earth Mother's pudenda. For instance, the summit of Mauna Kea on the island of Hawai'i, is where the goddess Papa, or Mother Earth, joins with the god Wakea, or Father Sky. It was this joining of the two realms, imagined as divine beings, which allowed the inhabitants of Hawai'i to enter this world, according to their creation stories. Similarly, the Tibetans' name for Mount Everest is Chomolungma, which translates roughly into "Mother Goddess of the world."

Many people around the world are fascinated by pyramids and struggle to understand why so many cultures around the world constructed such monumental monuments, with most simply believing that these structures imitate mountains, which some clearly do, but what happens when the cultures, like the Egyptians, had no such local mountains to imitate? Why, then, would they seek to build such structures just to imitate mountains?

If we connect the symbolism of mountains as creative pudenda peaks with pyramidal shapes, and we examine them from the perspective of their being symbols of Female Creative Power, we can see how magical such power was to our ancient ancestors.

For instance, the ancient Sumerian cuneiform sign for women was "munus" (see image below), which is where we literally get the word "moon," a triangle with a line indicating the vaginal labia. The shape of the symbol and the word itself demonstrate women's close association with the moon, thus their spiritual connection to the cosmos. The Akkadians, more androcentric, were already changing the word's implication and called women "sinništu,"[66] which eventually gives us sinister. The same sign—a one-sided pyramid with a "holy of holies" in it—indicated women for both cultures.

Sumerian and Akkadian sign
"munus" and "sinistra," meaning woman

Some cultures, like the Aztecs, Mayans, and Toltecs, had a more complex relationship with this female-pyramidal structure, with the Aztecs taking what must have originally been a menstrual ritual—blood from sacrificial victims dripping down the face of the pyramid onto Coyolxāuhqui, their goddess of the moon (aka women's monthly connection to the cosmos through menstruation)—to a violent extreme in order to evoke fear in their neighboring cultures.[67] This idea especially makes sense when we explore the Aztec beliefs surrounding the creation of Huitzilopochtli, their god of war, when we reckon with the problem of having two temples atop the Templo Mayor in Tenochtitlan that should belong to Tlaloc, the rain god, and Coatlicue, the earth goddess. Instead, at some point the war god replaced the earth mother goddess, an act I explore more deeply in another chapter.

---

[66] Sin was the Akkadian god of the dark side of the moon.
[67] I will discuss the Aztec transformation of what once was a menstrual ritual probably meant to return the moon to fullness once a month into an androcentric blood ritual to feed the sun in another chapter.

Aztec Templo Mayor with Coyolxāuhqui Stone Indicated
Pixabay & Joshuart3d

Many scholars believe the Aztecs acquired their love of blood rituals from the Maya or Olmecs. Since the Olmecs were one of the earliest[68] (circa 1800-900 BCE) major cultures in Mesoamerica, and undoubtedly influenced both the Maya and the Aztecs, we should examine their use of pyramids, as well. Probably most famous for their giant carved stone heads,[69] the Olmec name is Nahuatl for "the rubber people," since the Olmec were known to have widely traded in rubber, with which the Mesoamerican ballgame balls were made. Many scholars credit the Olmecs with inventing, or at least perfecting, the ballgame.

Despite the fact that the majority of Olmec deities were clearly androgenous, scholars have long insisted that the Olmec heads are of male leaders, even though many of the images could easily be of female leaders or female ball players. Similarly, the hollow ceramic pudgy baby figurines, which possess no obvious

---

[68] Debates still rage about whether or not to credit Olmecs with being the "mother culture" of Mesoamerica, especially since more recent studies have revealed even earlier cultures.

[69] The Olmec heads could be of either or both genders, since the only reason most scholars assume they are of male figures is because most of them wear ball player head gear. Assuming that women never played the ball game is not only sexist, but also uninformed, since we know women played ball games, such as what became called lacrosse, in North America before European Contact.

genitalia, are also considered by most to be male with no real explanation or proof offered.

This assumption that males were always the leaders, always the main deities, which plagues much of archeology, anthropology, and history, rears its ugly head repeatedly in examining archeological remains of the Olmecs. Part of the problem is the fact that biological remains disintegrate quickly in the Tabasco coastal plains area, where many Olmec cities once existed. While La Venta is often considered the center of Olmec society, San Andrés, which is three miles northeast of La Venta, is often considered a town built primarily for elites of the Olmec culture. While the Olmecs built mounds, ostensibly pyramid shaped, in La Venta, scholars disagree about their purposes in doing so.

Sketch of Olmec Pyramid at La Venta, Tabasco, Mexico
Author

There is considerable scholarly interest in determining which of the many pyramids that dot the world is the oldest. Are the monumental mounds built in Liaoning, China, which are attributed to the Neolithic Hongshan culture (circa 4700-2900 BCE) truly older than the pyramid structures upon which the elite of the Norte

87

Chico culture lived (circa 3000-1600 BCE) in ostensibly the first city of the Americas, Caral, Peru?

Sketch of Hongshan Pyramid, China
Author

Most scholars who puzzle over why there are so many pyramid structures around the world try to equate all of them with local mountains, believing the ancients merely wanted to get closer to the sky where it is assumed their deities lived, despite the fact that most of these same cultures also had deities who roamed the earth and who dwelled in the lower realm within the earth, which most archeologists associate with death because so many buried their dead there.

There has to be more to ancient ideas of divinity other than a Christian-biased hope to reach the heavens because not every culture sees the skies or "heavens" as the only attractive place to go in the afterlife.

The point of view that insists the pyramids are nothing more than imitations of mountains misses several important points.

First, mountain tops, presumed to be the closest humans could get to deities in the heavens, are places of worship in many cultures that live near them, but almost all of those cultures believe humans either emerged from the caves (aka vaginal openings) upon those mountains, or were created from the clay or alluvial soils with some god's blood (or other fluids) mixed in.

Scholars examining the monolith, called a *huanca*,[70] in front of the Major Temple in Caral (see image below) believe the stone and its alignment with the pyramid's steps is an indication of an astrological alignment, which is a real possibility, although most of the research lacks specificity about which astronomical features the monument lines up with. One of the first archeologists to excavate extensively at Caral, Ruth Shady, believes the buildings were constructed to "imitate the hills of nature, which are divine works. The heavens were also taken into account, for the axes of the buildings are oriented in relation to certain stars, such as the sun, the moon, and the Pleiades" (86). What she does not explain is why hills are considered "divine works," although myths from the local region often have a goddess dismembered, so her parts can become the mountains, streams, sky, and oceans.

As far as astronomical orientation is concerned, most things on earth line up with something in the night sky, but most ancient people were concerned mostly with the most visible satellite, the moon, which signals time periods in the year we now call months as well as seasons via its position at moonrise and moonset. I have only read one archeological astronomical analysis that differentiates between the moon's position at the solstices and equinoxes from the sun's position, and that was a discussion of the lunar alignments of Newark, Ohio's Octagon and Great Circle (Withrow).

Are androcentric scholars jumping to the conclusion that all alignments that are both lunar and solar must be mainly solar alignments because of a prejudice toward the sun, which is often assumed to be masculine[71]?

When we view the Caral Templo Mayor from behind the monolith, however, the alignment is clearly that of a phallus ready to penetrate the stone pudendum before it (see image below).

---

[70] A huanca is a long, upright stone that was either worshipped or worshipped at, since they were often given ritual offerings.

[71] There are, of course, many exceptions to the concept that the sun is masculine. The Japanese and the Cherokee both believe the sun is female.

Sketch of Caral, Peru's Templo Mayor with Monolith (huanca)
Author

What is equally fascinating about Caral, compared to other cultures that created pyramids, is that it was pre-ceramic, despite their appreciation for controlling fires for ritual purposes, according to archeologist Ruth Shady (88). Having groups of people create large structures from stone before knowing how to create ceramics is not unique, however, since the people who constructed the many megalithic temples at Göbekli Tepe (c. 10,000-7500 BCE) and subsequently built hills over them were also a pre-ceramic culture.

However, the primary evidence gathered around the world that death was associated with a form of afterlife that people had to be born into comes from the use of red powders or paints, such as ochre or cinnabar, to decorate the dead. Like the blood that accompanies birth, human beings assumed blood was required to for rebirth into the afterlife, so they supplied it in the forms of powders or paints with red or orangish-yellow tinges. This burial tradition is so common in ancient times that, like pyramids, it can almost be considered ubiquitous.

Many people also assumed that, since we came from "the mound of creation," we had to be reborn into the next life in a similar mound. That idea still lies within our modern subconscious because most cemeteries are built on hills, and many are named

"Mount" something or other, like the Mount Zion Cemetery near where I live. Places like Mount Zion, even though mythical, have long been associated with creation and power, such as the Mount Zion referred to in the Book of Isaiah in the Christian bible where in God dwells and where King David was crowned king of Israel and Judeah. This myth clearly harkens back to the Sumerian *qursu* ritual wherein the act of "mounting" or having coitus with Inanna's representative in a "holy of holies" sanctuary on earth transforms a mere man into a king. Interestingly enough, "the Old Testament" personifies" the city of Jerusalem...as a woman [who] is addressed or spoken of as 'the daughter of Zion'" (Zion). Jeremiah 3:14 prophesizes that the Jews exiled in Babylon will be restored to Zion where Yahweh will be their "husband," and that it is there that Yahweh "will be the God of all the families of Israel, and they will be my people" (Jeremiah 31:1). Clearly, Judeo-Christian traditions continue the idea of the important connections between reaching or connecting with divinity at the peaks of mountains or pyramids, adding in hinted suggestive sexual contexts along the way.

Perhaps less ancient than "mounds of creation," however, are the Ships of Death, which serve a dual purpose. Boats not only transport people and goods, but also symbolize women's sex,[72] our vaginal lips or vulva, in shape. This shape is most evident when you look at traditional reed boats still made in Peru and Iraq. This vaginal lip symbolism is why boats have been associated with and given female names since the times of the Sumerians whose reed boats looked exactly like vulvas. Thus, a boat burial can be seen not only as a ship transporting the dead to the afterlife, but also as the birthing vessel or vulva through which they must pass in order to be reborn into the next world, which is how they are depicted in the Egyptian *Book of the Dead*. Pharaohs needed these symbols of Female Creative and Protective Magic, in addition to the pyramids in which they were buried, to travel the river in the realm of the dead, the *Duat,* into the Field of Reeds, wherein they took on divine ancestor qualities.[73]

---

[72] I will explore these ideas further in the chapter on ship burials.

[73] Note that, prior to the discovery of procreation, males probably only ever became identified as ancestors worthy of "worship" if they became important

In fact, 14 boats buried in the Abydos desert alongside the burial spots for several Proto- or First Dynasty leaders "'were placed in their graves many years before Khasekhemwy's funerary enclosure was built—so that they were intended for the afterlife of a much earlier dynastic ruler, perhaps even Aha, the first Dynasty 1 (ca. 2920-2770 [BCE]) ruler of Egypt'" (New York University). In fact, the discovery of these afterlife transports has demonstrated how much more advanced the Egyptians were with boat building than had been previously thought, since the hulls are made of non-framed cedar wood lashed together with rope (NYU). What would make this find even more significant would be to find that the male leader in question was also buried with his mother as further assurance of his rebirth into the next realm.

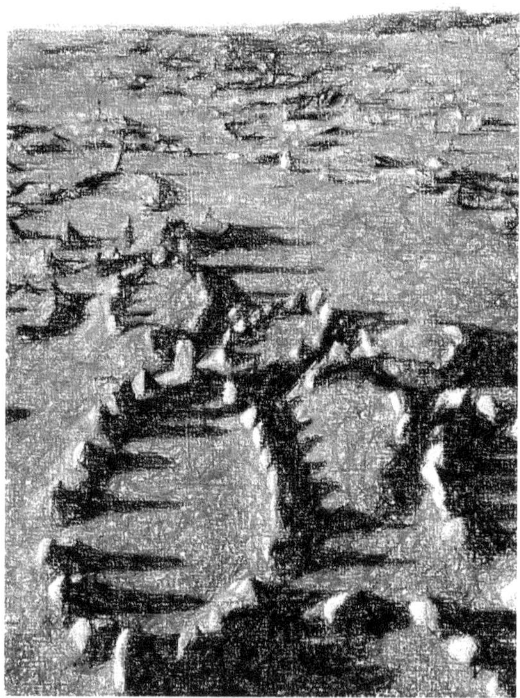

Sketch of Viking Boat Burials in Denmark
Author

community leaders; otherwise, the only blood ancestor everyone ever knew were their mothers.

Boats were especially associated with pharaonic burials in Egypt,[74] with the Bo people in China, with the Vikings in northern Europe, and are still used as the preferred coffin shape in the Philippines. Interestingly, while scholars like Euseblo Z. Dizon and Armand Salvador B. Mijares noted the similarities between the boat shaped burials among the historical (c. 1595-1850) Batanes living in the Philippines and the boat-shaped and actual boat-burials at sea practiced among the Vikings, they merely assume "the stone boat-shaped burial markers in Batanes reflect the material cultural remains of an ancient maritime people" (8), instead of connecting the coincidence in burial shape choices to anything other than the sea.

Yet many boat burials on land end up as mounds. One of the most famous Viking burial ships, Gokstad, was a very carefully crafted burial mound. The original mound was at least 45 meters high, but careful excavations found that a layer of blue clay silt was first put around the ship itself, up to its natural waterline (Biørnstad), probably to provide the keel-based boat with stability as it was laden with gifts for the deceased and to imitate the spiritual seas it was meant to sail upon. Above the blue clay was carefully laid turf, and above that thick layer of turf the soil that had been removed before the first layer went down was replaced. The effect was to create a nice hillock upon which people could pay their respects, if they knew there was a burial there, or could picnic. As Rebecca Cannell stated, "the mounds were meant to be seen up close and be a part of people's everyday lives," which was why they "were often placed by Viking Age roads" (Biørnstad). The nearness to roads can also be viewed as symbolic, since death was, for the ancients, just another road to travel, not an end in itself.

Also important to compare are stone circles, like Iniskim Umaapi in Canada and Stonehenge in England. While medicine wheels or sacred hoops, which stone circles in the Americas are often called, are often compared to stone circles like Stonehenge or Avesbury, scholars still struggle with understanding the symbolism of the use of stones in a circular pattern. Quoting Anthony Perks, "a doctor of obstetrics and gynecology at the University of British

---

[74] Khufu's ship, buried next to his pyramid at the Giza plateau, is a well known example.

Columbia," Starr Goode reports "that from an aerial view, 'Stonehenge's inner bluestone circle represents the labia minora and the giant outer sarsen stone circle is the labia majora. The altar stone is the clitoris and the open center is the birth canal'" (52).

It is important to note here that several "chiefly" mound burials in England, such as the Bush Barrow burial of a "stout" figure, were all **assumed** to be of high status males in the Victorian era. Unfortunately, the skeletal remains of most from most of these early excavations have disappeared, so cannot be studied further, which is a shame because the diamond shaped gold artifacts found with both the Bush Barrow interment and the Little Cressingham interment in Norfolk bear striking resemblances to each other. Each diamond shape is bordered by triangles, and those triangles are echoed in other ornaments found within the burials, which could very well indicate that the "chieftains" buried within the graves were female, something more recent scholars admit is possible, especially in light of new DNA evidence from other burials that has overturned several assumptions that such burials must be male.

Reluctantly, Stuart Needham, Andrew J. Lawson, and Ann Woodward admit that the Bush Barrow burial could be "a female who acquired a position normally attributed [by androcentric scholars] to males" (10), and they mention the burial's similarity to the Little Cressingham finds (20), but they, nonetheless, forge ahead with the assumption that both burials involved males.

Alan West, Curator of Archeology for Norwich Castle, notes that "these burials, and others like them have been interpreted as the graves of chieftains in a hierarchical hereditary society. However, there is an alternative...explanation. A continuation of hereditary power may have been prevented by the burial of so much personal wealth in the ground," concluding that "society may have been more egalitarian and kin-based, with individuals of families rising for a generation before disposing of their wealth and starting again."

Again, we have to overcome androcentric biases that assume all leaders were male, and that all people "honored" with extravagant burials that include gold artifacts or blades—useful tools no matter what gender the person is—had to be male.

Iniskim Umaapi is also an excellent example because it is as ancient as any of the stone circles in Europe, but it also has other symbolic elements that point to a gynocentric understanding of the universe. Quoting Gordon R. Freeman, who first dubbed Inishkim Umaapi "Canada's Stonehenge," Debbie Olsen reports that the stones of the medicine wheel were purposely placed, representing "the sun, the crescent moon, the morning star and constellations." This sacred hoop has 28 spokes, which indicate a connection with the days between full moons, but also connects to women's Female Creative Magic menstrual cycles. Like Inanna's rosette symbol, which has eight points, Inishkim Umaapi multiplies the power of the moon and Female Creative Magic by the four clearly marked quadrants in the wheel that represents in traditional Blackfoot culture, "the four cardinal directions, the four sacred medicines, and the four human needs"—emotional, physical, mental, and spiritual needs, according to Laura Sitting Eagle (Olsen). As with the Sumerian sacred sex act, *hashadu*, Inishkim Umaapi is seen as a liminal space wherein human beings can connect with the divine for healing.

While many wishful speculators believe the common place occurrence of pyramids, mounds, boat burials, and stone circles is because of an extensive interconnection between cultures, the simplest explanation is that **these practices are evidence of gynocentric views of life and death**, seen through the power of Female Creative Magic, with the necessity being that human beings need to connect with divine energies through these liminal spaces for healing or to re-enter the "mound of creation" (the human female pudenda) in order to pass on to the next world after death. Through these constructions, humans are "reborn" into the Next Realm.

We get the best visual example of this belief from the Māori of New Zealand. Similar to the beliefs of their cousin Polynesians in the Hawai'ian islands, the Māori believe in Rangi, the god of the skies, and Papa, or Mother Earth, whose joining created everything. However, Maui or Maaui, their hero god, attempts to stop Death, known as the goddess Hine-nui-te-poo, "by entering [her] through her vagina. In order to do this, he turns into a *moko huruhur* (a caterpillar or a hairy lizard). When he enters Hine, however, his bird companions laugh, thereby awakening Hine, who closes her legs and kills him" (Smith). According to legend, Maaui is unsuccessful at killing Hine, or death, because he is impure, having helped separate his parents from their incessant coitus, and who lassoed then beat the sun, Raa, who was moving too fast for there to be useful days and nights (Smith). However, his attempt to kill death by crawling through goddess' *mons venus* only to be killed when her vaginal lips clamp shut encapsulates the idea that humanity was born from Mother Earth (in some way), so must be reborn through reentering (or passing through) Mother Earth or Death in some way.

Even the basic kiva or ceremonial circle often used in many different cultures recreates the penetrative nature and inner secrets of female pudenda, demonstrating that the males who often controlled these spaces sought to assume the natural Female Creative Magic powers women possessed.

Both the basic structure visualized by the ancients who built the temples at Göbekli Tepe and the mounds which later covered each temple echo the importance of Female Creative Power. The structures symbolize a vaginal canal into the mystery of the interior temple, which was not just composed of spiral walls, but also held two upright monoliths carved to look like human beings, probably the two humans first envisioned by humanity as their original sacred cooperative couple, similar to Adam and Eve to Christians. The carved animals within the temples would then be the animals also created by Mother Earth, possibly with the help of Father Sky or Father Sea but just as likely alone, to accompany humans on the earthen plane of existence. These temples allowed the ancients to traverse the magical realm of the vagina and uterus as full grown adults to access the other sacred realms, possibly envisioned as a parallel realm wherein the divine resided, potentially as a sky realm, as a subterranean or as a water realm, or even a combination of those realms like the earth—a parallel universe.

While many scholars and interested commentators often dismiss the female symbols discovered in various mounds and temples as mere "fertility" signs, they miss the importance of the fact that ancient people would not have known that heterosexual intercourse was necessary for pregnancy to occur[75], so that female pudenda and the moon as full or crescent shapes are not just where human fertility came from, but, more importantly, where all human spiritual power came from[76]. These Female Creative Magic symbols were most likely how early humanity envisioned the source(s) of creation because the primary creators among them were women, not men[77] despite the last two and a half thousand years of androcentric propaganda.

Women's menstrual cycles not only imitated the moon's phases, but also allowed women to bleed for three to five days and not die—a very important magical element men attempted to imitate with circumcision, a clear form of sympathetic magic. The pain and blood would have appeared to indicate that women were "battling" with divine elements for many cultures, which explains why some viewed childbirth as a divine battle waged by women who are securing new humans from Another Realm, such as the Abrahamic religions' idea of the Guff, wherein children await birth. For some cultures, though, women also seemed to create other human beings through magical means all on their own. Ergo women were not only powerful because of their symbolic spiritual connections to a divine entity—the moon—but also because they themselves had enough magical power either to create other human beings on their own or to wrestle them to earth from a parallel universe.

For the ancients, circles, ovals, and triangles to be penetrated, pyramids, platforms, ziggurats, mastabas, and mounds to ascend, as well as boats that carry people and things from one

---

[75] Note that while the preceding two potential mythologies included acts of coitus in the descriptions, there are two possible ways to read that act—as a liminal activity that puts the participants in touch with the divine realm, and as incessant coitus because the "parental" gods are too hung up on their own pleasure, so ignore their children. The idea that such coitus "created" the other elements or divinities probably came long after humanity discovered the connections between coitus and procreation.

[76] I will discuss important differences between the idea of "fertility" and Female Sacred Power in another chapter.

[77] I will explore how women created most of the everyday necessities—from food to shelter, from agriculture to ceramics, from ceramics to metals—all while creating useful sciences and mathematics in later chapters.

place to another, and the ability to bleed for days without dying were all associated with the spiritual power of Female Creative Magic. All of these human creations—from giant stone structures to circumcision—were sympathetic magical attempts to tap into that gynocentric magic in order to assist a rebirth into the next realm, as well as to perform other acts of supernatural power, including healing, battling without dying, and possibly even through taking power over groups of people or of the resources of a whole region.

This book will discuss how the appropriation of Female Creative Magic that stems from a more gynocentric view of the world by androcentrists created the oppressive patriarchy that is now meeting its doom, having outlived its usefulness for the world of human beings.

Chapter Seven
What Was a Virgin, Really?
OR
The Origins of Divine Birth

Virginity, as a concept, is a conundrum on many levels, even today. So much so that in 2017, two Norwegian women, Nina Dolvik Brøchmann and Ellen Støkken Dahl, discussed, in their TEDXOslo Talk, "The Virginity Fraud," the myths surrounding the hymen as a means of checking the virginity of girls and young women, which is still practiced in many countries around the world, and the harm these myths do to women, such as in its most extreme resulting in "honor killings" because a woman's loss of virginity before legal marriage is supposed to bring shame on her family.

Where do such myths come from? Why is virginity, especially female virginity, still so important to so many people?

We have to emphasize that virginity, like gender, is a nebulous concept, but many people in patriarchal cultures value virginity in a woman above all other aspects of her character. As Gerda Lerner points out, "female [sexual] purity becomes a family asset, jealously guarded by the men in the family" (94).

Ask a woman if she is a virgin, though, and, depending on her sexual orientation, she might insist she is, only because she has never been penetrated by a male's phallus, even if she has had other objects—fingers or a dildo—put in her vagina for sexual stimulation. Is she still a virgin if someone performs cunnilingus on her?

Ask a man if he is a virgin, and he might actually be puzzled. What constitutes virginity for men? The concept is even more nebulous, since some men might insist that, since they never ejaculated inside a woman, having lost their erections before that point or preferring the "pull out method" of contraception, they are still virgins, even if they achieved vaginal penetration. Are they still virgins if they ejaculated before they even entered a woman's vagina? Are they still virgins if they have only participated in homosexual intercourse or fellatio?

The concept of "virginity" for ancient cultures must be more closely examined here because the Greek term, *parthenos*,[78] used most often to refer to cultural concepts regarding young females, has a primary definition of "young woman," any young woman. Yet the term often used when a female creates offspring without a male mate is <u>partheno</u>genesis, which literally means "virgin birth" to most people because here "partheno-" uses the Greek term's second definition, "sexually inexperienced," to imply that the creation of the young was without heterosexual intercourse. Because so much science relies on Greek or Latin for its terminology, that terminology is inevitably androcentrically biased.

A better Greek term for creation without sexual intercourse would be *kyoforía chorís synousía*, pregnancy without coitus, arguably a mouthful for anyone. Since parthenogenesis is the term most often used by scholars to mean a female who has not experienced penetrative sexual intercourse, but who has become pregnant anyway, we will first supply some clarity about how cultural prejudices have colored the ideas surrounding young women and their sexuality.

The confusion *parthenos* has caused among scholars and in scholarship regarding the ancients must be fully addressed because the ancient Greeks, the Mycenaeans, did not think like modern followers of Abrahamic religions about the concept of "virginity," so, for them *parthenos* would first indicate a young woman, not necessarily a woman who has not experienced sex (Breton Connelly 18). For Mycenaeans and possibly even the historical Greeks, *parthenos* referred to various goddesses, who are often oversimplified by modern scholars as "virgin" goddesses, were simply ones whose youth was frozen in time or whose youth was believed to be renewed annually, like Hera's was, by bathing annually in specific sacred springs. We must emphasize that such a concept of *parthenos* has nothing to do with sexuality or sexual experience. As Joan Breton Connelly points out, "after the onset of

---

[78] The word *parthenos* was most often applied to the goddess Athena and to her Parthenon, still standing in Athens, Greece today. Was she a literal virgin, the way we define virginity today? Or was she a perpetually youthful woman, as some scholars see her? There is growing evidence that the term simply indicated a young, nubile woman, rather than a sexual virgin.

menarche, the maiden assumes *parthenos* status [in which she is designated as being] in the prime of her youth and eligible for marriage" (27). For the ancients, the term *parthenos* meant the same thing nubile[79] meant 100 years ago. Unfortunately, nubile now means "sexually attractive" as its primary definition.

The fact that such definition distinctions do not occur to many scholars, including ones who consider themselves feminist scholars, further demonstrates how so many carry their programmed androcentric cultural biases into their scholarship. For instance, Marguerite Rigoglioso has two important books that address the concept of virginity, especially for the Greeks, but she never even defines what she means by virginity, let alone what the Greeks believed virginity was, but, instead, continues to carry forward the androcentric bias that *parthenos* indicates a sexual virgin, not a premenarchal female. Realize that, if sex was common and happened to children, too, lack of experience with sexuality would be rare in ancient times, so the modern idea that a virgin is someone without sexual experience[80] would not have applied in ancient times. Therefore, our ancestors differentiated women's ages and the responsibilities that came with each age, via their menstrual stages—premenstrual, menstrual, and postmenstrual.[81]

In fact, it only once occurs to Rigoglioso that the ancient people might not have connected heterosexual intercourse to conception when she discusses how Zeus swallowed Metis, Athena's mother, in order to prevent her from giving birth to a child that

---

[79] Even the word "nubile," which essentially means marriageable (eligible to get married) although it has taken on sexual overtones, is problematic, since its origin comes from the Latin term "nubere," which means to veil oneself before a bridegroom, a form of modesty women had to exhibit for a marriage ceremony and is why women often still wear bridal veils. Remember, hiding women from men's view is only considered "modest" in patriarchal cultures as an act of social oppression in order to keep them from actually competing on an even playing field with men. Only allowing husbands and blood relative males to see a woman unveiled is the ultimate form of male suppression of the feminine because allowing women to show themselves fully and in public threatens men, not women.

[80] Sexual inexperience is only important to patriarchies, which had few ways of determining paternity until recently.

[81] Women who menstruated irregularly or not at all could have been ostracized, just as some cis women now seek to ostracize trans women.

would overthrow him the same way he overthrew his father Cronus. Rigoglioso postulates "the possibility that the conception [of Athena] was understood to be unrelated to the sex act" between Metis and Zeus (Virgin 35). In other words, she argues that Metis created Athena on her own, parthenogenetically, while inside Zeus' stomach, but Rigoglioso does not press the idea further to reflect on the possibility that the people she discusses in either book did not associate coitus with pregnancy at all, although she comes close to that realization or admission when she discusses the Auseans of Libya and the Akan of Ghana.

When describing a ritual that included mock combat for young women among the Auseans, she states that "it is clear that this tribe was not interested in virginity in a moralistic sense as a measure of a girl's purity and 'value' in the brokering of marriage" because the tribe did not practice monogamous marriage. Instead, she believes the ritual was a way "of divining which maidens among them were fit to become priestesses of parthenogenetic conception in the manner of their parthenogenic goddess, Neith/Athena" (Rigoglioso, Cult 57, Virgin 40). Regrettably, her otherwise strong scholarship regarding the concept of "virgin birth" or parthenogenesis succumbs to her programmed androcentrism, even though she actively works against such programming, so that she does not see that the culture would never have valued a woman "morally" regarding sexual experience, but that their concept of "virginity" had likely been misinterpreted by the anthropologist whom Rigoglioso cites, Oric Bates, because of his own androcentric view of their traditions. It is this kind of unconscious androcentric bias that scholars carry with them, completely unaware that it is there.

Rigoglioso continually chooses to use the secondary definition of *parthenos*, such as when she discusses how Hera "annually bathed to restore her virginity in the Canathus spring at Nauplia" (Cult 70, Virgin 75). Had she used *parthenos'* primary definition, she would have recognized that Hera was restoring her youthful beauty, not her hymen or any other form of "virginity." But the example would also have failed to support her overall arguments about the importance of parthenogenesis to the ancient cultures she examines.

Lamentably, while Rigoglioso gives a great definition for "a classical matriarchal society in which childbearing was communal, monogamous marriage was not practiced, lineage was reckoned through the mother's line, and paternity was of little consequence" (Virgin 39), it never seems to occur to her that there was no monogamous marriage and no thoughts about paternity because the people in question had not discovered procreation, which takes the contribution of both genders to create offspring. Despite what birds teach us each spring about the benefits of having paired bonding to raise offspring, monogamy did not become a common practice for most human cultures until the Bronze Age, even though most scholars studying ancient cultures assume it was always practiced.[82] While it is most likely that the discovery of procreation circa 2400 BCE in Egypt through the domestication of the only monogamous animal ever domesticated, swans, was a precursor to the idea of legal monogamous marriage, many cultures did not adopt legal marital contracts until they were either influenced to do so because of contact with other cultures or discovered procreation themselves, and not always then.

Before humanity learned that males play a biological role in procreation in the early Bronze Age, there was little reason to bind a man to a specific House (built and run by women through which everyone traced their ancestral lineage) other than the one he was born into, so the creation of the role of "husband," literally meaning house-bonded or a man tied by marital or other ritual bonds to a specific House lineage, came later. This fact is brought home by the Mosuo of China, often considered the last real matriarchy in the world, because the Mosuo women do not marry. Instead, they practice "walking marriage," which means that the men only come to stay at night, but the men know they have no say in how the family is organized or run because they are not a "couple" in the

---

[82] Adovasio, Soffer & Page assumed that even Homo Ergaster (2 million years ago) practiced "at least temporary monogamy" (47) despite the fact that a small nuclear family would have found it much more difficult to survive on their own. Much more likely were small groups of humanoids living near each other, mating freely to develop social bonds, and helping each other whenever possible. Since humans, like elephants and bison, usually only produce one offspring per year, being herd animals makes more sense for survival than nuclear family units.

same sense in which most patriarchal cultures view marriage (The Mosuo). Instead, the matriarch's sisters and brothers help raise the children, treating their own children the same way they treat their siblings' children—as their own. While each woman can choose whom to allow into her bed, most Mosuo women stay sexually loyal to their chosen mates, practicing a form of unspoken, unpledged monogamy. The father of the matriarch's children stays with his own mother's family, helping his own siblings raise their children. The only taboo the Mosuo have when inviting unrelated people into their homes is that they avoid talking about sex when people of both genders are in the room (The Mosuo).

As we will see in Section 2, prepatriarchal cultures used a combination of household organizations and ancestral lineages to designate familial and cultural associations that were usually bound to specific households and ancestors. For instance, when Hatshepsut begins using the term pharaoh to describe her role—a term that will become commonplace and widely used by scholars millennia later to designate the rulers of Egypt, she was proving her right to rule through her female ancestral House lineage. After her rule, the backlash against women leaders[83] in Egypt pushes that culture more closely into becoming a patriarchy. Until that time, Egyptians were largely a previously gynocentric culture leaning toward androcentrism.

We now know that, for most ancient people, ancestry was traced, for practical reasons, through the women of each Household, something I will discuss more in depth when I discuss the transition from ancestor worship to Hero Worship for the Greeks in Section 2. Remember, too, that women were the original creators of their families' living quarters—whether the structures were tipis, longhouses, mud brick, or wattle and daub, which was still used in Elizabethan England, although house building, by then called construction and architecture, had been mostly appropriated by men by that time. Women were, literally, just to emphasize the

---

[83] Female rulers were much more common than most historians are willing to admit, but more on that later.

point, the homemakers, and often still are in various parts of the world.[84]

We know that human society changed over millennia in incremental ways, but we now know that one of the most profound shifts in society came circa 2400 BCE, <u>after the discovery of procreation</u>. Men had long leaned toward enacting on their violent tendencies like elk in rut, but the verifiable proof of procreation allowed them to begin spinning a new human story, one that lauded their might-makes-right views that allowed them to dominate not only the people in their own communities, but also, with the help of Bronze Age technologies, other cultures through warfare. Many scholars, like Gerda Lerner, claim that the brutality of hierarchy did not occur until women became chattel, property that could be bought and sold, believing that "hierarchy among men rested upon property relations and was reinforced by military might" (96). Her idea is intriguing but challenging to prove.

However, one element that could have started androcentrism's rise prior to the discovery of procreation could have been the concept of "brides by capture,"[85] which proliferated by the time we are into the Late Bronze Age, or the arrival of the Mycenaeans into Greece. Cultures like the cattle raiders Maria Gimbutas claims infiltrated Old Europe could have begun seeking women from other cultures because their own were dying in childbirth or were dying from diseases brought into human cultures from the herding of cattle, such as smallpox. Naturally, once a culture is mostly composed of males, birth rates drop, so these cultures would have needed more women to increase their populations in order to maintain their hold on the Eurasian

---

[84] If you refuse to believe women can build homes, I highly recommend you watch the videos on YouTube from Easy Crafts DIY, which feature young women building various styles of houses, either alone or with one other helper. The video, "Amazing Two Women Building the Most Beautiful House in Jungle," demonstrates how to make a basic wattle and daub house.

[85] Note: I realize this wording is problematic, since we have no way of proving that human couples were "marrying" let alone that they were monogamous during the Neolithic, Chalcolithic, or Early Bronze Ages. Marriage, which essentially comes to mean not allowing women to have sex with more than one man since few cultures ever legally limit how much sex a man can have, does not seem to be a legal concept until the Middle Bronze Age.

Steppes, which could have led to their invasions, indeed, could have led to the Mycenaean incursion into Greece.

Such cultures would **appear to be** patrilineal to archeologists from their grave practices because DNA isotope tests would show the men had remained "local," while the women were brought in from (i.e. bride by capture) outside to replenish their population numbers. However, while many eagerly buy into the theory that such apparent patrilineality meant these were patriarchal cultures, they fail to recognize at least two important distinctions: 1) greater female death from disease or childbirth diminishes a population, pressuring such cultures to bring in more women from outside their culture; and 2) the creative element that women brought with them—not only through childbirth, but also through technical skills, from weaving to metallurgy. Women could have been enticed by many means, not just by kidnapping and enslaving, to move from their original homelands to other lands for both reasons.

In this respect, Helen's story in *The Iliad* becomes vital to understand. Was she kidnapped by Paris? Or was she seduced by him? In other words, how would the ancient Greeks have viewed her willingness to leave one native husband for a foreign one? If the concept of "bride by capture" was truly the *modus operandi* of the day, would the ancient Greeks have faulted her for her "defection" to another country?

Additionally, regarding Lerner's theory that slavery of women preceded other forms, we have no actual proof that husbands could sell their wives as property before the early Bronze Age, so it is actually more likely that human beings found verifiable proof that men contribute biologically to procreation, making them fathers in deed, not just spiritually, and this elevated men's view of themselves, leading to regarding women as less important, less spiritually powerful, and less valuable as human beings. If some cultures were already attempting to attract or were actively kidnapping women from other cultures to augment their populations, the step to viewing women as chattel is a small step to make.

To make men superior in magic/religion, in politics, and in power required one important social change: the subjugation of women, which came in a variety of forms and degrees. As the knowledge of writing and procreation spread, laws were written specifically to prohibit women's rights—from being able to have sex with multiple men (polyandry) to the right, in some extreme cultures like that of Rome, of appearing in public spaces unaccompanied by men. Prior to this period, it appears as though most cultures kept men's tendencies toward toxic masculinity (using physical power to bully and degrade others) in check through social pressures like shame and humility. Once males, and the females who supported them, see an opportunity to put themselves in the social place of women who once wielded Female Creative Magic by assuming the role of creators, changing the narrative to male progenitors, instead of female ones, the patriarchy begins its seemingly inexorable rise.

Patriarchal cultures, such as the Abrahamic and Hindu religions, hype the importance of female virginity because virginity, defined today as the state of abstention from all sexually penetrative acts, especially penis into vagina penetration (aka coitus), was necessary in order for individual men to be certain that the women in question were actually giving birth to their children. Ergo female chastity and virginity become issues around which each patriarchy stakes its survival, which is why it is still an issue today in many countries around the world.

To paint men as being superior to women, everything female or feminine must be made to appear weak or evil, which explain the modern backlash against people who do not choose to adhere to gender stereotypes. Therefore, menstruation, that aspect of Female Creative Magic men longed for enough to cut off 40% of their most sensitive penile tissue in the "sacred" ritual of circumcision, which was sympathetic magic imitating menstruation in order to allow men to bleed from battle but not die, becomes unclean and dangerous. Women are taught to fear men and to hate sex, which becomes filthy and immoral, unless, of course, a man is doing it. Gang rape becomes a common form of social abuse to keep outspoken women, who are often defamed as harlots, whores, or "women of discord," oppressed. Women are blamed for the rapes, instead of men being held accountable for their actions, with some extreme cultures stoning raped women to death for their supposed immorality.

Most importantly, laws—the written word—are used in patriarchies to control what each and every woman can do with her own body, such as the "morality" laws imposed by Iran on women, which modern Iranian women have finally openly rebelled against.

Without these patriarchally oppressive acts, all to control women's bodies and who gets to use them, the patriarchy cannot exist.

However, there are clear cultural signals that the idea of what is often interpreted as "virginity" for ancient cultures was actually very different depending on individual cultures. For many, actual sexual activity is not a factor, but whether or not a girl or young woman has begun to menstruate is.

When cultures are not certain that it is coitus that causes pregnancy, their ideas about sexuality are much more fluid than patriarchal ideas, which clamp down hard on female sexuality, sometimes even forbidding women to interact with non-relative males at all, not even through eye contact.

Even in cultures that were not fully patriarchal, but merely androcentric heavy, like the traditional Lakota of the Americas, for instance, taboos were developed to prevent men from having any sort of social contact with women who were tangentially related, even by marriage. Charles A. Eastman and Luther Standing Bear, both raised as traditional Lakota warriors, wrote about mother-in-law avoidance, wherein men were required to avoid having any physical contact and were admonished to avoid direct eye contact with their wives' mothers. This traditional principle arose, not because mothers-in-law are notoriously mean people, but as part of the Lakota incest taboos because they viewed relations by marriage as "closely related," therefore social contact of any sort with relative women who were not their wives or mothers was taboo. Marla N. Powers explained the tradition "as a sign of respect," so that, "if a young man or young woman at home alone was visited by mother-in-law or father-in-law, the [visitor] would not enter the tipi because it always required a third party to communicate between parents and children-in-law" (80). Like other patriarchal cultures, the

Lakota even used gang rape as a way to control "wayward" women who otherwise defy control.

Navigating what "virginity" meant to various cultures can become hazardous, but far too many scholars apply the modern androcentric definition of virginity to ancient ideas, failing to recognize significant differences.

Remember, many cultures around the Mediterranean did not view the absence of sexual experience as important until later in their history. Instead, they saw sex as pleasurable and magical because it allowed human beings to feel, at least temporarily, like gods—euphoric and ecstatic, but, as androcentrism increasingly restricted women's rights, they began viewing sexual pleasure as solely a man's right. In fact, men, especially wealthy men, in ancient Greece enjoyed having sex with anyone with whom they wanted to have sex, including children. Many modern pedophiliacs adore ancient Greece and Rome because of this proclivity, with some even arguing that it is an older man's right and duty to "educate" young boys and girls into their sexual experiences, since the Greeks and Romans practiced such pederasty.

There are even tribal cultures, such as some of the tribes of Indigenous Australians, who believe it is men's responsibility and spiritual duty to "bring on" girls' menstrual cycles by having sex with them (Merlan 480), failing to realize that grown men having coitus with young girls will, more likely than not, cause them to bleed anyway, but in no way is that sexual exploitation responsible for the girls' naturally occurring menarche.

Therefore, the lack of sexual experience, probably even for the earlier Mycenaeans, was not a natural state, which is why the first definition of *parthenos* is "young woman," not "sexually inexperienced." There are many indications, especially <u>after</u> many cultures around the Mediterranean accepted the idea that males play a biological role in procreation, that the Greeks viewed sexuality as a male privilege, even a form of bonding with their fellow soldiers. After the discovery of procreation circa 2400 BCE, the ancient Greeks still had problems accepting that coitus makes babies, as evidenced by the fact that the Greek warriors in *The Iliad* did not desire to rape the women of Troy in order to impregnate them, but only to shame their husbands, who would be dead by that time, more by proving they could not protect their women. Because the Greeks also practiced homosexual intercourse, the idea that men were "planting seeds" during intercourse could have made no sense to many, since no males ever got pregnant.

What is more important here is the fact that, for most Mycenaeans, early Egyptians, and the Minoans, the idea of "virginity" indicated a girl who had yet to reach menarche, meaning that she had yet to begin menstruating. If we remember how important, how magical, menstrual bleeding was for ancient people, we can better understand why menarche was seen as much more important a demarcation for a girl reaching womanhood than sexual experience. For instance, modern scholars struggle over the use of lustral basins for the cultures that lived on Mediterranean islands near Crete, but all indications point to their use for girls' initiation ceremonies, marking their transition from child to adult, once they reach menarche. Because the cultures on Crete and nearby islands were nearly full on matriarchal, we must view this process from a gynocentric point of view to understand its importance.

Using menarche as the demarcation between child and adult becomes important when we interpret mythological tales from many ancient cultures, such as the Egyptian tale, the "Book of the Celestial Cow." There are many Egyptian creation stories,[86] most of which conflict with each other regarding details, but almost all of which refer to a water world from which rises a "mound of creation," symbolizing the female pudenda, the ultimate human creation device, from which will arise the use of mastabas and pyramids as rebirth engines for the sacred dead. In many versions of the creation tale, a lotus flower, symbolic of women's vulva labia, gives birth to Ra, so that the tales preserve clear symbols of Female Creative Magic. The fact that some of the creation tales include a "cosmic egg" also pays tribute to the discovery of procreation from studying the mating habits of the only monogamous animal ever domesticated, swans, while also seemingly moving creation away from the actual female animal and toward a more separate divine creation, especially male divinity.

---

[86] See Jennifer Houser Wegner's listing of such Egyptian creation tales on the Glencairn Museum website.

Through the "Book of the Celestial Cow," the Egyptians of the New Kingdom "reordered" the cosmos based on a more androcentric view (Guilhou 3) by claiming that Ra, the most common sun god, created the sky and imbued all creation with the *ba* or spiritual spark of the gods (possibly viewed as the material or corporeal part of the gods), and was responsible for more closely uniting "the king with the sun god," pushing the androcentric concept of divine kingship (Guilhou 4)—the idea that males were endowed by the gods to rule over others. Vital to the tale is the taming of Hathor, the Heavenly Cow goddess, by getting her drunk on beer—a special beer produced by pre-menarchical young women that is stained red to look like blood. What is important in this distinction of these "virgins" (as they are often called by scholars interpreting this tale) who prepare the beer is not the fact that they have not had sex, but the fact that they have not yet menstruated, meaning they have not taken on Female Creative Magic yet, so are asked by Ra to prepare blood-beer for the goddess, possibly because their lack of menstruation or Female Creative Magic means their magic would not taint the magic of the blood-beer.

It is important to emphasize that the concept of "virgin" as being a woman who has not had coitus becomes a moral value much later in human history, after the discovery of procreation. After the rise of androcentrism and the knowledge that coitus must occur for pregnancy, maintaining women's virginity becomes paramount in burgeoning patriarchal cultures, such as the Greeks and Romans. Virgin birth was not a new concept when Christians begin using it, since the ancient Greeks developed many variations of it with Zeus "visiting" many mortal women in various forms—from rain to swan—in order to divinely impregnate them. However, we must keep foremost in our minds that the ancient people really did believe women could generate other human beings through our own bodies, either as a battle between realms in order to bring a waiting soul in another realm to this earthly one, or through a method of conjuration that relied solely on Female Creative Magic.

It might astonish you to think that most people, prior to 2400 BCE, the most likely earliest date when humanity learned that men contribute biologically to pregnancy, believed all women were giving birth parthenogenetically—that is, that women were creating babies all by themselves. But the concept is only one step away from the idea of gods or divine forces as having sex with mortal women, who go on to bear semi-divine children.

After the Common Era, early Christians push the concept of "virgin birth" to explain Jesus' "heroic" and godlike status, following an ancient tradition wherein cultural heroes are claimed to have had extraordinary circumstances surrounding their births, a pattern Joseph Campbell picks up on in his comparative study of various myths in *Hero with a Thousand Faces*. In the basic Christian sense, Mary is depicted as a sexual virgin, clearly morally weighed and measured by both Yahweh, her god who chose her to bear his son (and, apparently, depending on the cultic views of different Christian sects, to bear all his less-than-divine siblings), and Joseph, her husband-to-be, who knew, according to the story told in the Book of Matthew, that he had never had sex with her, so wondered how she could be pregnant. Clearly, by the time the Book of Matthew was written in the first or second century ACE, the idea that males contribute something to pregnancy was firmly established among cultures around the Mediterranean. However, belief in Mary's "divine pregnancy" still demonstrates a lingering belief in women as sole creators of their offspring, not to mention puts the patriarchal fear that the children women give birth to might not have a discernable biological father.

Was the Female Creative Magic of birth <u>prior to the knowledge of procreation</u> considered parthenogenesis or "virgin" birthing? No, because the ancients likely did not think about the absence of sexual experiences as a virtue to aspire to because sex is not something nasty or shameful in gynocentric cultures. Ample evidence that women were expected to be full participants in sex as pleasure and as a liminal, thus sacred, act, comes from plentiful Sumerian sources, especially early poems about Inanna, the Sumerian goddess of sex and war. In fact, Inanna was all about sexual pleasures, since she never expresses a desire in any of the poems, hymns, and prayers for becoming a mother.

While real parthenogenesis, which occurs when females clone themselves in a process that is not yet fully understood, has yet to be detected among human beings, it has been detected in other animals, including vultures (Aridi). Rigoglioso argues that vultures were the Egyptian "symbol of parthenogenesis" because they were believed "to impregnate themselves by flying against the south win or opening their beaks to the east wind" (Virgin 33), which brings up the question: when vultures are depicted in ancient Egypt, were they a symbol of Female Magic?

There are, unfortunately, too many tales involving the wind (or other natural elements) as a divine entity impregnating women to discuss in this book, but we have to recognize that these tales probably originated after humanity learned it actually takes two humans to procreate in order to explain how women who have not openly mated with men became pregnant. The many tales wherein women become pregnant under mysterious circumstances are echoes, however, of a time when human beings did not know heterosexual coitus was required for pregnancy.

Just as the growing preference for androcentric ideas meant that both women and goddesses were supplanted in importance in many cultures, the sacred Female Creative Magical qualities of female sexuality and menstrual blood also diminished in importance, replaced by a sacred aura built around the idea that women who do not have sexual intercourse are more sacred, more pure, than other women—hence, that virginity as we consider it now replaces the Female Creative Magic wrought through coitus, menstruation, and our sacred ties to the moon, which have become tainted, unclean, and even associated with evil.

In order to understand just how rising androcentrism and patriarchies were preceded, naturally, by more gynocentric cultures, we have to be able to strip away our androcentric biases, first, by realizing that humanity did not always know heterosexual intercourse could lead to pregnancy, by understanding that sex was both a magical and a social activity for ancient peoples, and that Female Creative Magic, for the pre-procreation ancients, stemmed from women's ability to bleed but not die every month, demonstrating our ties to the moon, thus to the greater cosmos, to produce offspring on our own, and to heal or to transform others through sexual contact. While women are, of course, more than the sum of our body parts, we were also once regarded as far more magical than men because we were in sync with the magic of the universe, wielding Female Creative Magic.

Chapter Eight
Has Matriarchy Ever Existed?

In 1986, Gerda Lerner, noted historian from the University of Wisconsin-Madison, asserted that "one can truly speak of matriarchy only when women hold power *over* men, not alongside them" (emphasis hers, 31). In her important treatise, *The Creation of Patriarchy*, Lerner clearly believes that women have never held power over men but have certainly exercised power alongside men.

But does the definition of a matriarchy require that women suppress men in the same manner that men have oppressed women, often through violence and fear, as so many various patriarchal cultures have done in world history?

While it is certainly possible that women could oppress men in the same way men have oppressed women, especially since such oppression does not require superior physical strength, just weaponry and the ability to censor our natural human abilities to cooperate, none of the self-declared matriarchal cultures that currently exist, such as the Mosuo of China and the Hopi of the United States, do so.

Lerner overlooks the great possibility, nay, probability, that women have held powerful positions in many egalitarian cultures, even prior to the social creation of abusive hierarchically structured cultures, through their primary influence as mothers, but also through respect for their ancient Female Magic, which, in ancient times, tied them more closely to the divine entities that are believed to have created the world and our cosmos.

Matriarchy, by simple definition, must follow the leadership of females, presumably mothers, who lead their kin, clans, communities, and even cities. The Mosuo of China still practice this forms of matriarchy.

When I have asked colleagues, students, and acquaintances to think about their own relationships with their own mothers, the majority react in a similar way. Most have a love-hate relationship with their mothers. A few have antagonistic relationships with their mothers, who they often describe as overbearing or selfish. Some hold a grudging respect for their mothers, and, depending on the person's age and whether or not they themselves have been a parent, understand better some decisions their mothers made on their behalf. Others absolutely adore their mothers, who they describe as their primary source of self-belief—people who believed in their children as individuals who could accomplish anything, continually cheering their children on to whatever endeavor they attempted.

Realistically, we have to admit that mother-guilt, that special bond that many mothers learn to exploit to keep their children well-behaved and polite in public, is probably as old as maternity itself.

I remember reading one "feminist" writer who looked at the Stone Age Venus figures found throughout the world and being aghast at her and her students' reactions to the figures, which they insisted—as brainwashed by Western culture as anyone can be—were grotesque because of their obesity, firmly believing that the people who created them were mimicking or making fun of such callipygian women (Ruether 3).

As a woman who has hosted such a figure most of my life, I was deeply offended by these woman and her students' programmed reactions, but I understood that their programming made them see these female figures in this way.

Like many feminist scholars raised under Abrahamic religions, Rosemary Radford Ruether struggles with her feminist views as they bisect her religious views, creating a form of dissonance that she cannot reconcile. For instance, she discusses the poem often called the "Song of Songs" in the Judeo-Christian bible, comparing the poem with ancient Egyptian and Sumerian love poems, but clearly never having read the actual Egyptian or Sumerian love poems herself because she claims "it is unlikely that the Song is derived from the [more ancient] text[s]" because she believes the ancient texts all speak directly to how the sacred sex acts as a catalyst to rejuvenate nature in the spring, when, in fact, none of the more ancient poems she mentions makes any such claim (Ruether 89). Ruether is, in effect, willing to examine her culture's poetry, but does not make a concerted effort to truly understand the more ancient poetry she attempts to compare it to, so it is not terribly surprising that she has made little effort to understand the symbolism of the Venus figures carved by our ancient ancestors.

116

Much has been made of the fact that most of these Venus figures are featureless but given the fact that most are carved out of stone or ivory at a time before microliths were widely used for fine carving indicates that these figures were the best that could be created at the time, since no metal tools were yet available for precision carving.

The big question, then, was: why are so many of these female figures so plump?

ΩΩΩΩ

When I entered my 57th year of life still menstruating, my physician, who was concerned that I was so old and still regularly menstrual, had blood tests run to check my female hormone levels. The gynecologist who checked those levels was not worried that I had the hormone levels of a 20-year-old. She admitted that, yes, my ample fat cells were undoubtedly storing lots of estrogen, but when I pointed out my own research involving obesity and menopause indicating that the fat cells actually help produce estrogen, not just store them, she balked. "That's not what we were told in school," she said.

I drew her attention to an article from the *European Journal of Obstetrics, Gynecology and Reproductive Biology*, which indicates that "more than 70% of women with a long menstrual cycle had abdominal obesity" (Lay, Pereira & Miguel 225).

Throughout my research, I found no archeologist or anthropologist who discussed the Venus figurines who ever entertained the idea that obese women probably menstruate longer than thin women do, simply because of our ample fat supplies. Fat, after all, not only stores female hormones, but also actually produces them.

When I asked my gynecologist about the idea that obese men develop breasts because of their increased estrogen production, she looked at me blankly. Men's hormone levels were not covered in gynecology classes, nor were the associations with fat cells and hormone production.

Given the fact that so many cultures around the world derive their words for "leader" from the idea of "big bellies," it is easy to extrapolate backwards in time, especially given the plethora of corpulent female figurines.

Women were probably the original "big bellies," and the Venus figurines are a tribute to their power—not only were they people so revered in their groups because they were naturally larger and possibly seen as a sign of group prosperity once groups became more sedentary, especially, but they were undoubtedly viewed as closer to the divine because of their ability to menstruate long after most thinner women stopped. These women would have been the first guardians of the blood—tying the natural process of menstruation to lunar cycles and the life abundance of nature for a longer period than most women.

These figurines, then, demonstrate several things about our ancient ancestors.

First, they recognize women who menstruated longer than other women as spiritually enhanced, demonstrating how much the ancients revered women's ability to bleed and not die for three to five days every month. Much like the moon's 28-day cycle, women's cycles had similar lengths, so they began segregating themselves away, like the moon does every month, from others during their menses in order to prevent their more powerful menstrual fluids from weakening anyone else's personal magic. Men, desiring to take on this Female Magic as their own, began cutting their penises in forms of sympathetic magic, which is likely where the practice of circumcision comes from, since it removes up to 40% of the most sensitive penile tissue, so directly impacts men's enjoyment of coitus.

Second, the figurines possibly represent the first real leaders of most Neolithic communities—the "big bellies" who are so magical that they continue to menstruate—continuing to demonstrate that sacred connection between humanity and the cosmos. This idea is not a huge leap from what other scholars already hypothesize: that the dominant spiritual leaders in ancient communities became the dominant social and political leaders of those communities. We are simply acknowledging that women were very likely to be among those leaders because of this clear spiritual connection.

Third, the figurines could represent "goddesses," since we know that many ancient people revered their dead ancestors as divine spiritual forces, sometimes as outright gods. Even if these figurines are crude depictions of real women who lived and led so long ago, these same ancestors were probably also revered as gods, so that, like the Greeks who can still trace their family line back to the goddess Io, there were undoubtedly ancient people who traced their matrilineal roots back to these specific spiritual leaders transformed into goddesses.

Scholars with androcentric biases often jump on the female activities that sought to prevent women's superior magic from interfering with the weaker male magics to argue it is proof that women's activities have been separated from men's because of "traditional" (meaning their own androcentric tradition's biases) gender-based segregation. I find many of Jan Driessen's attempts to minimize the importance of women as leaders on Crete, for instance, amusing because he is much more willing to argue that images of athletic young males was for "homoerotic display for socially dominant males" (1), than they are erotic displays for the women in power.

Later, unfortunately, this gendered segregation became enforced by Abrahamic religions (which still influence so many scholars today) which change menstruation from a super power connecting women to the cosmos to evidence of women's uncleanness, thus deeming them unable to hold priestly offices—which is still the only reason[87] why women can never become ordained priests, so never a pope, for the Catholic Church. Similarly, because of an Islamic religious precept that does not allow women to lead prayers, women can only become imams for women-only mosques. While Judaism allows women to be ordained as rabbis, it is mostly only Reform and Reconstructionist temples that allow women to preach, since traditionally men are forbidden to hear women sing, let alone pray.

Because of these patriarchal religions, women have been oppressed and their voices suppressed for the most of the last 2000 years.

---

[87] Catholic Church doctrine, of course, says otherwise, making the claim that Jesus chose only men to be apostles, not women, clearly ignoring the many women who were apostles, even though they were written out of the chosen texts of the bible.

Given that facts demonstrate how many people around the world gave food and drink offerings to divine powers, food that was originally meant to feed the poor, but changed to feed the priests of various temples, it is also quite probable that, before they shifted their attention to men as closer to the divine (prior to an understanding that men had anything to do with procreation) women were seen as the natural leaders since they not only led, trained, cared for, and persuaded (possibly via mother-guilt and collective acts of shame) their offspring, but also led, guided, cared for, and persuaded their larger family groups, their *tiyospaye*, as the Lakota call them, which eventually became their clans, which became their tribes, which became villages, then cities.

Among more egalitarian cultures, gynocentric views, wherein matrilocal and matrifocal practices were followed and where women could lead, were in abundant evidence, even as late as the Bronze Age. Probably the most famous of these cultures occupied the island of Crete, wherein women were clearly priestesses and community leaders, often appearing in frescos as much larger than the men in attendance.

Even Gerda Lerner admits that "women's mothering and nurturing activities, associated with their self-sufficiency in food gathering and their sense of competence in many, varied life-essential skills, must have been experienced by men and women as a source of strength and, probably, magic power" (44).

While there is little evidence for oppressive matriarchies that suppressed men's ability to lead their communities among ancient cultures, there is ample evidence that demonstrates that women not only led in more egalitarian cultures, but also were quite successful in doing so.

The big question that we should really be asking is why patriarchies feel the need to continue to oppress women, when we have all learned that restricting women's sexuality has done nothing to prevent the problem of not knowing who a child's biological father is, as demonstrated by the new popularity of DNA testing, wherein teens often learn that the man they assumed was their biological father is not.

If matriarchies do not oppress men, why should patriarchies continue oppressing women?

Chapter Nine
Mere "Fertility" Versus Female Spiritual Power:
Why So Many Scholars Get It Wrong

You have undoubtedly heard an ancient work of art, such as figurines with prominent sexual organs and phallic or yoni shaped objects, called a "fertility symbol" by many people, quite often without any clear context, definition, or example to demonstrate what that label means.

First, we have to acknowledge that most androcentric scholars, after millennia of Abrahamic religions shaming human beings about sexuality in order to control women's bodies, thus enable the patriarchy to control most aspects of our lives, are reluctant to discuss sex, let alone sexuality, so most uncomfortable scholars eagerly dismiss symbols for sex and sexuality as mere fertility symbols, completely ignoring the fact that **the ancient people did not associate sex with pregnancy until after the discovery of procreation circa 2400 BCE**. If coitus was not what caused pregnancy for the ancients, symbols of human sex parts cannot, by themselves, be fertility items.

Why do so many scholars conflate sexuality with fertility?

One reason is because so many androcentric scholars assume that human beings have always known that sex causes pregnancy. To be fair and to be honest, I did not think about when or how human beings discovered the sex-pregnancy connection until I began research for this book.

We also have to admit that many sex-oriented rituals in many ancient cultures occurred in the spring, which is when many animals, especially birds, snakes, reptiles, insects, and small mammals[88], become sexually active. Many scholars assume that, because "the birds and bees" were doing it in spring, human beings believed that sex had to be performed in the spring in order to promote general fertility in the crop fields and among their livestock. One problem with this belief is the fact that some species

---

[88] See my chapter on the discovery of procreation to learn more about herd animal seasonal sex rhythms.

do not have coitus until the fall, such as when deer go into rut, so only associating springtime with coitus would be wrong.

Just to take one well-known sexual ritual, the *qursu* ritual practiced by the ancient Sumerians, however, there are other, probably more important reasons for why such rites had to be practiced in what would have been the ancient Mediterranean people's new year—sometime in between late February and early April, depending on the culture. Because spring was a time for the renewal of the earth, when rains come, when plants grow, when many flowers bloom, people practicing sympathetic magic would have capitalized on that time period to evoke more plenty, including such social practices as electing new city leaders. Therefore, it is quite likely that the ancient Sumerians, and possibly the Ubaidians before them, practiced this magical sex ritual in the springtime because male ejaculation was equated with the flow of water, such as via rain or river floods (e.g. the Nile River), which was vital to spring growth, **not** because they imagined the male was impregnating the female. Because of the differences implied from all this sexual activity, we have to be careful to differentiate between human fertility and human sexual activity that created sympathetic magics to promote the fertility of other living things like plants and animals.

I have lost track of the number of scholars who try to speculate about why so many female figurines with prominent sexual features, such as breasts and vulvas, have been found all around the world from Paleolithic, Neolithic, and Bronze Age times. Whether we are discussing the Venus of Hohle Fels from approximately 35,000 years ago, the Venus of Willendorf from 25,000 years ago, or the Venus of Laussel from 18,000 years ago, we have to recognize that many people found large breasted, corpulent women important enough to carve them, repeatedly, over eons.

Yet no scholar, that I have seen, has ever equated these corpulent women with the literal "big belly" leaders of ancient cultures, despite the fact that so many cultures, especially around the Mediterranean Sea, called their city leaders their town's Big Belly.

However, androcentric biases against admitting women were leaders and influencers of ancient cultures prior to 2400 BCE have meant that most of scholarly guesses about the importance of this plethora of female figurines are hardly scientifically based.

Some scholars simply dismiss these prolific female figures as "fertility" totems because they are unaware that women's sexuality had little to do with human fertility in the minds of the ancients. Yes, women could produce children, just as other living beings produce offspring, but the power of having babies was not the only aspect of Female Creative Magic that set women apart from men in ancient times. Nor did human women have sexual "seasons" like other herd animals, so there was very little reason to choose spring as human Sex Season.

Since many of the most famous corpulent female figures have no faces, some scholars speculate that they are not depictions of specific people, but general paleolithic pornography, while others speculate that women created the images in a "distorted" way because they could only see their own bodies from eye height down (Cook 41), completely ignoring the fact that few of the women would have been isolated or alone, so surely saw other women's bodies. Even Dorothy Cameron, who excitedly shared her theories about symbols in ancient art that demonstrated how people of Neolithic Eurasia had created "a belief system that regarded women as Creator," describes these female figurines as having "exaggerated breasts, buttocks, and oversized pubic triangles," apparently never having actually seen an obese woman's body before.

Replica of Venus of Willendorf
Author

However, so much of what we now know about these ancient peoples proves that communal cooperation, aka egalitarianism rather than men dominating women, was the primary form of social structure, so while there might have been randy men who fantasized about having sex with a fleshy woman, there were probably few if any social inhibitions about having consensual sex with anyone willing. But this idea that these female figurines are nothing but paleolithic pornography begs a question: why are so many of the women obese, yet so unobtainable that the men would fantasize about them? The answer, quite likely, would be that only the most treasured of women, elite women, if you will, could afford to be fed by everyone while performing very little labor to burn off her calories. So other scholars have speculated that these "fertility" symbols are actually depictions of powerful women—after all, creating human faces is challenging—so these could be depictions of actual women, women who were revered, women who led their communities.

Quite likely, women, like other matrons among herd mammals, organized their communities, directing them where to move and when based on the knowledge the matrons of the groups had developed over their lifetimes.

As Judy Grahn points out in her important book, *Blood, Bread, and Roses: How Menstruation Created the World*, ancient males believed women were more spiritually connected via their Female Creative Magic to the cosmos for at least three basic reasons:

1. Every healthy adult woman menstruates every month with a cycle that mirrors that of the moon, which is why we **still** call our periods our "monthlies" or "moon ties." Menstruation, clearly tied to the phases of the moon by the ancients, could also symbolize women's constant battle with supernatural elements in order to bring new life to the world. Each healthy menstrual cycle lasts anywhere from three to five days, without the women dying from that blood loss. This magic was something males were desperate to acquire, which is why they developed such

sympathetic magic rituals like circumcision and penis-piercing.

2. Many women gave birth, seemingly creating a whole other human being out of the spiritual and physical powers of their own bodies or possibly through battle with supernatural elements to bring human spirits to earth from another realm, such as the Abrahamic "guff," wherein babies' souls awaited for birth or in the Dream Time for native Australian cultures who still believe that every Spirit Child successfully born into this world chooses to be born and chooses the woman to be born to.

3. Vaginas are magical because they not only bring physical pleasure through heterosexual coitus, but also change size to accommodate both sex and giving birth, which is one reason snakes, whose jaws can also change size, are often associated with women. Snakes are also associated with women because they shed their skins in a way that is reminiscent of childbirth.

Add these verifiable facts to the fact that having sex is often quite a spiritual, often called a liminal (between worlds), experience, and you might begin to understand why the ancient Sumerians built temples to one of the most powerful deities ever conceived of by human beings: the goddess Inanna.

While she is often mislabeled as a goddess of "love," Inanna was unabashedly the goddess of sex—all kinds of sex; in fact, two of the sacred *me* she transmitted to humans have to do with different types of sex. And she was so powerful that she could change a person's gender, if she saw fit. Priests of all genders, including transgenderers, trained at her temple and learned how to heal through sex, a transformative[89] art that made a man out of Enkidu in the *Epic of Gilgamesh*, and made kings out of men through the

---

[89] Transformation is the main quality of magics for ancient people. Joseph Campbell lists several examples of transgendered shaman from around the world (*Way of the Animal Powers: Part 2 Mythologies of the Great Hunt*, 173-5), noting that "'among the Chukchi'" of North America, sacred shaman who changed genders completely adopted everything about their new gender, including changing language inflections (174).

sacred *qursu* ritual held every spring, and which is probably a ritual that is much, much older than when Sargon of Akkad made his daughter, Enheduanna, the High Priestess of Inanna, perform the sex ritual with him to sanctify his kingship.[90]

What ties sex to war, Inanna's other realm? Blood. Women bleed monthly, bleed after giving birth, and human beings bleed after being beaten, stabbed, and castrated during and after war. I challenge you to re-examine all the love-war goddesses that have been diminished in power by realizing how much blood loss ties women's Female Creative Magic to both menstruation/life creation and war.

Therefore, these ancient female figurines are serious stuff, but our ancient ancestors were not focusing on human fertility when they made them.

Why, then, were corpulent women, in particular, such revered figures to the ancients?

Menstruation, which was a physical demonstration of women's spiritual powers[91] to reach beyond this realm into the cosmos. Obese women tend to menstruate longer than thinner women do (Lay et al), so a woman who was old enough, loved enough to be fed well, and important enough not to have to move around a great deal, were probably also menstruating at 50, 55, or even 60 years of age. A woman who continued to menstruate longer than others would have been deemed more magical, more in touch with the cosmos and the spiritual world.

---

[90] The continuation of their practice of incest is one of many reasons why it is quite probable that these ancient humans did not connect heterosexual coitus with procreation.

[91] This fact is one many anthropologists struggle to understand. When Thomas Buckley encountered a Yurok (California American Indian) woman who claimed that menstruation was a time of women's greatest magical power, he was astounded, preferring to believe other anthropological reports that "menstrual blood itself was thought by the Yurok to be a dire poison" (191). He remained unconvinced of her interpretation, until he reviewed actual earlier anthropological notes that corroborate the idea that menstruation equals female spiritual power, so that women go into seclusion to prevent their more powerful magic affecting anyone else's (193).

Because women's menstrual cycles are tied to moon cycles, when we have figurines like the Venus of Laussel, who holds a moon-shaped horn that has thirteen marks on it, we know she is a depiction of that female-moon magical connection, perhaps representing the spiritual power of menstruation itself, what I call Female Creative Magic. As some scholars have pointed out, there can be 13 full moons in a year, so the 13 marks most likely makes note of that phenomenon (Achrati 5), Scholars know she was "once tinted with red ochre" (Venus), which is one of the universal elements that indicate both menstrual and birth blood, potent magical substances, especially for those who are about to pass into the next realm.

Until television convinced them that androcentric ideals of female beauty (thin, so weaker) were better than their own concepts of female beauty, many cultures around the world revered women with "some meat on their bones," as my father liked to put it.

When we examine, as we did earlier, the term many communities had for "leader" in their communities, many literally are interpreted as "big bellies." While some scholars argue that these "big bellies" only can refer to males, I believe they referred to these corpulent women whose evidence of their importance dates back at least 30 millennia.

Apparently, the ancients valued substantial women, like my father did, but these women were not necessarily glorified sexual beings indicating fertility; while they could have been Mothers, Matriarchs, Matrons of their communities; leaders, wise women, and commanding speakers, they undoubtedly also wielded strong spiritual powers through the retention of menstruation for a longer period than their thinner counterparts. Therefore, it is quite likely that many ancients would have believed their ties to the moon were greater than most because they continued to menstruate long after other women had stopped.

In many cases, these women were probably also worshipped, especially after death, since most ancient people practiced the deification of their most revered ancestors, praying to them for protection and assistance, and they would have all traced their ancestry back through their mothers, since males were not considered real "fathers" until procreation was discovered circa 2400 BCE.

Judy Grahn, Marija Gimbutas, and Gerda Lerner were all correct in their beliefs that women not only once led many different cultures, but also transformed our world. Women ruled, but most governed much more benignly than most men of history ever did, and they probably also ruled in conjunction with men more often than not, while men remained less sacred, less in touch with the spirit world than women did naturally. Unfortunately, so many scholars are so imbued with androcentric bias that they are challenged to see these figurines for what they really are: women of spiritual power because of their ongoing connections to the cosmos.

Chapter Ten
First Historical Industrial Revolution

There are many interesting discussions about when the first real human "industrial revolution" occurred, with some considering the invention of stone tools as the first real "industrial revolution" for humanoids, quite possibly instigated by Neanderthals in the Paleolithic era, which lasted up until about 11,000 years ago.

However, if we compare historical times to the European Industrial Revolution, which lasted approximately from 1760 to 1840 ACE, we can acknowledge that the first historical industrial revolution happened during the early Bronze Age, what is also sometimes called the Copper Age, in many areas of Eurasia and northern Africa, or, primarily, around the Mediterranean Sea.

The first major comparison comes from changes in the manufacturing of household goods, such as cloth and ceramics, but also extends into innovations in travel, from the use of the wheel on carts to the invention of the sail boat, which allows boats to sail further away from shore.

In the European Industrial Revolution, wealthy (or soon to be wealthy) men developed tools and machinery that replaced what women had been doing for millennia: spinning, weaving, sewing, clay manufacture, ceramic molding and painting (including the creation of innovative paints), kiln control, and house construction.

During the early Bronze Age, someone (we can only guess at their genders) also ramped up production of these same goods through several clever machine-like inventions, two of which introduced the use of wheels, the pottery wheel and, probably much later toward the Iron Age, the spinning wheel. The integration of the pottery wheel in ceramic manufacturing during the Bronze Age meant that women (who were usually the potters at this time) could throw, shape, and fire more pots, as well as increase the general size and uniformity of many pots, in a day, which allowed potters, for the first time, to make more than just the dishes they needed for their own homes. The surplus pots could be traded for other high quality goods, which helped to form the first "middle class" or monetarily-hierarchical business owners. The uniformity of the pots created on pottery wheels enabled people to store and transport their goods—from grains to oils and wine or beer—more easily, so that, eventually someone realized that pots with pointed bottoms could store more easily in the holds of ships—a form of pottery still used by the time we get to the historical Greeks, as evidenced by the plethora of amphorae located at the bottom of various seas from shipwrecked cargo. Perhaps more importantly, the use of the pottery wheel by ceramicists is often credited with leading to the idea of using wheels on sledges, turning them into carts, to improve the transportation of the goods over land. The wheel goes on to have far more implications on cultures, especially in warfare, which is discussed in another chapter.

Archeological evidence demonstrates that humans tinkered with ceramics for millennia, with some of the earliest ceramic sculptures, like the Venus of Dolní Věstonice, dating as far back as 25,000 BCE, prior to the application of ceramic technology for the creation of storage pots and cooking vessels. Around 16,000 BCE, Chinese potters began making ceramic pots, with ceramists from early Croatia making pots as early as 15,500 BCE (Ancient Pottery).

Because making figurines out of ceramics remained popular, many ceramicists learned to create clay molds, modeled after early rock molds carved for the use in both ceramics and metallurgy, fairly early on in the manufacture of ceramics, but the pottery wheel was "in widespread use by the beginning of the Early Bronze Age, around 2400 BCE" (Ancient Pottery).

Besides pottery making, the Bronze Age also saw the creation of mud bricks for building. Many of the ziggurats from Sumeria did not survive the ages because they were originally made with sun hardened mud bricks, but the development of ceramics introduced glazed tiles, which were added to the surfaces of buildings, leading to such spectacular architecture like Ishtar's Gate[92], ornately decorated with blue tiles and mythical creatures.

Clay, too, began to be used, especially by the Sumerians, for writing tablets, since it could be smoothed out and rewritten easily. The fact that thousands of written clay tablets have come down to

---

[92] Gates, of course, symbolize openings, and, since cities were also called "mother," the Ishtar Gate likely symbolizes sex, birth, or entering a new realm through either of those activities.

us through the millennia is largely attributed to the fact that the sun- or air-dried writing tablets were often heat-hardened by fires that destroyed the buildings in which they had been stored.

As with the discovery of metallurgy by women creating ceramics, which I discuss in the next chapter, many of these improvements to the processes of creating cloth for clothing and ceramics for storage of foodstuffs and other items, were probably invented by women, who were the most likely people to be creating the cloth and the ceramics at this time, at least based on what researchers have gleaned from studying pre-literature cultures around the world. Women were the likely clothing creators from early on in human development, learning over millennia to preserve whole hairy hides for warm blankets, as well as to scrape various forms of leather to form supple clothing that could be decorated with shells and other forms of beads—from porcupine quills, which were often died different colors to small pieces of polished amber, ostrich shell, seashell, animal teeth, animal bone, and even deer hooves. These neolithic developments accelerated the desire for different forms of clothing, so, since creating nets out of grasses was not far removed from learning to make fine threads out of various grasses like flax, it was quite likely women who invented linen cloth and the simple looms they created on which they wove various sizes of cloth.

Weaving cloth in Eurasia arose at about the same time as the use of clay to make ceramic figurines, according to archeological findings that indicate woven items, either bags or baskets, were used to store materials. It is possible that the idea of weaving such items occurred to people who created fish traps with woven sticks between upright logs tamped into lake and ocean beaches, but the archeological remains of grass woven textiles are scant because the fibers decay so readily in most conditions.

The weaving of grasses inevitably led to the discovery of flax strands, which are fibers that can be more easily woven into softer material we now call linen, but the process of producing flax fibers is time-consuming, laborious work, starting with pulling up the plants by the roots, removing the seeds, retting the stalks to begin the removal of the woodier parts of the plant, all before the breaking, scutching, and hetching of the stalks to remove the last of the woody material and to align the flax strands, so they can be spun using weighted spindles, like wool and cotton, into yarn or string for weaving (How Is Flax).

Since women were believed to be the primary plant gatherers of most cultures, they were probably the ones who discovered how to weave grasses into baskets and ropes, which was then applied to creating fish traps and fishing nets, while also expanding into the weaving of cloth for clothing and bags. The weaving of cloth boomed as an industry in the early Bronze Age because wearing cloth, instead of leather goods (which were also time consuming to make) became a mark of distinction, since fewer people had the time to produce the cloth to make clothing, so being able to do so was undoubtedly one of the first signs of wealth (at the least, it was a sign of technological innovation) among many people in the prior Neolithic Age.

Many scholars now admit that women created complex mathematics, like geometry, from learning how to weave designs into cloth as they wove it on their simple warp weighted looms, or, in the Americas, on back-strap looms (Edmonds). While much of the weaving on these looms was simple, it could be made more complex with various forms of counting warp threads and moving weft threads across them. Archeology reveals that loom weights have been found as far back as the Neolithic in Switzerland (Edmonds). When weavers began using bobbins or shuttles to pass the weft thread back and forth between the warp threads is not as clear. Susan Edmonds, a weaver herself, clarifies that the words for bobbin, spindle, kerkis, and shuttle are all used by ancient sources to indicate various devices that introduced different colored threads into the woven cloth, but clarifies that the shuttle is mainly the larger pointed tool that "may be shaped like a boat," can hold a smaller bobbin in it, but is the main device holding the primary color that is "thrown across the warp." Each of these different devices were introduced during the Bronze Age, which was the first time men seemed to take an interest in learning to weave (Edmonds), although it remained a primarily female-produced commodity with many women still producing their family's clothing themselves (Wright 409).

We know that the early Sumerians and Egyptians loved to wear woven clothing, especially linen, because it could be pressed into pleats and ruffles—something the wealthy men of both cultures loved wearing. In fact, the shift from only depicting gods wearing such skirts to depicting historical rulers wearing them signaled that the first ruler so depicted, King Naram-sin, desired "to be considered a god" or at least signaled his participation in the ancient *qursu* spring ritual wherein the king embodied the god Dumuzi as he had coitus with the High Priestess of Inanna's temple who stood in for the goddess Inanna herself (Wright 398). Either way, the wearing of "'flounced, fleecy, tiered, ruched, ruffled striped, plisse or *tayautee*'" clothing signaled a clear elevation of status—toward godlike—for individual human beings (Wright 398; see the image below).

Sketch of Sumerian men wearing pleated
and fringed skirts from the Standard of Ur
Author

Just as quality (thinner, more ornately painted) ceramics became symbols demonstrating a person's possible economic status or technological skill, clothing became the everyday visual way to distinguish those with means from those who struggled to get by. Among the Sumerians, for instance, wearing a very short kilt, even if it was cloth, meant you were a slave, while the wealthier you were the longer your wrapped dresses, robes, or skirts were. The same distinctions undoubtedly held in Egypt, as well.

Rita P. Wright argues that, among the Sumerians, "people in lower echelons were provided with cloth of courser varieties and at [reduced] levels of quality" (395), reporting that "robes and garments were woven of linen," which had been made into textiles as early as the Ubaid period, circa 4000 BCE (396), whereas "wool was made into saddle cloth, shaggy garments, headbands, headdresses, loincloths and menstruating garments, menstruating bandages and underwear" (395). Wool had been spun into yarn as early as 7000 BCE, but it was not until the early Bronze Age that sheep herding specifically for quality wools occurred (396), possibly because gathering wool from sheep prior to the development of copper or bronze sheers meant plucking the wool off the sheep by hand (397).

Undoubtedly, because women were the plant gatherers, they also experimented with pigments from plants for ceramics, paints and glazes, and cloth dying. Once wool and linen clothes became more common in Sumeria, colored cloth, which was more expensive that natural cloth, also became a social distinction with "yellow" cloth making up only 1% of the cloth for clothing, so "was exclusively worn by the king" (Wright 401). Black was apparently the next most coveted color with many gods and goddesses depicted in black or black and white garments. While Inanna's representative in the *qursu* ritual typically wore a white garment, potentially where wearing white to a mating ceremony comes from, the ruler acting as Dumuzi wore "a multicolored cape," according to the Enmerkar epic (Wright 401). Such multi-colored cloth came to be associated with great spiritual power, since Inanna in the *qursu* ritual began being clothed in more colorful garments, signaling this "more powerful...numinous power" belonging only to the goddess (Wright 401).

While few scholars address who coordinated and organized the production and distribution of such goods, Rita P. Wright cites records from Nippur demonstrating how many women from the temple of Inanna were in charge, with successive women named as "chief administrators," using their seals of authority "to authenticate records" (408). Women, often captives of war,[93] were

---

[93] Rita P. Wright says they were often listed as "'plunder, booty, captives, prisoners of war'" or designated as foreigners "from the 'eastern provinces'"

employed as "fullers, felt makers, leatherworkers,...weavers...[and] as spinners and in food-related activities" (406). These workers were typically paid in quantities of barley and/or beer, which was more like beer-soup than the beer consumed today, "depending upon the amount of time [the] individual had spent in [their respective] craft" (Wright 407). Wright even suggests that the "foreign" women brought ideas about "forms of dress, styles, and techniques...from their places of origin" to influence Sumerian clothing styles (410). Women further influenced the spread of crafts and style when wealthy women, especially royal women, exchanged gifts of goods to maintain close ties with people from other cities and lands, evidence for which comes from the women's personal letters (Wright 411-12). Such reciprocity was an ancient form of social obligation practiced, undoubtedly, since at least the Neolithic era.

While women undoubtedly discovered metals like copper as remnants from their ceramic firing processes during a time when every household made its own ceramic vessels, and while women clearly invented weaving, including inventing the spindle weights necessary for making thread and then weaving with a warp and weft using what are often called warp weighted loom weights and shuttles (Fashion Archives), even prior to the invention of the wooden frame loom, the fact that these processes became more wide spread in use and began to lead to distinctions in social hierarchy based solely on economic or technological resources provides evidence for women's growing continued scientific and mathematical processes that enabled them to make more goods more quickly.

Because of other innovations using clay and shaped stone, women learned to build homes out of stone foundations with better shaped stone or mud bricks, while still using weaving techniques to create wattle and daub walls. Their knowledge of grasses also probably led to thatched roofs, and, much earlier, reed boats.

However, it is quite likely that, as small communities became sedentary, and attracted more people, more planning became

---

(409). That such women could still level up to become administrators over their area of production demonstrates some consideration for skill and knowledge as important attributes.

necessary to create housing for everyone—most of which was still circled or squared around a community center, the most likely place for various community religious activities to occur. While small shrines grew into religious temples, houses, too, grew in proportion. Documents show that women were still enlisted for many of these building projects, including the building of palaces (Gelb 81).

More importantly, women probably oversaw the distribution of collected goods, either made in the temples, which eventually became closer to our modern concept of factories toward the end of ancient Mesopotamian history, with both private estate and temple made garments, "listed in relatively small quantities...exported in exchange for copper and tin-bronze," along with other products like "silver and scented oils" (Wright 411). Women could be independent merchants, such as an Akkadian woman "who financed agents that conducted business for her" (Wright 412).

Priestesses oversaw many of the commodity exchanges within the temple precincts themselves, such as when "in the eighteenth century [BCE] on a single day of purification 2770 people each gave a measure of bread and beer to the temple" (Makkay 5), foodstuffs that were probably either stored for a time within the temple precinct, or immediately redistributed to its workers (Gelb 75), with the blood sacrifice meats—goat, sheep, cattle—being reserved for the priestly classes. It is important to note that, sometimes, when the temple asked for "sacrifices or offerings" to support the individual temple or god's work, wealthier people often gave in the form of slaves, usually foreigners who had been captured and sold into slavery, but, occasionally, people who were essentially indentured because of debts, who could, when possible, work their way to freedom.

In this way, the temples gradually became great centers of social assistance, both in taking in the extremely poor and slaves, and in clothing and feeding them, although many did not work under the best of conditions judging by the number of people who attempted to run away (Wright 395). The temples and their priests also became wealthy as they accumulated land, stored donated food stuffs, and produced desired goods for trade (Makkay 5).

As with the later European Industrial Revolution, one key change, however, was that men began taking an interest in what had probably long been considered "women's work," so began to create ceramics and fabrics, too, truly making the Bronze Age Industrial Revolution very similar to the later European one wherein men almost completely took over women's crafts by putting them on an industrial scale of production.

So many wonderful things were invented during the Bronze Age—from various uses of clay for writing and ceramics to learning to create geometric patterns in weaving and building construction— that the early Bronze Age should be considered the world's first historical Industrial Revolution.

Chapter Eleven
Were Women the First Scientists?
Ceramic Experimentation Leads to Copper
Discovery

Imagine living in the Neolithic Age. You are your family group's potter. You are probably a woman who was taught by her mother how to choose clay, shape it, build a fire in a pit or even a small rock kiln in which you can fire the pottery to its super-dried and hardened state, so that ceramics can be used again and again for storage or food processing.

You and your daughter and your mother go outside the encampment to look for new clay to make more pots. You carry bags, possibly hide bags or woven flax bags or woven willow baskets made by you or for you by other women in the clan, and you each carry two or three different flint blades made for you by some of the flint nappers, possibly male, in the clan, although you are probably pretty good at knapping flint yourself. The flint is not necessarily for defense because you plan to use the sharpened rocks to dig clay out of the riverbanks. Flint is pretty tough stuff, good for digging in dirt, but you and your mother and daughter carry more than one each because they do break. You also carry a sharpened deer antler for digging, made for you by your favorite male lover, perhaps. Again, because you have to be able to survive on your own, if necessary, sharpening a deer antler would be something you, yourself, could do, though.

After locating some promising soil, you smell it and rub it between your fingers. Learning to use your sense of smell and touch to find the finest grain clay that will hold up during firing is a keen skill you rely on in order to make pots that can be used repeatedly.

You dig some of the clay loose with the antler and help fill a few bags, but then sit down to munch on some berries grandmother found along the bank. As you sit, relaxing in the sun and eating berries, you notice some unusually colored soil across the river. The soil, exposed by washing river waters, is greenish blue, and you think to yourself that that would be a lovely color for some pots. You know from experience that clay always changes color when it is fired in heat, but you enjoy experimenting with different soils in order to find the best pottery clay.

You suggest the group take a look before continuing to fill bags, so the three of you wade across the river. One section of the blue-green soil is soft enough to be considered clay-like, so you fill one of the bags with it, but you notice that some of the soil is composed of larger crystals. Your mother suggests that the crystals can be crushed to make temper for the pots—something she has taught you is important for cooking pots to keep them from cracking during heated use; the mussel shells you often use are scarce in this part of the country, so you also think such temper might make good ceramics. She shows you how easily the crystals can be crushed using a couple of river rocks. You are both interested in the garlicky smell the crystals give off (Bray 62), but your daughter thinks they stink, so she doesn't want to put any in her bag. You send her back across the river to finish gathering the more typical clay, while you and your mother gather a couple of bags of the soil and crystals to use in an experiment, a trial and error method you use quite regularly as you look for ways to make your pots stronger, lighter, and more attractive.

You imagine how the other women will trade you lovely things for a pot that is bluish green, instead of the usual tan or orangy-red.

The three of you enthusiastically make many pots with the new clay but wait a day to allow the pots to dry as much as possible before firing them. The next day, you dig a larger pit than usual, in which you build a fire to heat the surrounding soil. You quickly fill the pit with the dry clay pottery, add wood chips or charcoal and gradually build a bonfire over the greenware. Because you want the pots to heat evenly, you use reed tubes stuck into the soil on each side of the pit to blow air into the interior of the fire after the fire begins to subside. Besides your daughter and mother, other villagers eager to share the light of the embers in the growing night linger to watch and to take turns blowing air, which your people associate with life, into the embers. The kids, especially, think blowing into the pipes, which causes the embers to burn brighter, is fun, and you had promised the children who gathered the wood for the fire some of the ceramic toys you made. They get the common clay, not the blue-green clay, which seems so promising.

Since the other villagers are interested in the big production, someone suggests roasting meat next to the bonfire, and a clever hunter produces a deer to roast over the fire as it begins to die down, unknowingly adding heat to the fire through the deer's added fat.

Hours later, after the coals are out and the large flat rocks you have placed over the ceramic to separate it from the fire have cooled, you anxiously remove the debris to examine the ceramics beneath.

To your utter disappointment, the blue-green clay did not add those colors to the ceramics. In fact, three of the smaller pieces, which obviously got too hot, not only melted to lumps of unrecognizable ceramic, but also left behind small nuggets of shiny orangy rock, some of the first refined copper in the world (although you don't know that). Your mother is quite sympathetic, suggesting the soil might be best for face paints, instead of ceramics, and you disappointingly add the lumps of ceramic to your temper pile, hoping you can reuse it to temper other pots. After showing some of the villagers the lumps of shiny rock, no one can figure out what they are, so you dump them in the encampment scrap heap, not knowing that thousands of years later some industrious archeologist will dig them up and declare them evidence for an eneolithic or pre-chalcolithic (pre-copper) period.

Later, though, when you try another experiment with soils that appear even more green than before, you end up with a similar result, but a visitor from the south tells you his clan ceramicist had a similar result, so someone in his clan tried another experiment by placing the small shiny rocks on a flat rock during another ceramic firing. When the three or four small shiny rocks came out as one thinner layer of shiny metal, one of the ceramicists, in frustration, hit the material with a large rock hammer on top of another large hard rock. The people who made flint tools and the ceramicists were all interested in the fact that they could shape the slightly hot material with the rock hammer.

So you try a similar experiment, but some of the beads of copper that appear under some of the pots appear round, so your mother fishes one out while it's still hot and pierces it with a bone awl. The bone smokes a bit, but the awl pierces the copper bead. After the bead cools, you string it on a piece of thong leather, and tie it around your mother's neck. Others admire the piece and want copper beads, too, but you have one more experiment to try first.

Next, instead of putting the small bits of arsenical copper (the arsenic is what makes the soil smell like garlic; unfortunately, it is also toxic, but people will not learn this fact for millennia) into a ceramic bowl, you wonder if the material will take the shape of whatever it is in. First, you fashion a mold by pressing your bone awls lengthwise into a flat piece of clay then fire it. Then you place that ceramic mold beneath a pot with a hole in the bottom into which you place the scrap bits of metal you have accumulated so far. Some of the melted copper fills the awl grooves in the ceramic piece below. When the copper takes the shape of the awls, with some extending outside the molds, you are delighted, but concerned, so use one of your sharpest flint blades to trim the excess copper off of the awls. The excess copper is easier to trim off than you thought it would be, so you trim the others, and end up with three long, thin copper awls.

The leather workers are impressed with how well the awls pierce even green hide, so trade you more leather bags for each awl.

Meanwhile, your mother has used a stone mace to beat some of the copper metal flat enough that she can curl the metal into a longer bead. She smooths the ends of the curled copper bead with a sandstone. Using another piece of thong leather, she strings the newer, bigger bead and ties it around your neck, where everyone can admire it and can realize that they are talking to the magic-wielding woman who controls fire to harden earth into rock and melt rock into metal. Because you have also learned to use water to cool the metal more quickly, your reputation as a magic-wielding woman who can control at least four elements--fire, rock, air, and water— grows.

Now that you've figured out that the shiny rock is useful for making tools, your experimentation expands. Someone wonders how much soil you need to make a copper blade or axe, so others begin helping you collect enough green soil to make larger pebbles of copper. Along the way, you trade some of the beads you make for other items, including food, since you don't spend as much time foraging as you used to.

You use a freshly knapped flint axe head to create a flat open-faced ceramic mold for the new copper axe. Collecting enough malachite to smelt for the axe has taken weeks. Just as you have sometimes done for firing ceramics, you dig a small hole in the ground, inserting reed pipes to create blow holes along the sides. After starting the fire to heat the sides of the pit, you add charcoal to the pit, sprinkle the crushed malachite over the surface, then add more charcoal, covering the whole pit with a thick plug of turf. You and your daughter fan the heat inside the pit by blowing through the reeds for a few hours, thereby adding the spiritual element of air (and maybe some of your own spiritual power) to this powerful process. Unbeknownst to you, "the carbon dioxide produced by the burning charcoal bonds with the carbon in the ore producing carbon dioxide and copper metal. This chemical reaction is what changes the ore into metal and not the temperature in the furnace" (Ancient1580).

After collecting enough copper bits to possibly fill your ceramic mold, you heat the metal inside a larger ceramic vessel buried in a similar pit. Once the metal is molten, you use leather rope soaked in water to grasp the ceramic vessel and pore the metal into the mold. The technique needs refinement, since copper cools quickly once removed from the heat.

Overall, finding, processing, and polishing copper is more time-consuming for Neolithic humans than shaping flint is, so the early copper implements were probably considered much more sacred than flint, primarily because its creation required more control of earth elements like fire and rock, and more knowledge of ceramic creation. If ceramic creation was dominated by women, as many scholars believe it was, it is only natural that women would have been the first people to discover and to shape copper.

ΩΩΩ

The Missing Spirit

While this imaginary process that I have described here is probably not too far off from the events that probably led to the discovery of copper, what it does not capture are the spiritual beliefs and thinking processes that undoubtedly went along with ceramic creation. Anyone who could successfully use all four different earthly elements—fire, water, earth, and air—to create new shapes, new spirits, would have undoubtedly been considered a wise or magical person, so we do not venture too far off the ancient path when we acknowledge that such a person could have been seen as a magical person like a shaman, a witch, or a wizard—someone with a closer tie to the powers of the universe.

Many speculate, for instance, that the shaman, a 40+ year old female, buried at Dolní Věstonice sometime around 26,000 BP was not only the first person in Europe to invent a kiln and to create ceramic figurines, but also a shaman who put on spectacular wet clay and fire displays to awe her audience. Her knowledge of ceramics making was probably not passed on because both she and her apprentice died.

Humanity would wait another ten thousand years before ceramics became a steady human product in many places around the world, such as Japan and China, but, if knowledge of such activities made it into myths, nearly every early ceramicist would carry forward that mystique and community respect.

Many will scoff at the idea that ancient humans believed in magic as they learned to manipulate the physical world, but given the plethora of mythology that tells us otherwise we have to be prepared to acknowledge that early pottery sacrifices, which pre-date the ritual sacrifices of metallic objects, were a probable releasing of the natural spirits of fire, earth, air, and water within the ceramic objects—a release probably intended to give those powers back to the universe. The ritual might have also released the little bit of personal spirit the potter had imbued the ceramic with, as well, since the potter was the magician imbuing the pots with her spiritual powers. The better the pots, the more powerful the magician.

Imagine, then, that you are one of those early ceramicists experimenting (what else should we call trial and error work? These processes were fundamentally science at work.) with the amount of moisture that clay will tolerate when introduced to flames; with the kinds of clay to find one that does not explode or crumble in heat;

with the intensity of heat given off by different types of wood, grasses, or excrement. Through all this trial and error, the successes would have seemed random to a degree, so that the ceramicist would return to the types of clay, the techniques of drying, the techniques of introducing the shaped clay to heat, the methods of firing the pots to find what makes the ceramic hardest and what colors and designs are the most attractive, etc., that she felt was most successful, so that each successful piece of ceramic—whether a figurine or a bowl—would have seemed special, significant, and spiritually imbued.

All this experimentation took time, so the argument many scholars make that "specialization" in producing products only occurred after people were prosperous and sedentary, thus, they assume, finally had the time to dedicate to their crafts is a false narrative. Foraging life, while often mobile, gave most ancient people plenty of time to stay near particular sites, living off of the bounty of that area for a few months, which was also plenty of time for the more craft-oriented among them to try to make new things, beginning, of course, with flint knapping, which is also a time-consuming skill, especially with some forms of rock that are especially sharp, like obsidian.

I can imagine a mother, trying to keep her colicky child from crying or making too much noise, using the soil at hand to shape a clay figure to entertain the child. Who knows why such a gift would then be thrown into a fire to explode or to harden into a more permanent toy for the child? Perhaps the act of learning to harden such toys led to the discovery that wet ceramics can explode.

As many creation stories indicate, shaping animal or human images out of clay—even without turning them into ceramic—would have been seen as an act of creation, implying that the clay could be imbued with the "breath" of spirit (such as that blown into the air tubes that help the temperature of the firing remain even, but high) that allows the shaper to recreate a nearly identical likeness, possibly giving rise to the idea of Creation itself outside of the human body, giving rise to the idea that the creation of humans was by gods who manipulated earth, water, air, and fire. While women were undoubtedly seen as creators of new humans, the magic of ceramics would have allowed anyone, regardless of gender, to become a creator, an imitator of life.

So the ceramics contained spirits, the spirits of the water, the soil, the air, and the fire, not to mention a little of the shaman's spirit, which explains why so many cultures broke pots when they buried them with the dead. The fractured pot not only symbolizes a human spirit passing from the physical world into the spiritual realm, but also symbolizes the pot's own elemental and spiritual release.

Indeed, some of the early discoveries of copper slag in ceramicist's pit or bonfire kilns might also have been seen as a bright, shining bit of spirit separated from the clay by the fire. Not every ceramicist would have thought that thought, since we have evidence of small globules of copper tossed in midden heaps, discarded as useless.

While not every ceramicist would have been a woman, the likelihood that most were women is high, given that the women seemed to have been responsible, just as American Indian women were, for building their physical homes—first out of grasses, some out of hides, a few out of mammoth and whale bones, later out of rock and clay bricks—and for creating the household goods—from storage containers like baskets and parfleche boxes and bags to blankets, clothing, and floor matting.

Women were, by nature and by intellect, creators, fashioning life goods to make their own and other people's lives easier and more pleasing—a spiritual element many disregard as frivolous.

Note that I am not saying that men were not also creators, but, given the close ties of the idea of home and the hearth, which we will explore more later, women were probably once (and many still are) the creators of the actual physical place, not to mention the idea, of home.

Chapter Twelve
What Is Animal Husbandry?
Why Isn't It Called Animal Wifery?

The domestication of animals was a long, trial-and-error process for most areas of the world. As Rossel, Marshall, Peters, Pilgram, Adams, and O'Connor point out, "there is increasing emphasis on domestication as a nonlinear microevolutionary process influenced by the behavior of individual species, the nature of human environments, and management practices." In other words, domestication worked in stages, sometimes ending tragically, sometimes ending in success as human beings learned to live with, to protect, and to depend on animals native to their locations around the world to improve human lives. Whether we depended on the animal for protection, for food, or for labor, humanity's awareness of being able to control aspects of nature probably became heightened by our domestication of other animals, enabling us to view ourselves as above other animals[94].

Humans domesticated dogs in nearly every part of the world, some as far back as 12,000 or from some extremely generous estimates even 20,000 years ago. Dogs follow a pack mentality, so the earliest humans simply had to become the pack leaders for the dogs they domesticated in order to keep dogs in their camps for warmth, for security as early alert systems, as hunting partners, and as loyal companions. As many preliterate culture's stories indicate, dogs have long symbolized in art and literature the idea of fierce loyalty.

---

[94] We now know, however, that many animals often do things we associate with being "human," such as being able to use either sounds or symbols to communicate, having feelings, using tools, building art (puffer fishes are supreme, but some birds' nests are feats of engineering). The seemingly only truly unique thing humans do that no other animal does is weave clothing, but that skill, quite possibly, came from watching birds weave nests and spiders build webs. Even the idea of sewing could have been discovered by people examining or watching birds like the common tailorbird sew leaves together to make its nests. We should also acknowledge that most rituals and dances humans developed came from copying various animals, such as birds during mating season.

There a potentially hundreds of ways the domestication of dogs happened, possibly even happening earlier for some individual dogs and humans than we currently know, but the simplest way, still, is to take the pups of wild dogs and raise them among humans. Dangers still exist in the relationships humans have with dogs even among breeds that have long been domesticated, such as cocker spaniels which appear innocuous enough but are among the most frequent biters, but for most cultures in the world dogs symbolize loyalty and unquestioning love. The transient and semi-sedentary tribal nations of North American used dogs to pull travois in order to assist them in moving from place to place prior to the reintroduction of horses to the Western hemisphere. The travois is a simple device, essentially two long poles strapped to a human or animal's back that drag along the ground, allowing goods to be strapped between the poles, so they were probably first pulled by human beings, then by dogs, then later by horses, goats, and cattle not only for moving camp, but also for carrying goods for trade. Such packing devices work well over smoother land, but would have been untenable in mountainous terrain, although a resourceful group of people could have cut new travois as they moved from mountain to mountain as long as the mountains had trees.

Most scholars assume that early herd animal domestication was to maintain easier access to meat or milk from such herd animals, but, if they are raising the young first, they might have just been attracted to the liveliness exhibited by most lambs or kids, so kept the young for a form of entertainment after slaughtering the mothers, thus learning that these wild, nimble creatures, too, could be tamed. It was probably only after goats were domesticated that humans began using animal milk as food.

Other animals were naturally harder to domesticate because they would not have been as trusting of humans, although many species[95] are just as curious as dogs about other things. Many urban people are often fascinated, for instance, by how cattle will crowd over to a fence just to check out what is going on because most cattle are, by and large, curious creatures. After all, an adult bovine has

---

[95] Speaking from my experience growing up on a farm, cattle are extremely curious creatures, seemingly obliged to check out any new forms or features in order to ascertain whether to flea or to stay.

little to fear as long as they are near their herd because horns and hooves make excellent defense mechanisms, as tourists who ignore warnings and get too close to American bison learn, repeatedly.

By most estimates, goats and sheep were probably the first herd animals domesticated by early humans, starting roughly 10,000 years ago (Balter). While goats and sheep are both in the subfamily Caprinae, they are different species, and the inhabitants of Aşikli Höyük in central Turkey were among the first humans to progressively domesticate local sheep in an archeological snapshot of a period between 10,500 years ago to 9500 years ago, by which time "sheep represented nearly 90% of all animals at the site" with clear indications that female sheep (ewes) were being kept to increase the number of offspring, and males culled for eating or, possibly, for spiritual rituals" (Balter).[96]

The ancient Sumerians still used sheep livers to make predictions circa 3000 BCE, so such a practice was probably more widespread and common throughout the eastern Mediterranean than we know. The idea of "scapegoating," tying a community's transgressions against a god or gods to a goat that was then slaughtered, was not a new ritual for the Hebrews of the Judaic bible, but was undoubtedly a common social, possibly annual ritual throughout the region to allow individuals to be forgiven for their transgressions against others in order to ward off retribution by the gods.[97]

---

[96] Note that these humans did not realize that heterosexual intercourse caused pregnancy because sheep and goat gestation still takes several months during which time each ewe probably mated with several males. It was not until humans domesticated the only monogamous, short gestation animal ever domesticated, swans, that they would have made the connection between heterosexual intercourse and pregnancy. So the reason these people might have culled males was simple--they did not have babies—however, there are many cultures, still, around the Mediterranean that keep equal numbers of both genders in their sheep herds, so I have to wonder how these archaeologists came to the conclusion that the males were being slaughtered and the females kept for breeding.

[97] The ritual of scapegoating involved a community ritual in which the negative spiritual energies caused by doing immoral or unholy acts against others which could bring down the wrath of the gods on the community were believed to be expunged by casting them magically onto a goat, which was then sent out into the desert or similar wasteland to die. With the animal's death, the negative energies of the community died as well, bring hope for appeasement of the fickle gods and

Clearly, animal domestication occurred with various levels of trial and error for thousands of years with both humans and dogs learning to herd sheep and goats in such a way as to protect the herds from predators, to provide meat and milk for human consumption, hides and skins (later wool) for clothing and implements, possibly also for temporary forms of shelter, hooves for rattles, and horns for musical instruments and hoeing. In fact, the herder's signature crook becomes a symbol for kingship in many cultures, including Egypt with many gods being titled as a culture's "shepherd god," such as Dumuzi of ancient Sumeria. The city of Uruk's name literally means "the sheep pen" or "sheepfold"[98] because of its walls that made it look like a humongous sheep pen from a distance, which means it would have been an easy step to see its inhabitants as sheep needing protection, which is why Dumuzi was considered its patron shepherd god.

A theory I have seen no other scholar consider is the possibility that ancient human beings decided to gather in larger groups, originally, in order to mimic herding animals' annual mating and birthing moments, which is a very likely explanation for Neolithic sites like Göbekli Tepe, where the people gathered, built ritual sites, then later abandoned them, only to return to build new ones later. While scholars like Jan Driessen (Birth 1) argue that such commensality gatherings, wherein the people gather to eat and drink together to celebrate, he assumes, some divinity or another, we can compare this kind of gathering to what American Indian nations used to do pre-contact when they gathered people of

---

harmony and peace to the community. Much later, cultures in southeastern England and Wales practiced "sin eating," wherein the sins of a dead person are "cast into" a cake on the person's coffin, which a volunteer "sin eater" will then consume, so the deceased could pass untainted into the afterlife.

[98] According to Diane Wokstein and Samuel Noah Kramer, "In Sumerian the word for sheepfold, womb, vulva, loins, and lap is the same" (146). Kramer, clearly, believes the story of "Inanna and Enki" should be understood as Inanna's "wish to be 'fertilized' by the sexual as well as the magical, spiritual, and cultural powers of life" (147), still making the patriarchal tie between sex and procreation without proof that the Sumerians knew the connection when the tale was created. Even Elinor W. Gadon makes the same assumption, equating the cycle of Inanna stories with "the entire cycle of fertility from birth and death to rebirth," despite the fact that Inanna never gives birth (132).

different plains nations together in powwows, often engaging in sun dance rituals.[99] Just because these separate nations gathered to eat, drink, and dance together, does not mean they viewed the spiritual or magical elements of the world in exactly the same way.

Many scholars have speculated about why humanity essentially domesticated ourselves by learning to live in cities, with most only viewing the ancient communities through modern eyes, seeing only economic rationale or religious reasons. However, since human beings proved over several millennia that they could gather together, just as herd animals do and despite cultural differences, to find mates and to share birthing experiences, it is also highly probable that the first settled towns were formed for the same reasons. What androcentric scholars must learn to recognize is the fact that sex, food, and cooperative sharing, especially of information like how to make pregnancy and birth more successful, are the foundations of what led to "civilization."[100]

Relatively, it did not take much time before humans, at least those living around the eastern Mediterranean Sea, stopped seeing themselves as other animals among nature, but as human beings set apart from other creatures. Most cultures around the world used names for themselves that literally meant "the real people," and their words for other people meaning something akin to "those ignorant fools over there." Therefore, after humans learned to separate themselves from animals, it was only one more small step

---

[99] The Lakota/Dakota/Nakota had annual sun dance gatherings in order to allow individuals to make personal sacrifices to the spirits, but many misinterpret these sun dances as sun worship, when the name actually comes from the time of year—the summer solstice, when the sun stands in the sky the longest. This time of year was important because one aspect of sun dancing was a personal blood and endurance sacrifice that an individual endured—either from taking a thousand cuts to the skin, or from having a tether to the sun pole tied to their chests, so that they had to dance every day and night until the tether or their skin broke. How many other plains Indian nations practiced this same ritual?

[100] While scholars like Jan Driessen believe crises had to occur to create social changes (Birth 1), which makes sense up to a point, many still fail to realize that human beings are more cooperative, more nurturing, even toward people different than ourselves, than we are defensive or combative. Ironically, Driessen uses the biblical account of the golden calf, a clear indication that the mythical people in question preferred a nurturing goddess to a vengeful god, as an example of a crisis situation, as though the event was real history.

to separate themselves from every other human culture on the planet.

Understanding why humans did not more fully understand the procreation process until circa 2400 BCE is important to emphasize, though, especially since so many people will argue that humanity learned about procreation immediately after domesticating animals.

ΩΩΩΩ

The gestation cycle for both sheep and goats is approximately 150 days, or 21 weeks, approximately half of the time needed for humans, but still too long for early humans to have made a clear, verifiable connection between heterosexual intercourse and procreation. A primary difference between human procreation and sheep-goat procreation occurs because the ewes or does tend to go into an annual heat cycle in the autumn, so the animals' sexuality would have been seen as tied to a seasonal cycle, rather than being independently procreative like humans. Since human women were tied to moon cycles, menstruation, a decidedly different form of fertility cycling than the season-based procreation cycles of other animals, humans had no reason to believe heterosexual intercourse was tied to procreation even after they domesticated animals.

Like most herd animals, sheep and goats are matriarchal, with wild and domestic herds both controlled by a dominant ewe or nanny, further tying human females, who were probably some of the earliest nurturers of wild lambs and kids at the beginning of the domestication process, to the natural world as the more magical and more spiritually prominent gender because of their ability to "conjure" babies out of the workings of their own bodies.

While we now have archeological evidence that women crafted ceramic sippy cups for their babies (Whelan), we also know from anthropology that caprinae stomachs and bladders also make great food storage vessels because, even dried, they are relatively impermeable to liquids, so would have also made great substitute udders for baby calves. In fact, the storage of milk in rennet filled stomachs led to the creation of cheeses and yogurts (Fisberg & Machado). Since most archeologists and anthropologists claim that

151

most meat processing and cooking were performed by women for their families, clans, and community celebrations, it would have been women who discovered the properties of the stomachs and bladders, thus women would have been the first people to "invent" or discover these foods, and the first to come up with alternative feeding methods for baby animals they wanted to domesticate.

Peoples from the eastern end of the Mediterranean first developed gyro or kebab meats, which are traditionally made from lamb cooked over a rotisserie spit. While various chefs have been credited with the invention of the fast food kebab or gyro, the knowledge of cooking multiple meats together to combine their flavors as they roast is undoubtedly much, much older from people who regularly ate meat roasted over open fires for community celebrations.

Traditionally, in most preliterate cultures, a portion of the meat, usually the most fatty portion because it smells great as it cooks and helps keep the fire hot, was given, first, to the fire upon which it was cooked in order to honor the divinities or gods that provided the meat and protected the community. Smoke, it was often believed, carried those great cooking smells to the realm of the gods as it rose through the air. As hierarchies rose, it is quite possible that such fatty portions were also fed to the spiritual elites of each group, leading us to the "big" people who were identified as leaders of communities.[101]

Domestication of cattle was not as widespread initially, possibly only occurring in three places, the Indus Valley, the Fertile Crescent, and the western desert of Egypt, starting roughly 10,000 years ago,[102] but humans began to learn a few millennia after they

---

[101] Because obesity can prolong women's menstrual cycles, the obese female figurines found throughout the world were undoubtedly paying homage to such substantial women who were spiritual leaders in their communities, since their prolonged menses cycles would have demonstrated that they were more in tune with the cycles of the earth and the universe than other women, lending them an aura of more spiritual power. These obese women could have been the original "big bellies" referred to in many cultures, whose term for such powerful people eventually comes to mean "ruler" or "king."

[102] It makes sense that, once human beings learned to domesticate one form of herd animal, that they could figure out how to domesticate others more easily, which is why many people assume that most animal domestication started

domesticated cattle that these larger beasts could be used as beasts of burden, not just for meat, leather, and other byproducts (Pitt, Sevane, Nicolazzi, MacHugh, Park, Colli, Martinez, Bruford & Orozco-terWengel). However, humans probably did not learn how to utilize oxen, cattle trained specifically as beasts of burden for pulling carts, until after donkeys, who were eaten less frequently than sheep, goats, and cattle, were domesticated starting roughly 10,000 years ago in Africa (Rossel, Marshall, Peters, Pilgram, Adams, & O'Connor). So far, there is little evidence that allows us to determine whether donkeys or cattle were the first beasts of burden after dogs, but they were probably first used in the fields pulling plows, which had been being pulled by human effort alone once the first plow blades were created. The use of cattle to pull plows undoubtedly allowed farmers to expand their crop fields, leading to a surplus of food that could be stored, first in small communal granaries, then, later, in temple granaries, and, finally, in granaries controlled solely by rulers of a kingdom.

However, before going too far, we must note that there is a decided difference in the physical abilities of the two species, donkeys and cattle. Donkeys, while smaller than cattle, are more sure footed in rocky terrain, but cattle, which can grow much heavier than donkeys, can pull more weight. Sometime after these larger animals were domesticated, the travois became sledges (Uckelmann 399) or, in colder and snowier regions, sleighs, since it was some time, circa 4000 BCE, before wheels were invented for use in ceramics, then someone in present day Iraq decided to try to use the wheels to create two-wheeled vehicles by adding solid wood wheels to a simple rotating axle beneath the box of the sledge, creating the first cart.[103] Later, since all wild cattle, both males and females,[104] had horns, both were used to pull carts because, with a

---

approximately 10,000 years ago without qualifying which animals were domesticated when or where.

[103] Some of the earliest war carts recovered by archaeologists had solid wood wheels, so it took more experimentation before the lighter spoked wheels used on chariots were invented (Stiebing 61).

[104] It is important to note that many cultures—from Minoans and the Maltese to the residents of Çatalhöyük—displayed real or plaster cattle heads with horns in their homes and on their altars. While many people assume these bovine heads are all bulls, there is little or no proof that they are, since both genders of aurochs,

standard yoke, which has changed very little over the millennia, the horns are used against the yoke to slow a cart going downhill—a natural form of brake (Ford & Kreutzer). People probably learned to yoke cows, instead of bulls, first, since cows are more docile than bulls, so it would not have been until some experimentation with castration[105] that human beings learned that steers, castrated male cattle, actually make very good beasts of burden because they are not only larger, heavier, and stronger than the smaller cows, but also easier to handle when breeding season comes around, since they lack any interest in sex (Helmer, Blaise, Goruichon & Saňa-Seguí 89).

The primary advantage for a mobile family group for using cows instead of steers as oxen when they were on the move to new territory was that they could milk the cows every day, thus supplying an important source of protein and sugars to their diets. In the same way American pioneers discovered that cows made

---

the wild cattle domesticated by the people around the Mediterranean, had significantly sized horns. Many of the scholars who assume the heads must be bull heads overlook the symbolism of the shape of a cow's head, which is shaped almost exactly the same as women's reproductive organs—the muzzle being the vagina, the broad skull the womb, and the horns the fallopian tubes. Serious examination needs to be made of such findings in order to sort female magical strength from male brute strength symbolically.

[105] Why would humans experiment with castrating animals, you might ask, especially if they knew nothing about male contributions to procreation? Most foragers also hunted big game, sometimes on a seasonal basis. Such communities endeavored to make use of every part of the animals they killed, so that they would have regularly eaten the testicles of the males, whether they were goats, sheep, horses, cattle, camelids, boars, or even bears. Because humans can be observant, some would have noticed that eating the testicles actually gave people a temporary boost in vitality, even the aged among them. We now know that this boost is due to the temporary surge of testosterone we get from consuming animal testicles, but this kind of phenomenon undoubtedly also led to humans eating other body parts symbolically—the heart or liver for more courage, the tongue for better speaking abilities, etc. Even though most of these other sympathetic magic connections are not based on real biology, except for the iron we get from liver, if the people assume they work the same way testicles do, they could have felt a placebo effect. After witnessing many "calf fries" or "mountain oyster fries" in my lifetime, I have realized that many human beings still bear witness to the sense of vitality that consuming animal testes can give, although, reportedly, that boost is not more than we get from consuming some vegetables.

better oxen for crossing the western plains, ancient humans around the Mediterranean probably used female oxen for the same reasons.

What we can speculate, however, is that pulling travois probably preceded pulling sledges by several thousand years across the globe. Much is made, in fact, of the fact that the peoples indigenous to the Western hemisphere never invented the wheel, despite the fact that they, too, built monumental architecture from megalithic stone and developed innovative agricultural techniques, such as the *andenes*, terraced fields constructed to increase crop production in mountainous regions like the Andes, by controlling soil erosion, irrigation, and soil temperatures (Meghji) in order to feed millions, producing long lasting empiric cultures. Just as the Egyptians depicted themselves using sledges to move monumental stone from riverbanks to temples and pyramids, such devices were also probably used in both Central and South America, but their sledges were probably pulled by human effort, instead of animal.

As North American Indian nations demonstrated, dog travois were the most common form of transport for goods for thousands of years. In some areas of the world, goats were preferred over sheep as beasts of burden, although neither are native to the Western Hemisphere. In South America, the beasts of burden are their native camelids—the alpaca, llama, guanaco, and the vicuña—although the vicuna is so small that it does not make a very good beast of burden, and many doubt that the larger camelids were used to move stone. The small vicuña, though, has wool that is highly coveted and considered by some to be the finest animal fiber in the world. Many of the South American cultures used the camelids in their religious rituals, as well, usually as sacrifices to the gods.

It is important to note that sheep wool, the kinky curly hair we so often associate with spun wool, only arose, according to experts, after sheep from the eastern Mediterranean were moved north by herders, where they developed the curlier wool that more naturally mats for weaving (Bartosiewicz 335). Early humans probably discovered how wool can form felt by sticking it in their moccasin-like leather footwear in cold climates, such as that worn by Ötzi the Iceman who crossed the Alps circa 3350 BCE.

Cattle are also mentioned several times in *The Iliad*. In Homer's description of the new armor Hephaestus makes for

Achilles in T*he Iliad*, he describes a "herd of straight-horned cattle" (*The Iliad*, 573-86), the bull of which is being attacked by a pair of lions, probably lionesses who are the primary hunters in a pride, which cattle herders and their dogs try to drive off. Horned cattle become important all the way around the Mediterranean Sea because, as some scholars seem unaware, all cattle—both bulls and cows--at that time had horns. Yet an androcentric bias will lead countless scholars to **assume** that all cattle horns had to be from bulls.

The beginning of *The Iliad* requires the Mycenaeans to sacrifice a hecatomb of cattle to the god Apollo. A hecatomb traditionally takes 100 heads of cattle, with many scholars insisting that those cattle had to be intact bulls, although supposedly 12 exceptionally fine cows will suffice in times of great need (Jones 5). Since the cattle would have also served to feed the thousands of men fighting at Troy, and indeed, does feed the men who go with Odysseus to appease Apollo, such a sacrifice must have been dear.

The invention of the wheel for carts has long been lauded as a dramatic revolution for humanity, which it was. But the real revolution (which stems from a wheel term—to revolve) came after horses were domesticated, not just for food, but as additional beasts of burden. At first, horses probably were ridden, allowing much faster raids on other settlements, raids which, initially, were mostly for thievery or intimidation, not outright massacre.

If the early horse riders near and around the Mediterranean Sea and especially up into the Steppes regions of Eurasia were anything like the Plains Indians of North America, they used horses to steal other horses, and young men's reputations—and proof of their ability to support a wife and her family—were made via horse raids. The more horses a man had, the wealthier, and, presumably, more influential, he was.

We must acknowledge, however, that it would have taken years of breeding the best stock to create the larger, faster horses that came to be known as Arabians, which led to the even larger horses known as Thoroughbreds. In fact, Thoroughbreds are so inbred now, causing a loss of genetic diversity with 97% of the Thoroughbred population tracing its ancestry back to Northern

Dancer, that many Thoroughbreds suffer from genetic-caused diseases (McGivney, Han, Corduff, Katz, Tozaki, MacHugh, & Hill).

Initially, domesticated horses' sizes were increased, probably after castration became an animal husbandry tool, since castration at a specific time in a stallion's life, transforming him into a gentler gelding, also makes him heavier than intact stallions. Because people would have wanted larger horses to pull wagons, just like cattle did as trained oxen, it would not have been long before humans discovered that horses are considerably faster and have more stamina at speeds that outstrip cattle.

Within a few millennia of the domestication of cattle around the Mediterranean and in Eurasia, many cultures in these same regions began domesticating horses. Mary Bachvarova believes that early communities hunted horses for eating, but some might have begun herding them as food sources by 4800 BCE, although evidence for domestication, especially for riding, comes much later, around 3700-3000 BCE.

As wheel technology grew—from solid wood wheels to spoked wheels to the extremely light chariot wheels—horses proved to be even more valuable for "elites" because they became excellent war weapons that verged on legendary.

While Homer features chariot use prominently in *The Iliad*, many scholars disagree about whether or not his descriptions of their use are historically accurate because there was probably a lot of innovation about the structure of chariots between 1250 BCE, when the battle for Troy was supposed to have occurred and 750 BCE, the earliest time most scholars agree upon when *The Iliad* was probably first written down.

Achilles' horses, Balius and Xanthus, were sons of the west wind, Zephyrus, so, like Achilles, were part divine. These magical horses were originally gifts from Poseidon to Achilles' mortal father, King Peleus of Phthia when he married Thetis, an ocean goddess. So human-like were this pair of horses that they wept when Patroclus, Achilles' cousin and close companion, was killed in battle (Green, *The Iliad*, Book 17: 426-7 & 436-9).

In fact, Homer tells us that Troy is most famed for the horses the king breeds.

With fleet horses and light chariots, warriors could sweep onto a battlefield either to strike their foes as they pass or to dismount and fight face-to-face, with their chariots and drivers waiting to sweep them from the battlefield quickly, often to the detriment of common foot soldiers. Therefore, capturing a foe's horses and chariot, called "high-status symbols par excellence" by Josho Brouwers, was just as much of a victory as looting his armor and weapons.

ΩΩΩ

So Why Is It Called Animal Husbandry?

It is highly likely that women created most of the symbolism surrounding animals and plants that have been used in so many cultures around the world, symbols that they wove into textiles and painted on ceramic vessels and cave walls. Women also undoubtedly discovered unique foods like yogurt and cheeses through the processing and use of animal stomachs to carry milk. Similarly, it is also quite likely that women, who would have more experience nurturing other small, helpless beings like babies so could easily adapt those skills for small animals, initiated the domestication of animals. Far too many scholars are far too willing to relegate the domestication of large animals to men, simply because of androcentric biases.[106]

We know that female gods were also often worshipped as the Goddess of Animals and goddesses around the world are often depicted frequently with lions, snakes, dogs, birds, and horses, so that, clearly, women have long been associated with both wild and domestic animals.

Why, then, is the process of caring for animals as domesticated livestock not called Animal Wifery?

For the answer to that question, we must turn to linguistics and take off our 21st century glasses. The concept of a "wife" is not

---

[106] Adovasio, Soffer & Page, for instance, willingly divide labor along gender lines, despite originally arguing against such simplistic biases, when they claim women domesticated plants, but, surely, men domesticated "larger animals" (257). Why a male would be more likely to domesticate larger animals is never explained by these scholars, but they apparently assume that such animals can only be handled with brute force.

158

quite the same in every culture, and there is clear evidence that the idea of a "wife" as the female spouse of a heterosexual couple probably arose much more recently than most people think, probably within the last 4000 years. In fact, the term "husband" has had many different connotations, as well, but both "housewife" and "husband" are parallel terms that tie the two people in question to the House, which we should think of as a familial dynasty.

As with cultures that still practice matriarchal structures like matrilocal marriage and burials and matrilineal lines of descent, the House was a concept as well as a specific dwelling place.

Literally, women were the Home Makers. They tanned and stitched together hides for the first tipis. They interwove tree branches to create wikiups or wigwams, temporary domed dwellings often covered by tree bark, woven reed mats, or tanned hides. Women processed stone for perimeters and made mud bricks, which were a natural transition either to or from learning how to make ceramic vessels for storage and cooking. Small round huts with grass thatched roofs, often called rondavels, were made by women throughout ancient Eurasia and still in parts of Africa.

Undoubtedly, the Matriarch of a family was the one who directed where her sisters, daughters, nieces, and granddaughters could build their own homes, and she became the foundation of her family's House.

Therefore, a "housewife" as a concept arose out of the idea of a particular woman, bonded by blood to members of her family, as the one responsible for the safety of her family members, so would have been the person organizing and directing food gathering, processing, and storage, clothing processing, and organizing shelter for everyone of her blood line. This responsibility would have been shared among her eldest blood female kin.

The concept of a "husband," literally, a man "bound" to a "house" through some recognized connection, such as through committing his labor to or outright marrying a daughter of the Matriarch, was a man who dedicated his life to the prosperity of both his kin, being a descendant of his own Matriarch, and to the prosperity of his adopted kin through marriage or a concept similar to a marital "bond."

Traditionally, among the Lakota, for instance, a young man proposed marriage to a woman, but it was her family who decided if he was worthy, no matter how in love with him the woman might be, because he had to support not just her with the means to provide food, shelter, and clothing, but also members of her family in times of need, such as if her mother's sons (the wife's brothers) were killed in combat. In order to prove his worth, individual men in many cultures had to provide evidence of wealth or the ability to procure wealth for the bride's family, often called a Bride Gift by many scholars or a Bride Price by androcentrists. If a Lakota brought several horses he had stolen in a raid, for instance, he proved he was brave enough to face an enemy and strong enough to lead away a number of horses, ensuring his ability to provide safety and security for the woman to whom he was pledging his fealty and for her family—his "bonded" or adopted family.

Since men bonded themselves to a House in order to give their labor to that chosen family, as well as was expected to continue to fulfill obligations to his blood kin, and since many raiding cultures, such as those from the Steppes where horses prospered on the grassy plains, appear to put value in male interactions with both protecting and stealing such herds of horses and cattle, they were literally "husbanding," pledging, bonding, or otherwise securing the animals for their respective Houses.

While this concept might not have been widespread among ancient cultures originally, another factor might have played into the adopted use of the idea: castration.

Because castration leaves no discernable effect on most male skeletons, we have few methods for determining when castration began to be practiced on herd animals like cattle and horses, let alone on humans. One study that used horn cores and phalanx bones, however, possibly determined that castration was commonly practiced on cattle among some pre-pottery Neolithic cultures (circa 6000 BCE), namely at Tell Aswad in Syria and at Cafer Hŏyük in Turkey, in order to create calmer beasts of burden to pull logs, sledges, and to carry heavy items like bags of salt (Helmer, Blaise, Gourichon, & Saňa-Seguí 89). Since not every bull was castrated, even these people had little reason to suspect that males played a biological role in procreation.

However, if the act of castration was viewed post-procreation discovery as taking the male's "manhood" from him, there would have been no way he could sexually satisfy any prospective sexual partners. He also could not have fathered children.

While we cannot fully rule out bestiality as a practice among ancient cultures, since it could have had spiritual or predictive powers for some practitioners, it is doubtful that men were seen as "husband" of their animal herds in a sexual sense. However, if people began castrating bulls and stallions in order to control the breeding populations, once humans discovered procreation, the shepherds became the people who literally controlled those breeding interactions because they deliberately choose what two animals to "bond" in sexual union. Such purposeful matings are yet to be proven in the archeological record via DNA, so it is clearly an area that needs to be fully explored over the 10,000 years humans have lived with animals.

Controlling which animals could breed led to another revolution that continues to affect animal species around the world today. Long before human beings even understood what genetics might be (e.g. Mendel and his pea experiments in the late 1860s ACE), humans dabbled in genetically modifying our domesticated animals, breeding selectively for size, color, temperament, and other desirable attributes such as their fighting ability (e.g. roosters and dogs). Through such experimentation, humans have created not only the genetic diseases now rampant among Thoroughbred horses, but also derived the Teacup Poodle and the English Mastiff. Both the tiny poodle and the huge mastiff are the same species, *canis lupus familiaris*. Yet, ironically, dinosaurs with similar traits but minor distinct features each get their own scientific names; it is unfortunate that we cannot use DNA tests on such ancient species.

My own personal theory is that animals, since ancient humans also saw themselves as part of the natural world, were beings who were bound or bonded, eventually, to specific households, meaning they were clearly identified as belonging to a particular family or clan. Just as men were bonded to specific matrilineal households, so became "house bonds" or "husbands" for those houses and the women who ran them, so, too, did animals become "husbanded" to a female family line, which was then obligated to care for them.

Stories like those of Queen Pasiphaë, who was cursed by Poseidon after her husband refused to sacrifice a beautiful bull the god had sent him for that purpose, quite possibly make fun of the lingering idea that an animal like a bull can be "husbanded" to a woman's lineage. Pasiphaë falls in love with the beautiful bull, contrives to have sexual intercourse with it through a machine built by Daedalus, thereby conceiving the monster Minotaur, so that Daedalus must create a great labyrinth below the palace to imprison the beast. This story demonstrates that the ancient Greeks knew by this time that heterosexual coitus produces offspring and indicates that bestiality could also have still been a common practice, but it also proves that humans still did not fully understand how genetic traits were passed. There is a genetic disease called Occipital Horn Syndrome in which the mother passes on the disorder, usually to male children who can form horn-like bony protuberances on their skulls, usually on the back of the skull. The syndrome is often accompanied by other physical deformation, such as an extremely long neck, joint hyperflexibility, and coarse hair, which could exacerbate a child's appearance to be cow-like.

In addition, the tale could also demonstrate an ongoing male concern about the size of their own genitalia. Some men have long been jealous of the length bull penises can reach, with one habit remaining popular even today: canes made of preserved bull penises (yes, that's how long they can be).

Humans have had a long history, much of it unwritten and some of it written clearly by us, with other animals. Whether we have "husbanded" them well remains to be seen.

Chapter Thirteen
Human Adaptation of Bird Cultures:
The Advent of the Winged People & Gods

I am always amused by theorists who posit that humanity must have contacted "aliens" or that there had to be a super advanced human civilization, now completely disappeared, because of some of the odd ways human effigies have been depicted by our ancient ancestors, as though human beings should have known how to replicate human facial features, especially, right from the start of human-created art.

What the alien-theorists overlook is the influence on humanity, thus on the cultural symbols we developed, of various animals. Ancient humans, after all, viewed themselves as part of a greater world filled with living spirits, and they would have observed and mimicked many different animal behaviors for many different reasons, but especially for ritual purposes in order to align themselves with the imbued spirituality of their surroundings. This belief in the spiritual nature of nature is a basic tenet of pantheism, so it is reasonable to assume that our ancestors sought to align themselves with the spiritual power of various natural spirits in order to create sympathetic magic. For instance, many cultures had snake dances, wherein they imitated the movements of snakes, often to propitiate rain because rain runoff can look like snakes moving as it seeks lower elevations, and because so many snakes winter underground, and their emergence and frenzied mating, often in balls of snakes writhing around each other, seemed to bring spring rains.

Charles A. Eastman, a man raised as a traditional Lakota in the late 1800s, called wild creatures the "animal people," telling stories about beavers, mountain lions, and eagles as though they were human beings. One such group of animal people that influenced humans across the globe was the Aves class—birds. For Eastman and many American Indians like him, birds are the "winged people." Most pantheistic people viewed other animals as other people, and many especially admired the raptors, like the American Bald Eagle, because they could fly extremely high and had eyesight so keen they could spot small prey from those lofty heights, sometimes dropping down like lightning to snatch that prey—just like gods, many of whom were imagined to inhabit the skies or "heavens."

Even during today's American Indian powwows—formal dance competitions using traditional tribal regalia—there are still several dance categories that note the influence of birds in both the dance movements, such as the Fancy Dance, which imitates the mating and defense dances of prairie chickens, a type of plains grouse with showy tail feathers. Eagles[107] are still so revered that, if one eagle feather should fall to the floor from a fan or other piece of regalia, everything at the powwow stops until the feather can be retrieved in a respectful, often ritual, manner.

However, humans learned more than just costuming and dances for spiritual rituals from birds.

Just judging how some dancers imitate bird movements, such as strutting and swooping, ancient humans probably developed some idea about how to signal others during theatrical productions or even from a distance. Birds signal each other with flashes of wings, tail feathers, and from agile maneuvering in the air. We learned to do it with colored flags and call it semaphore.

However, bird signaling is much more complex than most people assume. For instance, many birds are known to imitate snakes, such as the black capped chickadee, when trying to deter predators from stealing their eggs or young (Lesley, "Black-Capped"). Some birds are clever enough to even act, pretending to be injured to lead possible predators away from their nests, like the common killdeer does. Furthermore, almost all birdsong is communication, whether it is a parent attempting to coax a hatchling to open its mouth for feeding, a parent attempting to coax a fledgling to take its first flight, or males attempting to woo females for mating. Such singing undoubtedly impressed early humans, who learned to imitate the sounds of various birds for their own signals, including pushing them to imitate birds through reeds, something that eventually gave us flutes—one of the almost

---

[107] American Indians are, by law, the only people who can legally handle eagle parts, including the feathers in the United States, which is a clear indication about how important these particular birds are to their spiritual beliefs.

ubiquitous musical instruments for early humans, besides drums, an instrument naturally shown to us through various species of woodpeckers.

Cormorant Drying Its Wings in the Sun
Beverly Strouse

Humans also imitate birds in many spiritual rituals, such as showing obeisance to divinity, much like the picture cormorant (above) as it dries its wings in the sun, which many goddess figures imitate in ancient art, like the famous Ishtar figurine below, wherein the goddess not only has wings and bird-like feet, but also holds her hands aloft[108] in the same manner as the cormorant above. This posture is common throughout the ancient world to signal prayer, an appeal to the gods or to the spirits dwelling within a natural feature, such as a rock, tree, or stream.

---

[108] Toby Wilkinson believes that figures holding their arms aloft, especially high up beside their heads, are imitating cattle horns, so might "have been taking part in a ritual 'cow dance'" (155).

Sketch of Ishtar, Babylon
Author

Our ancestors also learned how to make houses out of grasses, sticks, and mud-and-daub from observing birds' nests. Such lessons also taught us how to sew (like the tailorbird), weave baskets (like weaverbirds), to twist grass ropes (like orioles), to use mud to make wattle and daub houses (like swallows), to use fur to make felt (most birds will use hair to line their nests), which undoubtedly led to learning how to spin and to weave strands of hair to make yarn, to build reed mats and boats, and to shape clay into figurines and pottery. By watching birds weave nests and spiders weave webs, we also learned to weave cloth and to stitch seams. From observing which mosses some birds collect to line their nests for warmth and absorbency (Lesley, Black-Capped), humans would have also learned to use mosses to stuff woven cloth for bedding, as well as to staunch bleeding cuts, including use the moss as diaper material and menstrual or ritual blood absorbers. We know that the Maya and Aztecs used such fibers, which they often turned into paper, for burning ritual blood offerings at certain ceremonies.

Many birds peck on wood to shape the trees for their use, which includes hunting for insects, enlarging holes for nests, and

166

marking trees to allow sap to drip, such as the Yellow-bellied Sapsucker (Lesley, 7 Fascinating). These sticky syrups also attract hummingbirds, butterflies, and other insects to consume their syrups, while deterring reptiles that climb the trees to feed on eggs and young. Observing such patient and persistent birds, humans would have learned to shape wood and to access sap to use as glue and sweeteners.

Birds also demonstrate every spring how to attract and woo a sex partner. As Charlotte Perkins-Gilman put it, males who sought "the overmastering necessity upon [the male bird] that he secure the favor of the female has made [him] blossom like a butterfly" (8). Male bird plumage tends to be brighter, more showy than female plumage, which convinced many men in pre-literate cultures that it was the men who needed to dress in vibrantly colorful plumage and paints to attract their mates. Traditionally, the plains Indians, like the Lakota, wore great feathered bonnets (a term applied by racist colorless people, no doubt, to emasculate[109] the men wearing the elaborate headdresses), but few people know that each feather a man earned was through some heroic effort sanctioned by his clan, his moiety, or his tribe as a whole, so that a man who wore a fully feathered headdress was a man of note—like the male birds who worked so hard to attract mates. Such elaborate feathered headdresses were *de rigueur* for "big men" of many cultures around the world, from the Sumerians, to the Aztecs.

Birds' spring-time mating rituals, quite often, also demonstrate how important it is for a bonded pair to raise the young until they are old enough to take care of themselves. Such a spring ritual was undoubtedly the origin of the Sumerian *qursu* ritual, wherein designated couples had sex to imitate the birds and other animals' sex rites in order to propitiate the spring divinities that bring abundance in plant, animal, and human life. Of course, the

---

[109] Perkins-Gilman also points out that in 1911, when her book was first published, there is "no feminine analogue" for the word emasculate (11). Many different terms have been suggested over the last century, but none have proven useful because they do not capture the essence of the idea behind "to emasculate," which is to remove a man's power. I suggest the term **egynovi**, which combines the negative "e" from emasculate, emphasizes women "gyno" and adds the Latinate term for strength and life, "vi." **Egynovition**, then, is the removal of feminine power from anyone.

big mistake most scholars make is to assume that these sex rituals always related to "marriage"[110] and human fertility, when, in fact, they were all about the sexual union of two beings—that liminal moment when both beasts and animals are like gods and can join in the great spiritual connections between all beings. Sexual mating creates emotional pair bonds, most of which can last at least long enough for any resulting child to be able to learn self-care. Only by studying monogamous birds under controlled conditions, like domesticated swans, could humans have learned about the connection between heterosexual coitus and procreation.

However, ancient humans undoubtedly watched bonded mates raise broods of chicks, demonstrating how, at least among some species, two parents are better than one, although most tribal cultures around the world, even today, take a collective responsibility for raising the community's young. From observing birds feeding their young, ancient humans would also have noted the benefits of feeding their own young pre-masticated food before they develop the ability to eat solid food themselves, thus reducing the time period necessary to breastfeed a child.[111]

Our fascination with birds also leaks into other symbolic forms, such as wings on deities, feathered cloaks, and tufted hair styles, which probably also led to royalty wearing coned or domed caps in some cultures. The Hittites especially loved such domed caps, but it was Egypt that used the *hedjet*,[112] meaning "white one,"

---

[110] In fact, it is highly likely that religious people interested in controlling a self-appointed and abusive monarch might have encouraged common people to use a similar ceremony to create divinely sanctioned marriage, legally and spiritually binding themselves together for life, for themselves, thus siphoning away a bit of the monarch's god-given power over them.

[111] Too many androcentric scholars push the idea that prolonged breastfeeding is a natural form of birth control, but they avoid admitting that having heterosexual coitus while breastfeeding is still more likely than not to create another pregnancy; the ONLY deterrent inherent in breastfeeding is the excuse women can give for not wanting to participate in coitus during that time.

[112] The Hedjet could simply be an imitation of the beeswax cones that "royal" Egyptians wore, especially during the hottest summer months, in order to perfume themselves. Researchers analyzing the smells of ancient Egypt believe that, "in the royal palace,...the perfumed smell of rulers and their family members would have overpowered that of court officials and servants. That would have

as the symbolic crown for Upper Egypt, possibly styled after the White-Crested Helmetshrike or the Long-Crested Eagle, both native to Africa, but the crown, admittedly, looks more like the Southern Cassowary's keratin-coated crest. However, the cassowary is native to Australia, so it is highly unlikely that anyone in northern Africa during the time of early Egypt circa 3000 BCE ever saw one.

Even the crown for Lower Egypt (which included the swampy delta) the *deshret*, looks like a combination of two birds' plumage. The back swoop could be an imitation of the Hoopoe bird's crest, while the curly front swoop is similar to the tail feathers of the Long-Tailed Widowbird, although the curl is more likely an imitation of various tail feathers from ducks found closer to Egypt.

We know that the Egyptians, in particular, revered birds because so many of their gods took bird form: both Horus and Isis as hawks; Horus' son Qebehsenuef as a falcon; Geb as a goose; Thoth as the ibis; and Nekbet as a vulture.

Similarly, the Sumerians often pictured Inanna with bird wings and crowned with crescent shaped cattle horns to remind us of her ties with the moon and Female Magic, since the blood ties of menstruation and war identified her as the highly sexual war goddess that she was. Most Sumerian gods or divine beings, like the Apkallu (who had human bodies but are often depicted with bird heads; see image below) and the Lamassu (who had bull bodies with human faces), were winged, sometimes having four sets of wings, more like butterflies or dragonflies than birds, perhaps because the Sumerians and Akkadians realized that heavier beings would require more wings to actually fly.

---

perhaps denoted special ties to the gods among those in charge" (Bower). If true, then the perfume cone would have been an easily identifiable form of crown.

Apkallu Fertilizing a Plant, Symbolizing Anemophily (Wind Pollination)
Author

And the Hittites will, later, develop the winged disk, which not only becomes a late symbol of Horus for the Egyptians, but also Zoroastrianism, circa 600 BCE, will adopt the symbol to indicate the divine organizing power of Ahura Mazda, a self-created male deity after the fashion of Ra of Egypt.

Even the Aztecs associated their primary war god, Huitzilopochtli, with the hummingbird or eagle, so that he is often depicting with a sweeping feather headdress and feathers that indicate wings on his back.

Ancient humans would have encountered many different birds in nature, in temporary camps or villages, and in the towns and cities that developed near good river sources, good soil, and along plentiful stretches of seaside beaches, where humans tended to settle, at least for a time. Birds would have also flocked to freshly tilled soils to gather insects and worms, and again after seeds were sown to eat the seeds.

Observing bird behavior would have been fairly easy for most ancient humans as they fished, which attracts many sea birds, as they foraged, which would flush many birds from their nests, as they hoed fields for planting, which would attract birds to the seeds they planted as well as to the insects and grubs they disturbed, and as they reclaimed marshes and swamps for water, harvesting reeds and other vegetation, and building new soil for planting and, like the Aztecs, for building their cities.

Furthermore, it would only have been through making observations about domesticated (or at least captured and kept) monogamous birds like swans and quetzal birds, that human beings would have, finally, determined that males play a biological role in the creation of offspring.

While it is quite natural that human beings learned from and imitated many animals in order to learn how to live and how to propitiate the divine beings that they believed inhabited everything, birds, by far, had the greatest impact, even teaching us the importance of being good, protective parents for our young.

Chapter Fourteen
The Bull as a Gynocentric Symbol of Life

Ever try to count how many scholars assume that bulls must be a male symbol of power? Most do because it is rather simplistic thinking—bulls are male and tend to be strong, ergo all bulls must represent male power. For instance, according to Ute Günkel-Maschek, bull leaping "may have constituted one of the prerequisites for achieving status of social adulthood" in Crete (125), indicating that she believes only males interacted with bulls. Even Dorothy Cameron, who strongly believed bulls were a gynocentric symbol of women's generative powers believes that only males used the "back" cave walls "for male puberty rites, [since] imprints of smaller (boys?) footprints have been discovered" there.

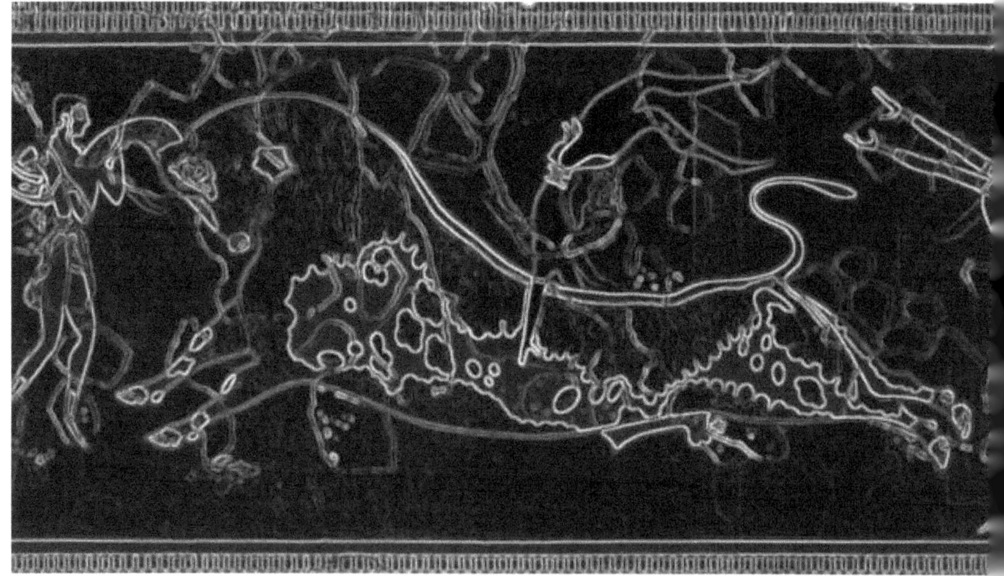

Sketch of the Bull Leaping Fresco from Crete

But do bulls always symbolize male power and were males the only people who were interested in bulls?

Ample evidence demonstrates that early humanity's interest in domesticating cattle began at least 10,000 years ago.

However, androcentric biases, which lead scholars to assume that males were the only people who dealt with animals, usually color the how, why, and who answers regarding domesticated cattle. Most androcentrists assume men domesticated cattle because they were assumed to be the only hunters in early cultures, but we now know that that biased belief is not based on facts.

Since men cannot breastfeed under normal circumstances, women were the people responsible for feeding and nurturing human babies. That much seems obvious, even to androcentrists.

However, when it comes to figuring out how aurochs, the ancestors of all domestic cattle, were first tamed presents a sticky situation.

Did some burly guys just build fences that kept the cattle in? Instead of mass slaughtering them after running them into such corrals, also known as "capture kites" because of their shape? If so, how would merely keeping cattle in pens domesticate them enough for humans to milk them?

As with most animal domestication, a relationship had to be built first. With dogs, the relationship was probably pretty easy, but the most likely scenario was that the mother was killed, so the pups could be raised by the humans. We still practice that process of domestication with every domesticated dog born today. If a puppy is not nurtured by caring humans, chances are great that it will not trust people as an adult either.

Since our ancient ancestors would have figured out the best way to domesticate any animal was by raising its young, instead of capturing adults, chances are great that that process also worked for cattle.

However, very young cattle still require their mother's milk. So how would our ancient ancestors have fed them in order to raise them as healthy domesticated cattle?

As the primary nurturers, women would have known the calves needed milk. Since human beings domesticated goats and sheep before they domesticated cattle, they could have fed a few calves goats' milk using goat bladders or stomachs as artificial udders,[113] then transitioning them to the same kind of grain-soup (think porridge) humans ate at about four months old, although most would be nibbling grass by then, too. By two years of age, cattle lose their milk teeth and grow permanent teeth because they are reaching adulthood, a time when they can begin to reproduce. By this age, too, most cattle, which naturally have horns no matter what gender they are, unless genetically modified to be polled (not have horns), will have grown their horns, too, which, unlike antlers, are a permanent fixture growing out of each cow's skull. Horns[114] were important elements in oxen because the horns act as brakes against the yokes to slow vehicles going downhill and allow a drover to use the oxen to push a vehicle backwards without harming the cattle (Ford & Kreutzer 10).

The tricky part of raising a calf is keeping the calf alive until it can forage for itself, but the same holds for any mammal baby, including humans.

The people who would be most practiced at doing that nurturing would be women, although there is every reason to believe men were capable of filling those nurturing roles, as well.

As noted elsewhere, the basic cattle skull, which people raising cattle would have undoubtedly seen many times after finding their carcasses in the wild or while butchering them for food, is shaped just like mammalian female reproductive organs with the snout of the animal symbolizing the vagina, the cranium the uterus, and the horns the fallopian tubes (Gadon 31). As Ahmed Achrati

------

[113] I grew up on a farm in south central Kansas where my father, who worked at a feed yard, brought home some of the calves born in the holding pens, which would have otherwise been trampled to death. Every morning, I was responsible for feeding up to fifteen calves via milk bucket, which I did in three shifts, since we only had five buckets. So you can trust me when I say that a baby calf will suck on anything that looks like it fits in their mouths, so using the stomach or bladder of another animal (even its butchered mother) would work as an udder substitute.

[114] While many androcentric scholars assume horns are from bulls, both cows and bulls had horns, and cows were often used as oxen. Many scholars claim gender can be determined by the size of the horns, which is highly problematic because studies have shown that domestication decreases "sexual dimorphism," so that it becomes much more challenging to tell males from females from horns or other bones (Helmer, Blaise, Gourichon, & Saňa-Seguí 82).

points out, "despite its phallic appearance, the horn has been used to symbolize the female genitalia" (6).

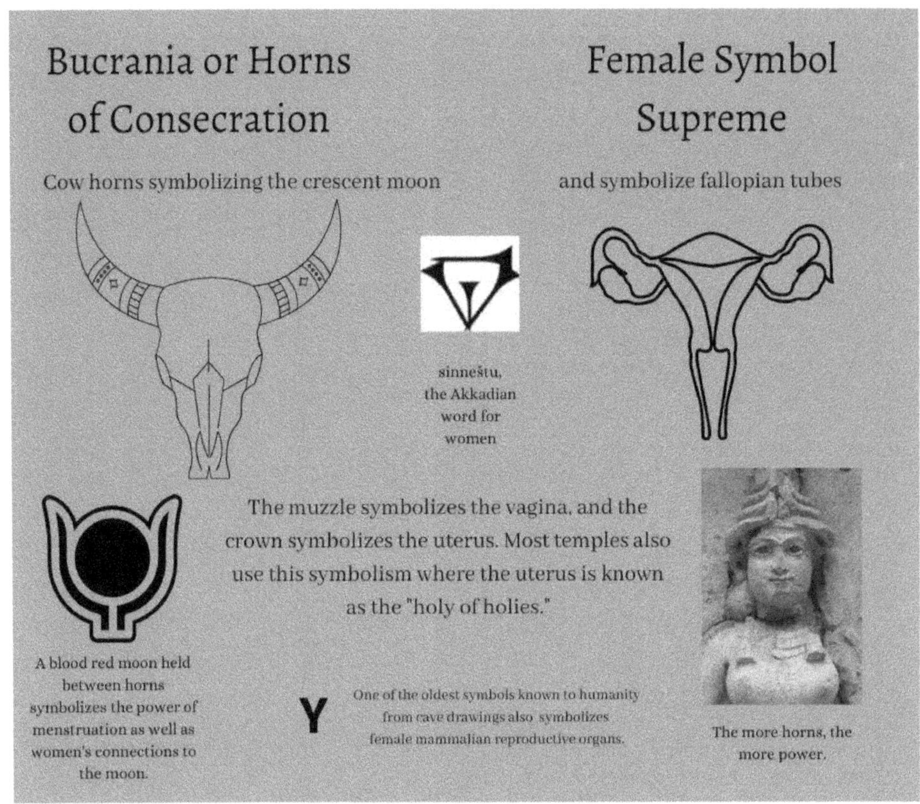

Bucrania: Symbols Across Time
Author

Couple the crescent moon shape of cattle horns with women's association with the phases of the moon, thus the magic of the moon (which is where we get the words "menstruation," "menarche," and "menopause," not to mention "monthlies," remember), with the nurturing nature not only of women, but also of female cows who will violently protect their offspring, if necessary, and cattle make the perfect gynocentric symbol, which is why horns became symbols of divine power, too.

The more horns a god had in Sumeria, the more powerful the god was. Goddesses, first, were given horns, probably based on cattle worshippers from the western deserts, in ancient Egypt, but even after Isis, who comes much later in the pantheon than Hathor,

175

goddesses are typically depicted with blood red moons[115] held between two crescent-shaped long cow horns. Androcentrists mistakenly tend to call these blood red moons suns, though, missing the intentional association of women with the moon, its cycles, and the magic period of a blood red moon during lunar eclipses. In Sumeria, Inanna is always depicted with four sets of horns—the most of any god—which also ties her to the powers of the four corners of the earth, which is appropriate since she was the Queen of both Earth and Heaven.

However, most ancient cultures were also more egalitarian than they were hierarchical, so the bull would have been seen as a combination of both the female symbolism in the head and the male symbolism with the penis and testicles. Once we realize that both genders' genitals are associated with bulls, the bull becomes a perfect marriage of both male and female sexual organs. Through this gynocentric lens, we begin to picture them differently, almost like the ankh, which is also a combination of male and female sex organs, and the yin-yang symbol, which includes both aspects within each other (so that yin has some yang and yang has some yin). We have to remember to distinguish between Sex Magic, though, and "fertility" symbolism, since the magic of sex is undoubtedly older than the idea of human "fertility," since the ancients would not have connected procreation with heterosexual intercourse until circa 2400 BCE.

Some people misinterpret such sexual coupling symbols— the bull, the ankh, yoni and lingam, yin and yang—as "fertility" symbols because they assume that our ancient ancestors understood the connection between heterosexual intercourse and pregnancy, but that knowledge was still beyond most ancient people, even after the domestication of animals like cattle because of too many sexual partners, too many attempts at coupling, too

---

[115] Why so many androcentrists believe the sun, Ra, is held in the horns of the goddesses is a mystery. It makes much more sense for the red disk between the Egyptian goddesses' horns to be "blood red moons," which is what a lunar eclipse looks like, typically. There are two such moons every year, although they are not always visible everywhere around the world. Odds are great, however, that the Egyptians would have seen many, many blood red moons—a double menstrual connection for the cosmos with women—over their long history.

many types of sexual interaction, and too long of time period between successful heterosexual coitus and birth.

Instead, these unified dual-sexed symbols represent Sex Magic, or, as I like to put it, the Magic of Ancient Sex. While male ejaculate (semen) was associated with rain and rivers, so that sacred acts of sex in the spring was a form of sympathetic magic that sought to bring spring rains, thus vegetation abundance, sex was a liminal, spiritual activity. Prior to the patriarchal attempts to make sexual intercourse a dirty, evil activity, sex was seen as spiritually connecting, as a healing method, as a way to get in touch with the divine, or to at the very least feel godlike for a while. And our ancient ancestors did not just limit themselves to heterosexual intercourse. As depicted in the Moche pots, sex was pleasurable however it happened. In fact, the skeleton versions of the Moche pots (see image below) are undoubtedly meant to symbolize how humans can reach beyond death into the Next Realm when enjoying coitus.

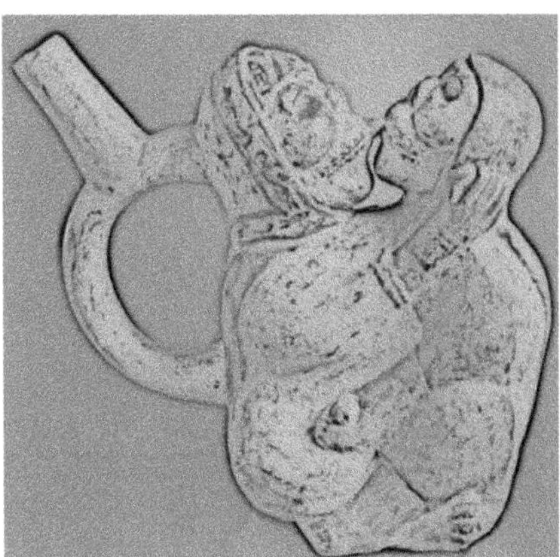

Sketch of a Moche Skeleton Sex Pot
Author

Sex, instead of being a reviled activity, was used by many cultures for social bonding, healing, and transformation, so was part of a balanced life, demonstrating that heterosexual women need men just like heterosexual men need women, but also that homosexual activity was also a spiritual, socially bonding activity, with sex as the unifying element that creates social bonds, social bonds upon which every community is built.

Like the Egyptian ankh, which is a symbol of Isis' vagina and Osiris' penis and testicles together, forming the Egyptian symbol for unified life, a bull, then, becomes the same sort of symbol, assuring men that, while women could survive without them, men are an integral part of human society, just as women are. Because bulls, in particular, are composed of the symbolic bucrania associated with Female Creative Magic and come with phalluses that clearly symbolize Male Magic, they should be viewed as symbols of egalitarianism—both genders working together equally.

So, remember, just because a bull is being used as a symbol, you should not assume it is a male symbol because it could be a symbol of the unity of both genders as society's glue for a balanced society.

Chapter Fifteen
Gendered Magic:
How the Ancient Mystics Viewed
Biological Processes
Through a Spiritual Lens

One of the greatest challenges I have found in discussing many of these ancient ideas with various individuals over the last few years has been the fact that most Christians no longer view spiritual or religious views as "magical" or, indeed, see miracles as "magical" at all, because magic has for so long now been considered "evil" and miracles seen as "god's-work," as though there is a major disconnect between magic and miracles or between magical beliefs and spiritual beliefs or between magical spells and prayers. This major disconnect points directly at the work of some practitioners of the Abrahamic religions, who have long sought to explain that a male god is all-powerful thus created and controls everything on his own, and that "magic" is somehow associated with witchcraft, thus evil and explained mostly through "the devil's-work." This opposition between good and evil becomes the defining factor for many who practice one of the Abrahamic religions (mainly Judaism, Christianity, and Islam), so that it, unfortunately, also colors how many scholars interpret images from our ancient past.

However, we have to admit that the disconnection probably began much earlier than the Abrahamic religions, and possibly began with early spiritual or magical practitioners themselves—the shamans of the ancient world. After all, if humanity's lot was to appease temperamental gods, at some point it would be natural to see the actions of those deities as negative. While many preliterate cultures viewed the conflict as between balance and chaos, not good versus evil, it is easy to see how that original conflict changes as androcentrism rises, thus making things associated with women negative, with males being on the side of "good" (aka "might-makes-right") and women being on the side of "evil" (if women have no overt physical strength, they must be manipulating outcomes using magic to succeed over men). To gain power, male shaman/priests had to differentiate between Male Magic and Female Magic, making Male Magic seem stronger, thus better, while demeaning Female Magic as weak and evil. Later, the Abrahamic religions change Female Magic symbols, like fruit and snakes, into "temptations of evil," and sex, which was once worshipped as divine, becomes a nasty thing, especially when women initiate or control it.

In fact, it is possible that some cultures viewed magic with a waxing and waning flow of interest. Gwendolyn Leick suggests, for instance, that the Akkadians expressed more of an interest in "magic texts, incantations, and [magical] rituals," especially that of reading omens, than the Sumerians who preceded them had (7). In India, Women's Magic, called *shakti*, was rediscovered in the 6th century ACE, returned to popularity, empowering both women and men through a practice of Tantrism, and later became a uniting philosophy that allowed Indians to rebel against their British colonizers (British Museum). Thus, it is fully possible that human beings have long flirted with the idea of being able to conduct magic repeatedly throughout our history, although androcentric forces seem to have clamped down on the idea that females had any good magical powers after each episode.

Unfortunately, most scholarship on magic has largely been a conversation between male scholars, who pretend they are legitimizing the ancient belief in magic by examining it as mostly desire manifested through ritual and writing (curses and spells), and a separate conversation between female scholars, who understand how magical simple bodily functions, like giving birth, were to the ancients. While the female scholars tend to assiduously try to accommodate the male theories in their discussions, the males tend, largely, to ignore the female discussions altogether as though they bring nothing to the conversation, something I discuss further in the very last chapter.

ΩΩΩ

Here's the spot where I admit that reading scholarship intimidates many lay readers and even educated readers who are not knowledgeable about the particular scholarly language being written. Please realize, however, that that intimidation caused by scholars using big words is actually on purpose. And that purpose for most scholars is much like **the same purpose** shamans (and some charlatans who speak "tongues" now) of old used to use strange language for prayers, spells, hymns, and, later, even curses.

Esoteric language, spoken only by a few people with specialized knowledge of the subject matter, is used by scholars in various fields and by shaman alike. Of course, they use different words, and sometimes different languages. However, put most simply, <u>the goal of esoteric language, whether it is religious or scholarly, is to create a hierarchical separation of the informed from the uninformed</u>, or, to use esoteric language, to differentiate between the cognoscenti (those in the know) and the oblivious (those who have little background in the topic).

A theoretical physicist, for instance, learns specific language regarding her field, which often, ironically, uses language borrowed from many ancient cultures to describe theories regarding physics. Similarly, a fundamentalist Christian who "speaks in tongues," can use a smattering of ancient language to confound those watching and listening, or simply makes up gibberish that sounds like ancient language. This kind of linguistic borrowing is as old as humanity because it was through this kind of borrowing of portions of some words to adapt them to mean new ideas that keeps so many languages "alive." But the primary purpose behind using such languages as Latin, as it is used in science, for instance, or even Sumerian or Hebrew used by glossolalia (people who speak in tongues while in religious ecstatic states) is not because those languages are more "magical," but because fewer ordinary people know them, thus setting the theorists and shamans (or charlatans) apart from ordinary folk.

We must remember that our ancient ancestors might not have been conscious of the efforts by those "in control" of messages from the "spirit realm" to make themselves exclusive spiritual mediums through the use of esoteric languages, so that other people would have to come to these people who claimed to speak to the divine to have supernatural signs read or to be healed of illnesses caused by paranormal agents. We will never know if any knew that some people were purposely confounding them in order to manipulate them.

However, ancient human beings most likely approached all of life's events through a lens colored by beliefs in **spiritual magic as a form of energy that flowed through everything**. This magical energy was, of course, believed to be of divine origin, meaning that gods or beings greater than themselves had set up the universe in such a way that spiritual magic was available for all, but that individuals had to access it through liminal activities—activities that allowed them (they believed) to touch the magic through a spiritual connection or through liminal people, such as the priestesses who practiced *hashadu*[116] for healing or transformation in Inanna's temple in Uruk. Enacting spiritual magic was akin to opening a door or a water channel that allowed the human being on earth to touch, to experience, and even to use the magical energy the gods regularly used (they believed) in those liminal moments that made the experiences so profound. While drug use probably created deeper ecstatic moments, many important liminal activities included sex—actual heterosexual or homosexual intercourse, but also oral sex, cunnilingus and fellatio, which were specified as actual sacred *me* powers passed to humans by the gods, according to Sumerians—the ecstasy of which allowed the participants to feel as though they were "reaching through" the mundane, ordinary world into the spiritual realm.

Most ancient people believed in many different kinds of magic, but there were two main categories: natural magic and human-enacted magic. Natural magic explained most natural

---

[116] That heterosexual intercourse could be used for healing purposes is a belief that lasted a long time in Sumer, even after the Akkadians took over. In the poem, "Enki and Ninhursag," Enki's rape of his own offspring angers his wife, Ninhursag, so that she plants his "seed" in the earth, where it grows into plants, which Enki then hornily devours in very sexual terms (thus the writer equates eating with raping), but Ninhursag had cursed the seeds, so that Enki ends up poisoned. In order to heal him, she "places him in her vagina" (Lieck 35), which is another sexual act, not procreation, but a healing act through sex. While Lieck considers this act a "motif of male womb-envy" (37), she also believes the tale is attempting humor by having "a male made pregnant by his own spilled seed" (39) because it had to be humiliating to be "at the mercy of the Wife and having to borrow her sexual equipment" (39). Like so many androcentric scholars before her, Lieck misses the clear symbolism of the *hashadu* sex healing rite, preferring to believe, instead, that the tale is a cautionary one about how men who desire sex too much become "vulnerable to the power of women" (40-41).

phenomenon, from the magic of water finding its way up through rock or across a landscape unaided by human beings, thus viewing springs, rivers, and flood waters as magical beings imbued with spiritual energy, to the magic of women creating new human beings seemingly solely out of their own spiritual will power and their own physical bodies. The second example of natural magic, pregnancy and childbirth, could be initiated by humans, however, because many cultures believed women were able to "conjure" babies out of their own spiritual powers, out of bodily fluids by clotting their menstrual blood, or through propitiating divine spirits[117] with sacrifices and libations.

While natural magic for the ancients had a dangerous side, such as flood waters drowning people or women dying in childbirth, most ancients believed in good actions and intentions as being more important, even judging by later sacred texts like the *Egyptian Book of the Dead*, which emphasizes that a person's negative actions and intentions are what will make her/his/xir heart weigh down the Judgment Scales after the person dies, thus spell the end to their soul forever.

In fact, it was not that long ago that many human beings considered diseases to be manifestations of evil spirits set upon individuals to torment them. Even in the 21st century, some Catholics have relied increasingly on exorcisms to drive out demons. In the Americas alone, more than eight million people joined "the cult of Saint Death, or Santa Muerte," with some drug traffickers actually offering human sacrifices to this saint in their prayers "'to avoid arrest and to make money'" (Hernandez), increasing the violence between the drug cartels, the populations of

---

[117] Such divine spirits, for many cultures around the world, included deceased ancestors as well as gods. Some ancestors, we know, were believed to have become divinities themselves after death, sometimes because of the powerful way they had lived their lives, and sometimes because of something magical or mysterious that occurred during their lives or after their deaths. Catholicism preserves this ancient ancestor-as-divine belief when the church canonizes a deceased person as a saint. Some tribal cultures, however, believed women had to entice the souls of Spirit Children into their wombs before they could successfully give birth. Remember, fetal and infant mortality was high in the ancient world.

Latin American countries, and the military forces in those countries.

Little evidence exists, thus far, to demonstrate if, when, or why humanity might have begun to envision "evil" as a manifestation of the gods or as a spiritual reality prior to 2400 BCE when humanity was able to verify that males play a biological role in procreation. When did women begin to believe problems with pregnancy and childbirth was caused by negative spiritual activity?

Evidence abounds, however, that humanity dealt with the grief of death by manifesting rebirth engines, *wiedergebut maschine*, if you will, in the shape of female pudenda or "birth mounds" like the burial pyramids and boats I explore in another chapter, to ensure that chosen human beings could be successfully reborn into the Next Life or Next Realm of existence. Death, to the earliest ancients, then, was not a negative or evil event, but merely a transition, a form of birth or, as Christians call it, rebirth, from one realm to the next.

Somewhere along the line, however, possibly sometime around the transition from the Mycenaean culture to the Greek culture most of us are more familiar with, humans began to believe in specific malignant spirits. For instance, the Mycenaeans appealed to the gods who guard the realms where the deceased go, such as the gods Hades and Persephone,[118] but the later Greeks began to view the Nether World as "chthonic," a place more of misery than of the Elysian Fields. As androcentrism rose, the kingdom of Hades was becoming, for many Greeks, a permanent place of dread, with threats of being tossed into the Tartarus much like Christian fears of being sent to Hell, so that the afterlife changed from a place to await rebirth to a permanent repository for the dead. This change is possibly tied to the fact that the Greeks began writing curses and leaving them with their deceased family members' remains, who were expected to act upon their wishes to exact

---

[118] One often overlooked reason Persephone must become Hades' bride and rule with him in the realm of the dead is because she, like Osiris for the Egyptians, symbolizes spring growth—a time when the earth itself seems reborn. Just as the earth revives in the spring, those who died were expected, also, to be reborn at some point into a new life or at least into a new spirit realm, which gave birth to the idea of resurrection.

retribution on their enemies for them (Quah). For many post-Homeric Greeks, casting a curse meant writing it out on "a thin sheet of lead" that was then rolled up "and pierc[ed] with a nail" before it was placed in a corpse's hands and buried (Collins 2). These activities seemed to increase post-Homer because lines from many Homeric verses were used in various forms of Greek magic, especially curses, so that "late authors, such as Philostratus, conceived of Homer as a necromancer" (Collins 104). By the time Christianity rises and begins to dominate the world, the Next Realm, Hades' kingdom, has gone from the place everyone travels to, at least temporarily, after death before being reborn, to an evil realm full of sinners and unending pain, the Christian Hell.

The temptation is to dismiss these spiritual views of demons and hell as just superstition, but we have to approach spiritual ideas from the way our ancient ancestors would have, wherein they spent most of their lives in an effort to appease various divine powers, and saw events as spiritual manifestations of their fears, of their hopes, and even of their desires. Most ancient cultures were much closer to pantheistic beliefs than many Christians are today, so let us discuss pantheistic spiritual views first before delving into two main forms of gendered magic, which can help readers understand how shifting from gynocentrism to androcentrism changed human beings' perspectives on human-divine connections through spirituality and transformed religions from being the epitome of communal cooperation to being used to justify conflict.

αααα

Pantheism

Pantheism is a relatively simple concept for what can be a very complex spiritual point of view. "Pan" simply means "all" or "all encompassing." "Theism" indicates "gods," "divinities," or "spiritual beings."

Essentially, a pantheist believes that everything—rocks, plants, animals—are imbued with spiritual energy. Samuel Noah Kramer describes the basic Sumerian understanding of the world, their universe, as a combination of "the great realms of heaven, earth, sea, and air; the major astral bodies, sun, moon, and planets;

185

such atmospheric forces as wind, storm, and tempest; and finally, on earth, such natural entities as river, mountain, and plain, such cultural entities as city and state, dike and ditch, field and farm, and even such implements as the pickax, brick mold, and plow—each was deemed to be under the charge of one or another anthropomorphic, but superhuman, being who guided its activities in accordance with established rules and regulations" (113-4). This pantheistic view of the world in which the ancients lived, allowed them to blame someone outside themselves when things went wrong, but primarily allowed them to explain how the cosmos, as they saw it, worked. By creating divine beings who were in charge of the various aspects of life, the ancient Sumerian people relinquished their responsibility for many outcomes, unwittingly allowing leaders to create an order of importance for everything from deities and people, to animals and things. To the Sumerians, humans were created to serve the gods. Serving the gods by doing the work the more divine beings no longer wanted to do was the sole purpose for the existence of human beings (Kramer 123). We have to wonder, of course, why they fashioned their religion in this way, since early Sumerians, who voted on the leaders of their city and on their council members, were more egalitarian than later Sumerians, who were eventually overrun by ambitious men who desired to rule the world. We have to wonder how many changes occurred in spiritual beliefs between the early Sumerians and the later Sumerians for these drastic social changes to occur.

While the Sumerians and other ancient people created various unseen divinities to explain why things were the way they were, modern humanity can use physics to acknowledge that everything we can see in the universe is composed of invisible atoms, which are highly energetic but require special equipment to see, even when they compose solid objects. We cannot view these incredibly small moments of being (aka atoms) in an energetic state because our eyes simply do not allow us to see microscopically. In fact, we need electron microscopes to view atomic structures. Perhaps these are the invisible forces our ancestors sensed must be in everything?

Much of what we can witness with our naked eyes, however, would have seemed miraculous, magical, divinely created, or highly spiritual to many of our ancestors.

Take, as an example, the basic chemical reactions that create beer, something ancient Mesopotamians called "'the divine drink'" (Brews Cruise). We cannot see the yeast involved in naturally occurring alcoholic beverages, but someone around 4000 BCE (quite possibly long before, but probably rediscovered by the Sumerians) discovered that barley bread left soaking in water for days formed a tasty drink that gave people a bit of a buzz. What had to follow was a flurry of scientific (cooking/chemical) experimentation because, approximately 1000 years later, the Babylonians recorded over 20 different types of beer that they had learned to produce, still using grains, just as beer continues to be made today. Ancient beer, which was more like bread soup, had to be sipped through straws, if the imbibers wanted to drink less grain or soggy bread and more alcohol, which was something both Sumerians and Egyptians did (see images below) (Brews Cruise).

Upper sketch: Sumerians sipping beer through straws
Bottom sketch: Egyptians sipping beer through straws
Author

What is important to understand here is the fact that, while ancient humans learned a simple process for making alcoholic beverages in many parts of the world, no one understood that yeast, a microbial form of life, was the secret to the fermentation process until Antoni van Leeuwenhoek (1680 ACE) developed a compound microscope that allowed him to view the yeast, but he still mistook the globules he saw as a form of starch, not as the tiny, living fungus Charles Cagniard de la Tour finally identified as yeast. one-celled organisms, in 1835 ACE. Forty years later, Louis Pasteur finally demonstrated that these microscopic organisms turn sugar into alcohol by consuming the sugar molecules and then excreting the alcohol (Alba-Lois & Segal-Kischinevsky). So, yes, when we drink alcohol, we are drinking another organism's urine, essentially.

What the ancients viewed as "magic" or "god-created" can now be explained scientifically, just as we can now fairly competently explain how heterosexual intercourse causes pregnancy. Birth, the act of a woman pulling a new human out of her own body, was another natural event that seemed magical to our ancient ancestors.

As we move forward, we have to remember that the ancients did not have all the scientific knowledge we have today, but, instead, viewed everything as being derived from some form of divine power or spiritual energy.

ΔΔΔ

Two Gendered Forms of Magic

Inevitably, there were at least two kinds of gendered magic for most ancient people (pre 2400 BCE): Female and Male. There was at least one more form of gendered magic for those who were transgendered or had no perceivable gender, but there is very little specific evidence at this time about how that magic manifested itself circa 2400 BCE. The only thing we can say with certainty is that many prepatriarchal cultures viewed such people as liminal—between this world and the next, thus more magical or spiritually powerful than ordinary human beings.

Female Magic included Creative Magics, Protective Magics, and Power Magics, most of which were tied to women's menstrual cycles and menstrual blood, which were their clearest connections to divine powers, especially those of the moon (aka Lunar Magic), since its phases mirrored women's monthly cycles. For the Aztecs, for instance, women were warriors, which is why they bled every month from their spiritual battles, who were attempting to spiritually wrestle other human beings from Another Realm into this one.

Male Magic included mostly Creative Magics and Protective Magics, most of which were tied to phallic magics. Many cultures later linked Male Magic to Solar Magic, a form of Creative Magic, as well, a clear indication that men wanted to counter women's magical ties to Lunar Magic, and that form of magic developed further, after the discovery of procreation, so that men, too, develop Male Power Magic.

Where most modern androcentrists misinterpret these magics when they view pre-2400 BCE symbols, either in art, architecture, or mechanical devices or in written hymns, poems, stories, or glyphs, is that they assume that any symbols indicating gendered magic signal either levels of social power or are merely instances of "fertility" symbolism. While gendered magic could have, and probably did have, an impact on each individual's social status, we have to continually remind ourselves that **gendered magic had little to do with ideas of human fertility**[119] until

---

[119] I feel compelled to point out a major flaw in androcentrists' beliefs about the power of lactation to prevent pregnancy, which so many still seem to believe is a scientifically accurate claim to make. There is no verifiable evidence to support the belief that women no longer ovulate while breastfeeding. Factors androcentrists do not consider in these sweeping claims about how prolonged lactation must have acted as birth control for the ancients in order to explain reduced population numbers are: 1) that many women "find the act of lactating so emotionally fulfilling" they are not interested in sex, so that the idea that lactation is a form of birth control preventing pregnancy when coitus occurs is absolutely not accurate; and 2) some cultures have "taboos against sexual intercourse after parturition," so that the reduction in coitus post-partum is what is reducing pregnancies (Thomason, Hytten, & Black 337). Ergo, very little hormone change is detected in women who lactate, so that **coitus while lactating can still cause pregnancy**. The damage being done because of the false belief about lactation as a form of natural birth control is incalculable, therefore unconscionable among my scholarly colleagues. Instead, they should factor in other issues with population, such as the high number of miscarriages

after humanity discovered the actual biology of procreation circa 2400 BCE with two important exceptions:

> 1) ejaculation was often tied with rain or river floods, which usually provided the water necessary for plants to grow in the spring, so many spring rituals are tied to either a ritual act of coitus or with male masturbation with the successful ejaculation an indicator that spring rains or river flooding necessary for crop growth would, well, come; and
>
> 2) penises visibly "grow," in much the same way plants grow in the spring—seemingly suddenly and without much warning, so the phallus has long been equated with crops, mostly grasses.[120]

Who is likely to forget Andrew Marvell's famous line in his poem, "To His Coy Mistress," when he proclaims, "My vegetable love should grow." In these two important ways, males have long been associated with both the water necessary for plants to flourish, a sign of abundance or crop and animal "fertility," and with the visible and rapid growth of most plants, especially in the early spring, so that symbols of male genitalia are most likely symbols for this kind of rapid growth and the rain or river floods that ejaculation was believed to bring forth.

Two primary symbols of Male Creative Magic for most ancient cultures, then, comes through the flow of semen (ejaculation) and through the growth of plants—with neither really having anything to do with human fertility.

However, the fruit that the plants bore would have been Female Creative Magic because plants create offspring the same way women did—seemingly completely on their own.

---

and infant-toddler deaths, both of which would have been traumatic to most women.

[120] In "Enki and the World Order," Enki's prodigious penis "'sticks out of the marshland,'" because, like rapidly growing grasses in the spring, "the sight of the lovely young girl in the marshes has produced an immediate effect on the god" (Lieck 33). As the tale progresses, however, Enki's insatiable lust ends up nearly killing him (after he rapes his own daughters and granddaughters), with many scholars finding it humorous that "a male made [himself] pregnant by his own spilled seed," the offspring of which grow into plants that end up poisoning him (Lieck 39).

We need to begin with deeper descriptions and explanations of Male Magics for several reasons. One primary reason is because so many men get super defensive when women are, in any way, seen as superior to men; therefore, we need to admit that most ancient cultures believed in some forms of Male Magic, so we can discuss a few of the forms it took before humanity discovered that men actually play a biological role in procreation circa 2400 BCE, which then allowed men to claim a new form of Male Power Magic, as they attempted to wrestle away spiritual power from women.

YYY

Male Creative Magic revolves, unsurprisingly, around the penis, a magical instrument by any standards measuring magic. We can call it Phallic Power, but the ancient Sumerian word for a woman's lover's heart, *ša*, also means penis (Assante 40), so it is possible that Sumerians believed the way to win a women's heart was through a man's penis. In fact, many cultures who worshipped goddesses, like the Philistines, "honored" the goddess by leaving clay phalluses at her shrines (Maeir). Penises, then, were known to Create Magic when used in sexually liminal ways, and most women would probably agree that a good orgasm caused by coitus is magical, indeed.

As I mention elsewhere, stone megaliths were undoubtedly one aspect of Female Creative and Protective Magics because so many are arranged like women's reproductive organs—a path leading into a sacred circle, like Stonehenge, for instance, can be seen as the vagina, the circle itself can be seen as the uterus, and the altar within is the center of Female Magic, also known in many temples as the "holy of holies" where creation magic takes place. However, Stonehenge, in particular, also symbolizes the joining of Male Power and Protective Magics and Female Creative and Protective Magics because each standing stone is erect, like a turgid penis, and the ones covered by lintels or capstones, symbolize vaginas, which not only gives and gets so much sexual pleasure, but is also the entrance to the womb, from which we are all born. Thus, human beings who celebrated various rituals within the safety of the womb of Stonehenge and its accompanying Woodhenge were also, undoubtedly, celebrating the magic of sex, both heterosexual and homosexual forms, as well as celebrating life and death, which was one of the major transformations in human life as people pass into the Next Realm.

However, we have to be honest, open, and thorough in our examination of Phallic Power because many scholars and lay people alike look to symbols like the penis and testicle images, alternately called *fascina* or *priapi*, left behind by many Roman soldiers across western Eurasia and interpret them only as representing male physical power or "good luck" tokens.

Depictions of human penises are old, as old as 28,000 years ago when the Hohle Fels phallus was carved out of dark grey rock by someone in ancient Germany. Despite the many scholars who wish to view these phalluses as signs of power, by most accounts, the people of those very ancient times only experienced sporadic conflicts with each other. They never waged all out war. Their more passive lifestyles were probably caused by many circumstances, such as because they did everything on foot, since horses had not been domesticated yet, but also probably because cooperation was still *modus operandi*. There is little archeological evidence to support the common belief that war is a natural human thing.[121] There is ample evidence, however, that the ancients thought penises and vaginas were magical, highly spiritual human organs.

Both penises and vulvas, sometimes even the whole female (e.g. horned animal skulls) or male (phalluses with testicles) reproductive systems, are depicted by our ancient ancestors for many reasons.

Most simplistically, vulvas were magical doorways that could grow to accommodate enlarged phalluses but were also Protective

---

[121] In discussing a study on domesticated foxes, Brian Hare and Vanessa Wood demonstrate what happened to the brains of the domesticated foxes, which showed decreasing levels of corticosteroids, also known as stress hormones, so that "after fifty generations, the friendly foxes had five times more serotonin—a neurotransmitter associated with lower predatory and defensive aggression—in their brains than the regular foxes" (25). Humans, who domesticated themselves approximately 30,000 years ago when outside threats were minimal, have a superior capacity for cooperation, but the patriarchy, which has clear demarcations of who is "in" and who is "not," is divisive and has been enabling the rich to oppress those with less through division and fear (Hare & Wood xxx).

Magic in the form of boats that carried sailors from one port to another and the deceased from one realm to another. The Sumerian goddess Inanna famously traveled over her sky realm in the Boat of Heaven, which was possibly what the Sumerians called the Milky Way, which, in a totally dark sky, looks like the prow of a boat pushing through foamy waters.

Hoes and plows, like penises, could penetrate the earth in order to create channels wherein water flowed, such as when a man wanted to feed a fetus in the womb with his man milk. As androcentrism rose in both Sumeria and Egypt, the male leaders were increasingly depicted wielding hoes to break open dikes or to carve canals because controlling the waters of their respective fields meant crops would grow, so that the people could thrive. Both cultures also had spring rituals involving penises and ejaculation that were meant to ensure that river flood waters would come or be controlled. I will discuss some of these ideas in more detail soon, but some are also discussed in depth in Section 2.

Together, penises and vulvas, like the lingam-yoni made famous by Hindus and Buddhists, were symbols of Protective Magics through time from ancient Sumeria to ancient Rome.

Clay plaques depicting heterosexual couples having coitus were mass produced using molds by ancient Sumerians because they were popular for warding off negative spiritual energies. It is possible that the Sumerians believed such spiritual entities did not like seeing people enjoy themselves, but it is just as possible that coitus was so spiritually powerful for both people that its act was enough to create a sacred, spiritual safe space inside a home (Assante 28). Evidence abounds that the ancient Sumerians valued sex on its own, without tying it to marriage, originally, especially from the many erotic poems, such as "Inanna and Dumuzi," the various forms of which demonstrate that they copulate because "sexual intimacy, spontaneously desired by both partners, is practically a prerequisite for the decision to marry" (Lieck 71). Thus, coitus was good magic; if the first coupling did not go well, the Sumerian couple probably never committed to each other.

It is important to remember that the ancient Sumerians would not have associated the sex with procreation, so, even if the woman became pregnant, they would have believed magic and the woman's own spiritual power created the child, not the sex.

Similarly, the ancient Roman soldiers, several thousand years later, carved their *fascina* or wore bronze amulets made to look like flying penises and testicles supposedly for good luck and, sometimes, for good health (Silver). The good luck for soldiers was most likely hoped for in terms of Sympathetic Magic that tapped into women's menstrual magic—bleeding but not dying. If the soldier got injured in battle, he believed he would not die as long as he wore the amulet, which is why so many of these personal amulets show clearly circumcised penises. In fact, it is possible that the men probably endured circumcision before entering their military service to enact that sympathetic Female Protective Magic that they knew allowed women to bleed but not die every month. So the *fascina* are not just good luck tokens; they are Female Protective Magic subsumed into Male Protective Magic.

One fascinating recent find, now being acclaimed as the "oldest narrative scene" in the world, is an 11,000 year old carving found in southeast Turkey at Sayburç (Pandey). Archeologists interpreting the scene imply that the scene, which has a man in three-dimensional relief, facing the viewer, apparently holding his right hand on his genitals while his left arm crosses his abdomen, shows the man protecting himself from the onslaught of the two leopards or lionesses facing toward him from each side. One archeologist believes the scene depicts a scene from a mythical tale, so that the man in question is possibly a hero of some type Pandey), which is a possibility.

However, if we look at the Sayburç scene through gynocentric eyes, we see a whole other possible interpretation. If the man is ejaculating, he could be ritually performing the act to bring spring rains. The animals which appear to be "attacking" the man and another indistinct figure, two lions and a bull, in fact, indicate the presence of divinity. Goddesses from the eastern Mediterranean are often depicted as flanked by lions, and the "bull," as I discuss in another chapter, has its head turned in an unnatural way, so that we see the cow horns as crescent moons on either side of the animal's head, which would also indicate both the presence of the goddess, via her ties to the moon and to cattle, but could also indicate the joint idea of a balanced life as represented by the "bull's" penis, which is not clearly evident in the image, and the female reproductive organs as represented by the cow skull's shape. What the image could be depicting, then, is the equality of the genders—each gender's symbolism being represented in several symbols. However, if the lions are truly attacking the man, this scene could also symbolize how women use their magical, spiritual powers to keep men under control.

A more interesting interpretation, however, is that the scene depicts a time between the ascension of the Taurus constellation, which is visible from November to March in the Northern Hemisphere but is best seen in January, and the appearance of the constellation Leo, which is visible from March to May. Taurus, on the left, is pointing toward a person who could be dancing, another spiritual activity, but the people who carved the narrative into the rock emphasized the ejaculating man, which seems to emphasize the parallel between men's semen flowing and the spring rains brought by the presence of Leo in the skies.

Not all penis imagery includes the testicles, though, so another form of sympathetic magic the Roman males might have believed is tied to consuming the testicles of other animals, such as bull or boar testicles. The sympathetic magic comes in because even modern men, still, believe they can improve their own sexual performances by consuming such testicles, often called a "calf fry" or "mountain oysters." In a manner similar to eating the heart of an enemy, males believe they can activate their own gonads' production of testosterone and/or semen via the consumption of the testes of animals that are considered "virile," even though there is currently no scientific evidence that they get any more of a testosterone boost from eating testicles than they do from eating ginger or leafy green vegetables.

Admittedly, penises exude ideas of power and physical strength because, like the Hulk, they tend to grow in length and girth when excited. Penises are the most visible body organ[122] to

---

[122] Vaginas also expand but are not visible. While some would argue that biceps are just as visible and can radically change in size when flexed, both genders have them, and ancient women were just as strong as ancient men pound for pound. For our current era, some men's (and women's) obsession with body building

demonstrate such magical growth for any mammal, and so many masculine values, even today, revolve around the question of whether one man's penis is bigger (thus better) than another man's. Because of this possibly long-held masculine value, many men develop inferiority complexes simply because they feel their penises are too small,[123] so let's start by examining averages.

Scientists finally had to physically begin measuring men's penis sizes because, naturally, men who self-reported their penis sizes tended to exaggerate a bit, well, actually a lot.

According to Bruce M. King, who compared several penis measuring studies, "the mean penis lengths were found to be 4.01 inches." Adrienne Santos-Longhurst reports that 95% of all penises "fall into the average range," which means a turgid girth of 4.59 inches compared to a flaccid girth of 3.66 inches—a growth of .93 inches—and a turgid length of 5.1 inches versus a flaccid length of 3.6 inches—a growth of 1.5 inches. Despite this "normal" range, with fewer than 5% of all men having penises that are shorter than 3.9 inches in length when erect, 45% of "people with penises wish they had a larger penis" (Santos-Longhurst). King reported that the average penis length most men desire to have is 7.7 inches, but estimates that fewer than 12% of all men have penises longer than 6.3 inches.

However, we must not lose sight of the fact that many penises grow, visibly, in size when turgid. Some do grow more visibly (aka "growers") between their flaccid and erect sizes, say from a three-inch size to slightly over six inches, which would be a growth of three

---

demonstrates a desire to impress people via size, not substance, which, along with the current obsession over super heroes, is merely an echo of androcentric Hero Worship begun by the Greeks. See that chapter in Section 2.

[123] For example, there is concerted effort online to answer the question of whether being a "grower" is better than being a "shower." A "grower" is a man with a smaller flaccid penis, which can nearly double in size when erect. "Showers," which seem to have reached mythical status, are men who have longer flaccid penises, so much so that their phalluses show through their clothing, but they experience very little size change when erect. While many men seem to assume men whose phalluses show through their clothing are more likely to attract women, studies in countries like India seem to indicate otherwise. Perhaps "showers" are for men what women with enhanced breast sizes are for women? A healthy skeptic remains skeptical about the actual size and appeal of the organs in person.

inches. Most men with flaccid penises over 5 inches long do not see much visible growth (aka "showers") when turgid, and, ironically, if a penis is too long, the man will always struggle to maintain an erection, since the cost to his body's blood flow could easily make him pass out.

Nevertheless, the ancients would have viewed penile growth upon erection, which happens to no other bodily organ so visibly, as magical. The closest second magical body part on the human body that shows visible growth flaccid to firm muscle are the biceps, so we can understand why so many men like to show their "guns" off—as though the growth of those two muscles are proof of the penile growth or size. The reverence people held for phalluses was clearly evident in the number of male gods depicted with erect penises—from Egypt's Min to Greek's Priapus, after which the medical condition of priapism is named.

Thus, men's contributions to sex and its social bonding capabilities, not to mention sex's healing abilities, were important. Remember, sex was believed to be so magical that it could transform beasts into men (as it does Enkidu in the *Epic of Gilgamesh*) and men into kings, as well as heal the sick.

However, as men age, erectile dysfunction inevitably occurs, so our ancient ancestors had tests to determine whether or not elderly male rulers still had the capacity to provide for the people. These tests often involved successful ejaculation.

Successful ejaculation by the male ruler of Egypt became a responsibility of the God's Wife, one of the highest offices women regularly held throughout ancient Egyptian history. Her job was to help the pharaoh ejaculate to ensure that the Nile would flood its banks—a form of sympathetic magic meaning that the male had to "overflow" in order for the river to overflow, too, enabling crops to grow and the people to flourish. Unfortunately, many scholars reduce the Male Magic, especially Phallic Magic, to a mere "fertility" power, overlooking this ritual's origins in Sex Magic, a form of magic that nothing to do with making babies, but was used for transformations, so was a Power Magic.

For most ancient people, the sexual activity many creatures exhibit during the spring would have demonstrated what human beings needed to do in order to ensure plenty of refreshing water for human consumption, for crops, and for livestock. Whether we are talking quetzal birds in Mexico or vultures in Egypt, bird mating rituals were often thought to bring rain or flood waters, so human leaders often had to perform spring rituals similar to what the birds and other animals[124] were doing to ensure rain or flood waters for the crops humans depended on for survival. So, during spring mating season, human males, especially religious or community leaders, often rushed to ejaculate as often as they could, too, in order to keep the earth and its processes in balance.

Many of the male gods who were first introduced into the more gynocentric cultures at the eastern end of the Mediterranean were gods of water. Lieck quotes the "Disputation Between Wood and Reed," a Sumerian poem, circa 2400 BCE, that relates the creation of life on earth with the sky god An ejaculating into the earth goddess Ki, with Lieck noting that the implied heterosexual intercourse describes "his 'fluid' [being] poured into her body" (18). While Lieck acknowledges that the Sumerian sign used to indicate fluid in the poem "can stand for water as well as for male semen," she makes the assumption that the Sumerians knew semen contained sperm when she attempts to connect conscious procreation to the act of symbolic falling rain (18). Like many androcentrists, she fails to recognize that sperm were unknown to humanity until **much** later in our history, even though she recognizes, earlier in her book, that "Nammu stands for the female sex as the one apparently able to create spontaneously, as expressed in a hymn to the temple of Eridu; '*E. engura*, womb of abundance,'" admitting that "the 'Single Mother' matrix [was] replaced by a [heterosexual] couple to represent the primary constituents of the universe" (14). In other words, once male gods began to replace female gods in importance, goddesses who could create on their own were replaced by male gods who created on their own.

---

[124] One of my favorite Egyptian animals used to symbolize the importance of the flooding of the River Nile is the frog (and not just because my high school had a frog as a mascot). Frogs hibernate during droughts, reemerging once rains or river flooding replenish the soil in which they are buried, so they symbolize both rebirth/regeneration power of springtime, but also fecundity because their songs of woo fill springtime lands around water sources. The next time you view Egyptian art, look for the frogs; they liked to sneak them into tomb paintings, especially.

Other Sumerian gods existed to explain the existence and life-giving process of canals and irrigation systems. For instance, the Akkadian god, Enki, who was often considered the god of fresh waters, was imagined as controlling his sister (Ninhursag, symbolically reedbeds that surround water), daughter (Ninsar, symbolically the river bank), granddaughter (Ninkurra, symbolically the dike and/or channel opening that connects irrigation canals to the river), and great granddaughter (Uttu, symbolically the irrigation canal) in such a way that he literally raped them.[125] The rapes—probably couched in sacred sexual terms—were symbols of Enki's spiritual/magical control over how water, something in his divine control, reaches crops in the fields. His forcing these goddesses into coitus probably symbolized humanity's ability to dig irrigation canals with hoes and plows to transform the landscape. Therefore, ejaculated semen was seen a magical element itself, but it was mostly associated with water and milk, such as when the goddess Inanna invokes her lover Dumuzi to make his milk "thick and sweet."

Occasionally, though, we find scholars, like Lieck, who assume that the ancients knew semen contains sperm, even though sperm were not actually visible to human beings until well after the invention of the compound microscope in 1590 ACE when Antony Van Leeuwenhoak first used the microscope to view creatures he called animacules sometime around 1661 ACE (Wills). Lieck, for instance, repeatedly asserts that the Sumerians knew about procreation through various interpretations of poetry, despite the fact that human beings are "fabricated [by the gods] and [were] not the result of divine intercourse, although Enki is recognized as the virtual father" (26). Despite what many androcentric scholars attempt to assert, the title of "father" was actually a formal social title, not a biological acknowledgement of men's contributions to procreation until after 2400 BCE.

Penises were such revered sexual instruments and bringers of rain that they were even depicted in cave art, although their exaggerated lengths often make people wonder if what we are supposed to see are tails, not phalluses. Even the ancients seemed

---

[125] Such acts of incest were common in ancient stories before humanity learned that it is heterosexual intercourse that creates pregnancies.

to have penis envy, but it is important to note that almost every depiction of such penises shows them turgid and erect—their most magical form, of course.[126]

Add to the magical size change the fact that the penis can be quite stimulating in a pleasurable way, and it is not difficult to imagine how most men who were good lovers would have been viewed as magical humans, indeed, possibly being "endowed" with a higher social standing than less able lovers. If the person buried with the 6500 year old "Varna gold penis sheath" is proven to be male, he could have been given this homage to his penis by a very grateful lover upon burial (Curry). What I have not seen the Bulgarian archeologists admit, yet, is that the metal working discovered in this late Neolithic or Chalcolithic cemetery was probably discovered and worked by women through their work with ceramics. While it is distinctly possible that a male lover also adored this particular person's penis, so gave this great gift for use in the afterlife, we must recognize that no one puts their own grave goods in their burials, so we must exercise caution in assuming that the person buried with such objects was richer than anyone else, when they could simply have been so beloved that they were given special honors after death.

Does modern male obsession with penis size reflect a loss of Male Magic? Or is it a residual effect of males trying to prove their Phallic Magic was hugely important?

According to "12 studies that measured the penises of 11,531 men," the *Hindustan Times* reported that "women [are] more interested in a man's personality than the size of his manhood, but it's the men who can't seem to get over the fact that bigger is not necessarily better" (Men More Obsessed).

Possibly one of the worst aspects of androcentrism and patriarchal values is the negative impact masculine values,

---

[126] According to Devdutt Pattanaik, "Unmarried girls who worship Shiva-linga on Mondays hope like Parvati they too will be able to transform a hermit[, a solitary unmarried man,] into a householder, and get a good husband, one who is as accommodating, loving and benevolent as Shiva. Monday is associated with the moon, the graha (celestial body in astrology) associated with emotions and love."

especially those surrounding the ideas that "bigger is better" and "might makes right," have done to men's psyches. Men are constantly being pushed to "measure up" to other men's (and some women's) artificial standards for perfection, such as by having a larger than average penis or being tough, a quality which is often strangely likened to their weakest and most vulnerable physical feature—their testicles—since having an aggressive demeanor is often called "having the balls" necessary to act.

Speaking of testicles, we have to acknowledge that the ancients probably also noticed that they move, too, seeming to extend far from the body when a man is physically warm and moving up closer to the body when the man is cold. Such changes would have, undoubtedly, been viewed as magical, as well.

ΩΩΩ

While ancient men obviously were credited with having some magical abilities, given the magic of their primary gendered organs and their ability to ejaculate semen, ancient women were undoubtedly seen as more magically powerful[127] because theirs was more extensive, encompassing Creative, Protective, and Power Magics, and more clearly tied to the cosmos through the menstrual connections to the moon's cycles or Lunar Magic.

Female Creative Magic stems from women's ability to create other new humans through the spiritual and physical agency of their own bodies, just as animals and plants seem to do. Women are often associated with plants and animals specifically because of this ability to create offspring.

However, given, unfortunately, elements that are counterproductive to successful pregnancies, however, not every woman would have been able to carry fetuses to term, just as many plants' first fruits remain seedless and, often, inedible. Ute Günkel-

---

[127] The reverence for Female Magic, called shakti, for Hindus who follow Tantrism was reawakened in the 6th century ACE. According to Imma Ramos, curator of Medieval South Asia Collection at the British Museum, "the ultimate aim of Tantric practice is to become a deity like Kali—to fully internalize female power. Tantric images re-envisioned women as independent practitioners who were capable of achieving self-deification" (Tantric).

Maschek, for instance, makes many unsubstantiated fertility connections in images painted on walls in Akrotiri between women and fig branches (366), which usually bear fruit with no seeds the first few times they bear fruit, and can even produce, when mature, figs "parthenocarpically, which means they produce seedless fruit without the process of pollination" (Gill) because, even though fig trees contain both the female and male sexual parts, their "flowers still require...stimulus, such as a bee's buzzing wings or the wind, to pollinate properly," so that artificial stimulation of the blooms is often required to get fruit with seeds that is edible (Madore). Günkel-Maschek sweeps sexual rites together with marital rites with no clear evidence for matrimony (364),[128] since the female figures are all alone in most of the paintings she discusses. Instead, what is probably occurring are rites of Female Magic involving menstruation and sex, which I explain in more detail in the chapter on the Cretans.

We also know that many infants died within a few years after birth from disease and malnutrition, as well as from congenital birth defects, many possibly brought on from the women having coitus with men who were too closely related to them genetically. Evidence exists that human beings did not connect birth defects with such sexual relations until well after procreation was discovered circa 2400 BCE, since the first laws against incest did not occur until Hammurabi's famous code was written in stone circa 1780 BCE, 620 years later.

Some cultures, such as the Aztecs, went so far as to see Female Creative Magic as women fighting to bring the souls of new humans into this Earthly Realm from the Spirit Realm, such as the Abrahamic Religions' sacred space called the Guff, wherein the souls of children await birth, and the native Australian cultures'

---

128 Günkel-Maschek paraphrases another scholar who claims the women's "thick waists and their large, heavy breasts indicate that they were represented primarily in their capacity as fertile **wives** and mothers of children" (emphasis mine, 364), despite the lack of males associated with the women at all. If Minoan women and men did practice group marriage, of course, the males might not need to be present, but there is little actual evidence from the early finds of this civilization that they practiced monogamous marriage.

Spirit Centers, from which women must entice the Spirit Children into their wombs in order to give birth.

Clearly, the ancients viewed the process of giving birth as not only spiritual, but also highly sacred, in that most women had to appeal to or wrestle with the divine forces in order to become mothers. Thus, women who succeeded in giving birth to healthy babies would have been viewed as spiritually powerful in wielding their Female Creative Magic. In some cultures, like the Aztecs, women who died in childbirth were viewed as sacred warriors who had fought a spiritual battle and lost. Unfortunately, the patriarchy came to view women's only real role as producing children, so the magical hierarchy that probably affected ancient women's self-esteem has been blown out of proportion with many modern women feeling as though they are something less than real women if they have never successfully given birth. Perhaps our ancient ancestors felt they had not been spiritually blessed if they were never able to give birth successfully, but some scholars, like Judy Grahn, argue that menstruation, not birth, was the most power form of Female Magic.

Part of Female Creative Magic comes from menstruation, which not only helps prepare women for childbirth, but in and of itself would have been seen as evidence of women's Female Creative Magic because of menstruation's parallel ties to the moon phases (which is where we get the word monthly as well as the root word for menses), because of women's ability to bleed for three to five days straight and not die, and because of women's spiritual battle to bring children to this earthly realm. The fact that women fought this battle, so that they bled but did not die, was a monthly testament to their greater spiritual powers—a form of power so strong that many women in many cultures chose to segregate themselves away[129] during their menses in order to keep their stronger magic from messing up the weaker Male Magic. We see evidence of Female Creative Magic in many different forms, from animal horns adorning gods' heads to demonstrate ties with the crescent moon,

---

[129] This voluntary Magical segregation is changed by the patriarchy which tries to make menstruation seem unclean and unnatural, so that some women in some Abrahamic cultures are not even allowed to bathe or to cook during their menstrual cycles.

to red ochre or cinnabar sprinkled on the dead[130] like birth blood to allow them to be reborn into the Next Realm.

∆∆∆

Men sought to recreate Female Creative Magic for themselves through sympathetic magic rituals, such as circumcision and penile piercing. If a man survived having his genitals bleed, he took on Female Healing Magic, so hoped to survive other times he might bleed, such as during combat. Many of the *fascina* amulets worn by Roman soldiers clearly depict circumcised penises, further demonstrating their belief that, if they could survive circumcision, they could surely survive battle.

It is also possible that some men sought spiritual and social power through castration[131], such as was historically written about in ancient China wherein men agreed to become eunuchs in order to gain access to sacred spaces only the royal family normally entered like the royal palace and its grounds. DNA studies, which show that only one man for every 21 women were reproducing in the early Bronze Age, would suggest that castration was far more common in the ancient world than was previously understood, but such a claim is extremely difficult to prove since skeletal remains provide no indication of whether or not a man was a castrate.

Castration was probably also used by Sargon of Akkad on the thousands of men whose technical skills he needed in order to keep

---

[130] Interestingly, among the dead at Catalhoyuk, pigments of blue (azurite) and green (malachite) were used mostly "in burials of females and subadults" as early as 11,000 BCE, with these colors "associated with concepts of growth, fertility and ripeness," according to Schotsmans, Busacca, Lin, Vasic, Lingle, Veropoulidou, Mazzucato, Tibbetts, Haddow, Somel, Toksoy-Köksal, Knüsel, and Milella. While male remains were usually colored red. The color differentiation could also indicate that women were associated with sky deities, like the moon, so might have been seen as having, naturally, a more divine afterlife, and men were born, died, and reborn again on the earth.

[131] For most, castration would not just have involved severing the testes from the body, but also included removing the penis altogether, too. According to some evidence from ancient Sumeria, such men could be accepted into the goddess Inanna's temple for service and taught to heal through sacred sex—possibly as the first transgendered women known to the world.

his captives from rebelling.[132] Since the Sumerian goddess Inanna was also the goddess who decided each person's gender, Sargon could have used her as an excuse to transform males into females because we do have evidence that some of these transformed males served in Inanna's temple as priests who wore both male and female clothing at the same time (Grahn 24). So far, these examples are some of the few we have that our ancestors recognized and still honored people who changed genders, either willingly or unwillingly.

Female Protective Powers were broad, but the most obvious place where they occur were in sacred spaces like temples, sacred circles, and rebirthing "machines" (*wiedergebut maschine*) like the Egyptian pyramids, Sumerian ziggurats, and boat burials that are common for many cultures. Almost every temple built in the eastern Mediterranean had similar attributes, whether constructed for large population participation outdoors or for more secluded (thus more hierarchical) worship indoors: long hallways or steps led up to successively smaller spaces, wherein only the most sacred people could enter or participate—the "holy of holies," the "mound of creation," or the place where the earthly realm was believed to touch the spiritual realm—the womb that connects this world to the Spirit World. Some depictions of such spaces have images of beings flying into the sacred space from the spiritual realm, such as the "Ayia Triada 'Girl on a Swing' model" in the Iraklion Museum, which seems to depict a young goddess swinging into an epiphanic scene from south central Crete (Cucuzza 201).

Most depictions of Female Protective Powers, however, revolve around either female pudendum or vulvas. While we can easily imagine pyramids, ziggurats, stone circles, and boat burials as female pudenda, we cannot ignore mountain top sanctuaries,

---

[132] William H. Stiebing notes that "tens of thousands of Mesopotamians and neighboring peoples must have been killed in battles during the Akkadian Period, but many thousands more were captured and enslaved. These slaves were...a form of wealth the king could give to supporters or temple establishments. Most, however, were retained by the state and put to work in quarries or mines extracting stone or ores to be used for stelae and statues to glorify the king or to make products for later distribution. The ruler's ability to create and dispose of these various forms of wealth increased his power and enlarged his already paramount status" (76).

such as a revered vulva-shaped chasm on Mount Juktas from the Protopalatial period on the Greek island of Kythera. Peak sanctuaries, like the one on Juktas, were important not only for being closer to the contact point between the earthly realm and the sky realm, but also because such places allowed a wider view of the world, enabling worshippers to potentially unify their rituals via bonfires with people of nearby islands (Georgiadis 297). At the outdoor temple on Juktas, a chasm shaped something like a vulva (see image below) is clearly visible with an interior stone of a bluish color, possibly a betyl, or a sacred stone that possibly was understood to be an earthly representation of an important spiritual being or deity, possibly one associated with water or birth. According to Mercourios Georgiadis, "the buildings erected at Juktas...are aligned to [the chasm], and all the finds around the chasm suggest that it was the focal point of the rituals that took place" (299).

Sketch of Juktas chasm on Kythera, Greece.
Note the interior blue stone, possibly a betyl,
a rock designated as the presence of a deity.
Georgiadis

Such Female Protective Powers extended all the way into medieval period when many churches in northwestern Europe were built where *sheila na gigs*, female figures depicted holding their vulvas open, conveyed the idea that the people were being welcomed into the protective womb of Mother Church where they were supposed to be "reborn."

Female Power Magic also took several different forms, but by far the most misunderstood form that continues to put off androcentric scholars today is that of sacred sex rituals—from the healing properties of Sumerian *hashadu* rituals to the transformative powers of the Sumerian *qursu* ritual, which transformed mere mortal men into divinely sanctioned rulers. Far too many androcentric scholars still believe that sex with "taboo" women in a temple means the women were harlots, whores, or prostitutes of some kind, when, in fact, they were trained in the healing and transformative arts of sex. Misinterpreting the idea of "taboo," these scholars prefer to see "women who are supposed to be untouchable" instead of "women who are only supposed to be touched by certain people at certain times because of their sacred abilities."

What no other scholar has ever admitted is the fact that the ancient symbol for women, the triangle with a bisecting line at one corner, a clear symbol for women's vulvas is extremely ancient. A recent find in South Africa confirms the fact that ancient humans revered women's vulvas and associated them with those wonderful spiritual conduits as far back, possibly, as 143,000 years ago, which is the oldest possible date on the rock carving found off the cape coast (Rock Stars). Despite the archeologists' disclaimer that they could not interpret the meaning of this symbol carved on the rock they rescue, this same symbol continues to appear as the actual written word for women among the Sumerians and Akkadians, between 3100-2000 BCE.

ααα

After humanity learned that males had to be biologically contributing to making babies, circa 2400 BCE, a slow, but determined shift in perspective occurs over the next 1500 years, wherein males become viewed as the only humans wielding Creative Magic with females becoming seen as merely fields to be plowed and vessels to carry the miniature humans (homunculi) that the males came to believe they were "planting" inside women's wombs. As human women became degraded in importance, so, too, were female gods, so that by the time we get to the Greeks, who so many people prefer to see as the "original" start of European or Western culture, the male gods[133] are circumventing the need to have women give birth by doing the creating themselves, leading other nearby cultures to assume that we never needed female gods for creation at all, an idea that directly leads to the self-proclaimed rise of the patriarchy primarily through the Abrahamic religions, which have so greatly affected "the West."

As a result, Female Creative Magic, Female Protective Magic, and Female Power Magic were eliminated from our human consciousnesses as much as possible, leaving only Male Power Magic and Male Creative Magic in their stead, with some males still struggling hard to maintain their power over Female Creative Magic in order to maintain their often brutal hierarchy over women. Such a hierarchy requires conflict in order to actively suppress women's rights. Gynocentric ideas, however, while still allowing for some hierarchy involved less conflict and more cooperation through nurturance, guidance, and collaboration, proving that, while Roman soldiers might have worn *fascina* for luck so they would not bleed to death in battle, gynocentric cultures largely focused on nurturance and protection through cooperation, so focused less on a need for luck and more on social bonding through sex.

---

[133] Zeus, for example, gives birth to nearly all of his children himself.

Chapter Sixteen
Why Ships Are Called She

Sometimes sexism is so engrained but so clearly bigoted that we have to wonder how such biases began. Examples of such sexist bias include the traditional meteorological method of calling super storms by female names, a tradition which ended in 1978, and the traditional association of boats with women or of giving them female names.

This last bias was often dismissed as acceptable because it was believed only men own and captain boats, clearly ignoring all female captaincies that we know about in history, even from ancient history, such as Artemisia of Caria, a female admiral who captained a boat in support of Xerxes' attempt to invade Greece.

The reality of the origin of the boat bias, however, is much simpler, and, perhaps, not as sexist as it seems at first glance.

Whether we are talking about a dugout canoe or a reed boat on the Ganges, the Euphrates, the Nile, or Lake Titicaca, early boats had a basic shape—like vaginal labia or the vulva. The primary reason the Sumerian goddess Inanna called the vessel on which she sailed the star filled skies the Boat of Heaven was because the boat symbolized her sexual labia in many of the poems, hymns, and prayers to Inanna. Physically, the Boat of Heaven could be the Sumerians way of referring to the Milky Way, which appears in truly dark skies like the prow of a boat plowing through foamy seas.

Through heterosexual intercourse with the goddess' representative on earth, male leaders became kings, so it was long believed that Inanna's sex organs produced godlike attributes in the human males who courted her. The ancient Sumerians, possibly even the ancient Ubadians who preceded them, practiced a New Year's ritual called the *qursu* ritual, which allowed Inanna, the keeper of the divine knowledge from the gods, to transmit that knowledge to humanity through their male leaders via heterosexual intercourse on an annual basis. It was her "boat," her vulva, which put the men in touch with the gods by transporting them to the heavens, demonstrating not only the transformative spiritual nature of coitus, but also the idea that the enjoyment of sexual relations were what put humans in closest parallel with the gods who were believed to be greater beings than humans themselves.

However, we also learn in the story of "Inanna and Enki," which was composed after male contributions to pregnancy were discovered circa 2400 BCE during the rise of the Akkadian culture, that Inanna's Boat of Heaven is one means by which Inanna protects herself when Enki[134] tries to stop her from delivering the sacred *me*, divine knowledge imparted to humanity through Inanna's sex organs, to her favorite city, Uruk, so we also have to discuss how the vulva and vulva shapes were also symbolic of the womb's protective powers.

The word "vulva" wonderfully encapsulates the fact that early humanity reflected heavily on women's sexual powers a lot. The "v"s in the word not only represent the triangular pudenda symbols we discuss so heavily in this book, like pyramids, ziggurats, and the chemical symbol for transformation, but also encapsulate the basic ends of most boats.

You might rethink voyaging in Volvo[135] cars when you learn that the word "vulva" comes from an earlier Latinate word, *völva*, which not only indicated the female sexual organ, but also encapsulated the idea of re**volv**ing or turning, changing, re**volu**tionizing (Vulva), although the root word could come from Proto-Germanic word "*waluz*," meaning "staff," since the seeresses called *völva* or *seiðr*, famous from early German and Roman

---

[134] As the god Enki rises in power and significance with the Akkadian empire over ancient Sumeria, he appropriates many of Inanna's previous powers. Tales about Enki, like "Enki and Ninmah," work to prove to Akkadians that males knew all along that they were the real wielders of Creative Magic, changing Inanna's sexual Female Creative Magic into Male Creative Power as they lead their people to believe, instead of sex itself being a sacred, liminal activity that puts humans in contact with divine powers, such as through the *qursu* ritual, but that "mankind has a share in the divine because of sexual reproduction; just as the gods in the beginnings of time made love and gave birth, so do mortals; but the essential substance came from the male" (29). This androcentric propaganda worked to reduce women's and female gods' status over time.

[135] Johannes Kepler, in his science fiction story, The Dream, calls the planet Earth Volva, so named by his moon inhabitants in order to emphasize that the Earth revolves around the sun, not vice versa, which was commonly believed in his time (Popova). Read Popova's well written article in order to learn how much science had not yet revealed by Kepler's time (1571-1630).

sources, were powerful magical women who wielded wands or staffs (Simek), such as the woman buried circa 1550 ACE in the bed of a horse-drawn carriage along with a metal wand and a bag of henbane seeds (A Seeress). Despite the fact that the automobile company adopted the Greek god of war Ares' sign (Reeves), which is used almost universally to indicate males, as its company logo, the origins for the name of the company has more revealing roots. Originally, the term probably indicated that people could ride in Volvo cars as safe as in their mothers' wombs.

Prior to Latin, the vulva's Proto-Indo-European root, *wel-*, referred to curved, enclosing objects, like a boat (Vulva).

Imagine early sailors climbing into their canoe or reed boat. Some might assume such sailors always thought about sex, but it is more probable that they were hoping they were climbing into a protective womb that would keep them safe, something like entering the protective womb of Mother Church was for medieval Christians.

Some of the earliest evidence for boats comes from the pre-Sumerian Ubaid culture, as early as the sixth millennium BCE with what Robert Carter calls "advanced boat-building and sailing technologies" (52). Archeologists found a ceramic model of a reed boat at As-Sabiyah (Kuwait) that dates from circa 5000 BCE (Carter 54), as well as a ceramic disk painted with a two-masted boat (Carter 55), marking the earliest known indication of sailing.

Boats became especially important during the Bronze Age, however, because they got bigger, sturdier, so could navigate deeper waters and carry more cargo—from invading armies to trade goods. Improved boats allowed sailing further from shore.

Ships play a vital role in the believability of Homer's epic poem, *The Iliad*, because all those Greeks warriors who laid siege to the city of Troy had to get across the Aegean somehow. Such was the Greek obsession with the power of boats that Homer takes the time to "catalog" nearly all 1000 boats, listing their origins, captains, and cultures of the warriors each carried in Book 2's "Catalogue of Ships."

Some scholars speculate that the tradition of carving a female figure on the prow of a sailing vessel appealed "to the gods of the ocean" to allow safe passage (Harris). However, if we go back far enough, we discover that the ships mentioned in *The Iliad* were likely to have had eyes painted on their prows. While some scholars debate why such eyes were there, "Greek literature shows that the eyes of ships primarily served to mark the presence of a supernatural consciousness that guided the ship and helped it to avoid hazards" (Nowak 2).

Since our ancient ancestors believed everything was imbued with spirit, each boat built was undoubtedly also seen as a separate, conscious entity, with nearly every single vessel being considered female. Sea nymphs were believed by Greeks and Romans to be real spirit entities that saved sailors from drowning, so with discoveries of marble eyes that had once been affixed to such boats, we can confidently understand that the ancient sailors believed they were putting their trust in a sentient vessel that determined if they would survive or die on the seas. Such beliefs would have led to the often-heard statements made by sailors, "she'll hold together." A quote often credited to Aubrey de Selincourt, author of *Cruises in Small Yachts and Big Canoes* in 1883, is that "a sailing vessel is alive in a way that no ship with mechanical power ever be," demonstrating how long the belief that sailing ships were, indeed, living beings has held onto human consciousness.

In fact, some scholars speculate that the simple round eyes often sold as amulets for warding off back luck or "evil eyes" cast upon us by people of ill intent were born out of the practice of painting eyes or adhering stone eyes to boats (Hall). Whether the eye is the Eye of Horus or the Eye of Fatima, the human eye is also shaped like a boat, so we end up with layers of Protective Magic from both the vulva shape of the boat and the eyes seeing vessels to safety.

While some might believe that human women were superstitiously not allowed on sailing vessels because women might distract the sailors at a moment when they needed to keep a watchful eye on the seas, the reality might be a superstitious belief that a female sentient vessel might get jealous of a human female on board, so sink the whole crew out of spite.

Perhaps the idea that boats were inevitably females who had to be controlled by a number of men fed into the rising androcentrism that we witness during the Eurasian Bronze Age.

While boats originally carried people, especially traders and explorers, from place to place on peaceful missions, once we get wooden ships with sails and masts that allow further and faster exploration, innovations that occurred during the Bronze Age, a new form of hierarchical oppressive power arises. Being able to command warriors was one form of Greek power; but being able to command a whole fleet of mobile warriors was clearly seen as a desirable form of power because such mobility enabled different cultures to dominate areas of the Mediterranean, Aegean, and Black Seas through brutality.

Thus, while boats originated as means to navigate rivers and shorelines safely, reminding later sailors of women's sexuality in the process, ships became weapons of toxic masculinity that only sought to glorify men of wealth and social standing and to dominate women and poorer men, who became slaves—captives bought, sold, and traded among powerful men.

Chapter Seventeen
Female Scholars Raised in Androcentric Cultures Might not Recognize Their Own Androcentrism

I continually hope, as I read articles and books written by female scholars, that I will read work written by "woke" people who not only understand how patriarchal systems are grossly unfair to women, but also recognize that the androcentric bias we are all fed on a daily basis is a cultural preference imposed on people around the world because some people continue to want everyone to assume males are superior to females in some way. As I point out in my chapter, "The Translation Problem," where I examine the androcentric bias evident in six translations of *The Iliad*, we cannot assume that, just because a woman has written scholarship or translated literature into English, she is not carrying this androcentric bias with her into her work.

Yet far too many people who presume to argue against many, if not most feminist arguments, also presume that, if a woman says or implies something bigoted toward or about women, those women must be accurate in their statements or beliefs. Women, they naively believe, cannot possibly be bigoted against other women, despite the fact that we can be just as socially programmed as some men are.

So, before we continue further, realize that **women are just as influenced by cultural biases, mores, and beliefs as men are**.

In fact, female scholars, because of social pressures to believe men more readily than we believe women, often prefer to believe male scholarly views, even when those views are obviously biased or insubstantial.

For instance, Gerda Lerner wrote an excellent foray into the idea that the patriarchy was created in the Bronze Age, just as I am doing in this book, in her own book, *The Creation of the Patriarchy*, which she published in 1986. However, she chose to believe many unsubstantiated claims by many of her male colleagues. An important belief she embraced without question, which became a central idea to her arguments in her book in her chapter "The Woman Slave," is the idea that women were enslaved before men ever were, and she couched that belief in scholarship which made it sound as though her ideas were facts from ancient Mesopotamian history.

Lerner makes the sweeping claim that "prior to the invention of slavery was the subordination of women" (77), then spends the rest of the chapter citing sources which appear to support her claim. One of the primary sources she cites to support her claim is by Ignace Jay Gelb, a Polish-American historian and Assyriologist, entitled "Prisoners of War in Early Mesopotamia," which was published in early 1973. Gelb, initially, makes a similar sweeping generalization without evidence: "at first, men were killed, and only women and children were taken captive." Later, however, Gelb becomes much more tentative about his proposed ideas than Lerner makes them sound, stating that it is a "possibility that in addition to killing enemy warriors in battle, the Mesopotamian rulers may have been in the habit of gathering enemy males and probably others and putting them to the sword at some place within the territory of the conquered cities" (74).

Both scholars, Lerner and Gelb, are making assumptions about ancient people based on little factual information, and, without seeming to realize it, were heavily influenced by their contemporary patriarchal cultures.

In fact, scholars have long argued over two important issues regarding the forced capture of women:

1) were women sought after because of their child-bearing abilities, for sexual reasons, or for their technical expertise, such as growing, harvesting, processing, and weaving flax into linen?

2) Were female children routinely killed in order to control population numbers?

These two issues highlight an important disconnection of logic. Why would people who supposedly sought to control their population numbers, so routinely murder female infants, then want women for reproduction?

Contemplating this confusion leads to another important question: if women were not actually sought after for reproductive purposes, what were they sought for?

For Lerner and Gelb to be correct in their assumptions that only women and children were originally enslaved upon capture by an enemy culture, the captured women and children were probably not desired for reproductive purposes, even though they could have been sought for sexual purposes. This distinction brings us back to what the ancients undoubtedly thought—that there was **no specific connection between sexual activity and reproduction**.

If we examine how other cultures treated people who were captured or rescued from other cultures, however, we have to question the whole premise of slavery being a basic human institution. American Indian cultures, for instance, regularly adopted survivors of battles who were not their own people. In fact, early on during the European conquest of North America, so many Europeans found the lifestyles and the way Indians treated others so attractive that they left the colonies in large enough numbers for many English colonies to pass laws forbidding "going native." Like many tales about women and girls captured by American Indians put forward by popular newspapers and publishers of pot boiler novels, however, European Americans tend to romanticize such narratives to mythological levels, something that was also quite popular in ancient Greek stories, as well. Because of this tendency to exaggerate exciting circumstances, we must exercise caution.

Lerner goes further with her ideas, claiming that "the impact on the conquered of the rape of conquered women...**dishonored** the women and by implication served as symbolic castration of their men" because "men in patriarchal societies who cannot protect the **sexual purity** of their wives, sisters, and children are truly impotent and dishonored" (emphasis mine, 80). If, as she claims, the men were killed outright, why would they **also** need to be symbolically castrated in this way in order to dishonor them? Furthermore, if the ancients truly did not know that heterosexual intercourse led to pregnancy, why would they ever care about "the sexual purity" (a purely patriarchal idea) of their women?

Lerner's main logical flaw is believing, as many of her colleagues of the time wanted her to believe, that the patriarchy was in full swing (as opposed to being androcentric heavy) in ancient Mesopotamia circa 2300 BCE when Gelb's first piece of evidence to support his claims was written (Gelb 73), which, as I have discussed elsewhere, tends to be an assumption far too many scholars make about cultures thriving during the Bronze Age around the Mediterranean Sea.

The lack of clear logic by many scholars demonstrates how carrying cultural biases into scholarship with us, especially if we are completely unaware of those biases we are carrying, can undoubtedly taint our conclusions about ancient cultures.

For much of written history, women have been lauded as wellsprings of culture because of their disproportionate influence on children for so long that many women do not realize that it is often they, themselves, who are enforcing androcentric bigotry in their children. Too many mothers today still worry that their daughters will not be successful if they are not attractive enough, failing to recognize how mentally and emotionally straining such pressure is on girls, not to mention sexually confusing for all of us. Dressing children in sexually provocative clothing or even in typical "princess" fashions limits their ability to see themselves as full human beings worthy of being more than just sex objects who have to flaunt their physical bodies (possibly doing more than flaunt) for power or influence.

One giant obstacle to doing serious scholarly interpretation of ancient literature, archeological discoveries, or historical periods is **the assumption** by most scholars that this social construct we call "the patriarchy" has **always existed**, so much so that some anthropologists are trying to determine how extinct human beings must have viewed fatherhood[136] (Gray & Anderson), when the

---

[136] Gray and Anderson admit that, among primates, males who are not the biological fathers of infants, but who choose to help a female with the care of her infants, are more likely to be able to mate with her when she is, once again, ready to do so (12). Both scholars play rather loose and fast with the term "monogamous," however, when discussing mating habits, seemingly satisfied

larger, more believable probability is that ancient human beings had no way to determine that males contributed biologically to the creation of children until circa 2400 BCE when they finally domesticated swans, so that ancient humans **had no concept of biological fatherhood**. Note that I am not arguing that males cannot act in nurturing ways toward children. What I am saying is that ancient humans had little ability to differentiate what the catalyst was for pregnancy, so they adopted many different ways to explain how women became pregnant.

In fact, most pre-literate cultures developed spiritual beliefs systems to account for the fact that some children looked similar to some men in their tribes, with many believing that men's spirits—all the men's spirits—in the group worked their influence on the child either in the womb or after birth to shape the child to resemble males in the group. This belief meant that all the men of the community were responsible for looking after the children. This idea still holds in those cultures that see raising children as a community responsibility.

However, we have to acknowledge that most pre-literate people lived mostly in small groups, so the people were already related and shared similar features, so the mere fact that some children favored one or another male was simply **not enough evidence** to prove that men had a biological role in procreation. Jean Auel, who wrote the *Clan of the Cave Bear* series of fictional novels, struggled with this idea. In her first novel, she assumed that a Cro-Magnon woman pregnant with a Neanderthal's child would notice the similarities between how he looked and how the child looked, but her story is predicated on the idea that the woman knew what she looked like and that she looked different than the Neanderthals, despite the fact that she had been rescued by them when she was about five years old. While this novel was first published in 1980, and while Auel had tried to conduct research to

---

with "relatively monogamous" as good enough a sign that males understand which offspring are theirs (12), never considering how natural it is for many adults of most species to care for infants in order to ensure survival of their kind. The fact that adult males of different species have still been spotted caring for the young of other species should point to the fact Gray and Anderson are trying to endorse—that males can be natural nurturers.

make her story believable, she clearly carried a bias forward into her story that intelligent human beings would have figured procreation out just based on appearances. Novelists have to imagine other scenarios every day, so if even a novelist has a difficult time imagining what it would be like to be an ancient human being who knows nothing about procreation, imagine the struggles scholars raised in androcentric cultures have.

As a result of our current knowledge and social programming, we have to recognize that just monitoring the sexual intercourse of long-term gestational animals even after domestication would have provided no proof of a male biological contribution either because of complicating factors such as multiple sex sessions and multiple sexual partners, not to mention instances of homosexual activity. Our ancestors would have wondered, if a male ejaculating into a woman caused pregnancy, then why did men who had anal intercourse with each other not get pregnant, too?

However, ancient humans also have believed a number of nonsexual things about how pregnancy occurs—from the wind blowing up a woman's skirt and implanting a spirit inside her to the woman swallowing something magical—from a seed to a whole pomegranate—that caused her to become pregnant. The Aztecs imagined their Earth Goddess becoming pregnant because she tucked a tuft of hummingbird feathers in her belt. The Sumerians have a tale where a male god gets pregnant from swallowing seeds. One Navajo story tells of how Changing Woman became pregnant by spreading her legs before the sun, and her sister became pregnant from lying in a stream. Europeans have tales about women who get pregnant by looking at rabbits on the full moon, from swallowing ice or snowflakes or flowers, or from cinders that fall on their laps. And we would be disingenuous to leave out the number of different ways Zeus transformed himself to impregnate women— as a swan, as a bull, and even as golden rain.

The idea that human spirituality can influence how children look as they grow probably led to such practices as cradle board shaping of skulls to form more pointed skull caps, but also lent to the idea that certain tattoo marks on the skin, sharpened or purposely discolored teeth, piercing or scarifying various body parts were physical attempts to alter our spirits within. However, these physical changes created distinctly human features that became recognizable signposts distinguishing one person as coming from a particular tribe or clan. Among some traditional Lakota, for instance, a particular geometric pattern had to be tattooed around one wrist before death in order for the deceased to pass successfully through a hole in the constellation we now call the Big Dipper to reach the next world. Before Western culture redefined beauty for them, many cultures, such as the Amazigh of northern Africa (Ahdad), continued such beauty practices as tattooing faces until rather recently, and renewed interest in such body art in other parts of the world indicates that many people are still attracted to this artful, tribal, and spiritual aspect of some current subcultures.

In fact, among some Australian Aboriginal cultures, neither women nor men contribute to the creation of babies. Instead, "spirit children" decide what woman they want to give birth to them, although some males believed they have a direct influence as to what woman gets chosen, taking direct credit for "finding" the spirit child in some way (Merlan 475). Even among the "women who distinguish the man who 'made their baby'" via heterosexual coitus, some believe the true father of the child is of her choosing (Merlan 477), so that spiritual connections are more important than genetic ones.

In short, early humanity had no reason to believe that males contributed anything biological to procreation, so they did not associate heterosexual intercourse with procreation until much later in our history than most scholars are willing to believe. As I prove in this book, the ancient people around the eastern end of the Mediterranean were only able to scientifically prove males have a role in procreation after the domestication of the only monogamous animal ever domesticated by humans—swans—circa 2400 BCE. After that, humans had a better, more scientifically verifiable method to prove that males play a role in procreation, so that we should all know by now that some form of heterosexual intercourse, even in a petri dish, is necessary for humans to create offspring.[137]

---

[137] Even after learning this biological reality in school, many cultures around the world, such as the Aboriginal nations of Australia, continue to believe that sex has little to nothing to do with pregnancy. In interviewing people about the earliest age where they learned heterosexual coitus causes pregnancy, I met two women

n. 216, Part of what blinds modern scholars to this fundamental lack of human knowledge in ancient times stems from the fact that human beings enjoy a number of different sex acts, such as those depicted on the Moche sex pots, which meant that isolating heterosexual intercourse as the cause of pregnancy from all of these other varieties of sexuality would have been a struggle for most ancient people, including the Myceneans, a fact I discuss in my analysis of *The Iliad*, which takes place long after 2400 BCE, so the Mycenaeans should have known that heterosexual intercourse alone can cause pregnancy, but the poem indicates that the Greek warriors might not yet have realized this fact, which means the knowledge still held problematic issues that did not fully convince the people of Homer's day, circa 750 BCE, either.

The biological discovery of procreation (meaning using both female and male genomes to make offspring, as opposed to parthenogenesis, which is when females essentially clone themselves) made through observing swans also led to the proliferation of the egg as a symbol of birth and rebirth in some cultures, such as Hinduism and ancient Greek beliefs that Zeus mated with Leda as a swan, so that the egg as an image competed with the "mound of creation" or women's *mons venus* as the source of humanity[138]. Prior to the discovery of a link between heterosexual intercourse and procreation, semen was seen as "milk" males produced to feed babies in the womb, or like water flowing from a spring, which is why kings' semen was often linked to the flowing of rivers and to the fertility of the land along those rivers—not because it could produce offspring, but because it flowed. See any of the ancient Sumerian Inanna and Dumuzi poems for evidence because Inanna wants to make Dumuzi's milk (aka his semen) flow "thick and sweet," demonstrating no desire to become pregnant (Assante 39). In short, Inanna as the Sumerian goddess cannot be used to

---

who had not learned this fact until they were teenagers because of sheltered childhoods!

[138] Interestingly, the mating of Zeus as a swan with Leda was depicted on a fresco found in Pompeii, and inspired paintings by Leonardo da Vinci, Michelangelo, and Tintoretto, among others. And Gustave Courbet knew the symbolism of the "mounds of creation" when he painted "L'Origine du Monde," which means the "The Origin of the World," as an anonymous female with her legs spread showing her pudenda and vagina lips.

make "fertility" connections in the poetry because the sex-pregnancy connection did not yet exist for the people, despite the many scholars who misname Inanna as a fertility goddess. She was the goddess of sex, evoking all of sexuality's powers of transformational magic, something I will discuss in more depth in the chapter on her.

Yet this idea that **the patriarchy has not always existed and was not a full on thing at the beginning of the Bronze Age** occurs to few scholars, even though J.J. Bachofen, who began writing about what he called "mother right" in the mid-1800s, firmly believed that matriarchal cultures had to precede patriarchal ones, an idea that must be seriously considered because human beings undoubtedly believed women were sole creators of human offspring prior to discovering procreation.[139] Bachofen notes in his Introduction to his book, *Mother Right*, that the Greek historian Herodotus reported that the Lycians named their children "exclusively after their mothers," kept only genealogical records that listed the maternal line of descent, and that "the status of children was defined solely in accordance with that of the mother" (70). Bachofen went on to argue that such an inheritance system had to be the original inheritance system because so many other cultures, what he calls "pre-Hellenic peoples," practiced matrilineality, as well (70). Bachofen has been alternately lauded and suppressed by various scholars and publication companies since his works were first published, mostly because he does not explore one undeniable fact thoroughly enough: we always know who the mother of a child is, but males have to control women's bodies in order to be relatively certain of paternity.[140]

---

[139] See my discussion of parthenogenesis and virgin birth in another chapter.

[140] Unless, of course, the mother abandons the child anonymously, we truly do know who every child's mother is. Mother abandonment, however, does play a strong role in many early pieces of literature, with the mother often putting her child or multiple children in reed baskets, which are then placed on a river, allowing the gods of the river and seas to decide the child or children's fate. Note, too, that DNA advances have led to a flood of discoveries by children that the man who they thought was their biological father is, in fact, not their biological father, so this millennia old problem for men still exists, despite attempts to control women's bodies.

In the end, the inability to accept the idea that humanity has not always known that men have a biological role in procreation demonstrates androcentric bias so deep that many thinkers cannot overcome their cultural programming to admit the truth.

Even though some feminist scholars, like Marija Gimbutas, Lithuanian archeologist and anthropologist who was once laughed off the podium at a professional conference for daring to present evidence of preliterate goddess cultures in Old Europe, have continued to chip away at the androcentric bias clouding most cultural scholarship, that bigotry still exists, even among feminist scholars.

I have a lot of respect for other scholars who are able to publish works that open our eyes about how undervalued women have been in most androcentric cultures or have created works that demonstrate that women have, indeed, risen to positions of eminent power in some cultures.

Kara Cooney,[141] for instance, in her book, *When Women Ruled the World*, tries to assert that having women in power was a common and necessary thing in ancient Egypt, yet she still assumes that the patriarchy existed during the first dynasty of Egypt—prior to the domestication of swans, which was the only way humans could verify that males play a biological (versus merely spiritual) role in procreation. She allows the growing androcentric biases against women, which are demonstrated in documents written by later, more bigoted Egyptians, to cloud the fact that Merneith, one of the few clearly female rulers to be listed on an Egyptian king's list, was probably a far more influential person in the earliest Egyptian dynasty than even Cooney admits or, more tragically, understands.[142]

Cooney, like most scholars still viewing evidence through an androcentric haze, theorizes in her introduction to *When Women Ruled* that the reason why women are not respected as leaders in America is because of having less "sexual value" than men. She

---

[141] Realize I am not focusing on Cooney here to pick on her, but she is a great example for illustrating my point here. Personally, I like Kara a lot, and even follow her on social media.

[142] To be fair, Cooney admitted at a speech she gave at the Nelson-Atkins Museum of Art in February of 2020 that she is not an expert on Egypt's first dynasty.

admits that "sexual value is social value to human beings, whether we openly admit it or not" (19), but goes on with faulty logic to assume that because men can procreate as long as they have viable sperm and the ability to ejaculate no matter how old they are they are considered valuable sexually, but, because women lose their reproductive abilities after menopause, "a woman in her fifties, sixties, or seventies can provide no such sexual value to society through reproductive ability." Strangely for a self-proclaimed feminist, Cooney seems to believe "no female leader can take on more sexual partners because of her power" (20), which completely misses the point about why sexuality was and still is valuable in human society.

Like many androcentrists who try to control women's sexuality and anthropologists and archeologists who conflate sexual symbols with fertility symbols, **despite the fact that pre-literate humans did not connect the act of heterosexual intercourse with human fertility**, Cooney seems to assume the main (or only?) purpose for sex is procreation. However, sexual intercourse is not just valuable for reproductive purposes even today, and its value for ancient people went beyond mere sexual satisfaction into the liminal—a magical power so strong that it not only healed the sick, but also transformed beasts into men and men into kings because of its ability to connect us to other worlds beyond this one, which I will discuss in more detail elsewhere in this book.[143]

Sadly, even in Cooney's most recent book, *The Good Kings*, in which she clearly tries to break up humanity's romance with "elite" or powerful men, Cooney again demonstrates that she clearly believes the patriarchy was in full swing in ancient Egypt:

------

[143] I examine the West's first epic, *Gilgamesh* and the ancient Sumerian sexual ritual, *qursu*, via which a man was able to copulate with the representative of the goddess Inanna to become a "king"—the closest thing to a living god among humans for the ancients because of divine sanctioning through the sharing of sacred secrets with the god's representative on earth; the *qursu* ritual is oversimplified by many androcentrically-biased scholars as *hieros gamos*, even though the ritual has nothing to do with marriage. See also Julia Assante's "Sex, Magic and the Liminal Body in the Erotic Art and Texts of the Old Babylonian Period" in *Sex and Gender in the Ancient Near East*, 2002: 27-51.

"In Egypt, a few women were crowned king, but only when there was some kind of crisis. Those women who did manage to tear through the papyrus ceiling had to submit to a variety of compromises, from remaining unpartnered and accepting constant and unsolicited advice from men to having no legacy after they were erased from the history books. Then, as now, the patriarchy jealously guarded the halls of power, saying that only a man could do the job, that the people needed him, that they would be unsafe without him" (Good Kings, Chapter One).

Cooney not only ignores all the females named on the King's List as well as all the male rulers who were also scratched out of Egyptian history, but she goes on to claim, without evidence, that "patriarchy has been the unifying political religion of every polity since the agricultural revolution" (Good Kings, Chapter One).

The advent of agriculture, which was not the sweeping one-time game changer most scholars attempt to make us believe it was, did not give human beings any more insights about the mysteries of procreation than we had prior when we observed animals as predators hunting them for sustenance. We had to have a controlled environment in order to prove beyond the shadow of a doubt that males had anything to do with procreation, and the controls necessary did not occur anywhere in the world prior to approximately 2400 BCE when the Egyptians finally domesticated the only short-gestation, monogamous animal ever domesticated, swans.

The reality about the important role of sexuality and sexual intercourse plays in our society as well as to the ancients is much easier to grasp than many scholars let on, so why do they so assiduously avoid talking about the importance of sexuality?

Many men who hold powerful positions in society are seen as sexually attractive by women and men who fall under their sway. In fact, if we accurately report why so many women continue to support Donald Trump, which Cooney tries to explain in *When Women Ruled* (19), despite his open misogynistic attitudes and actions toward women, we would find that most of the women who support him do so for just two reasons: 1. They think he's sexy, remembering, obviously, the younger Trump, but his age and girth is compensated for by these women by his wealth and perceived power; and 2. Because they have been brainwashed to believe that women must be punished for getting "caught" having had sex, so they hope he will make abortion illegal again. Note that these negative beliefs about women never applies to these conservative women themselves, but only to other women, which is a clear sign they have been brainwashed, since they fail to recognize themselves as anything like the women whom they hold in contempt.

Women who hold power, on the other hand, frighten people who believe males are superior, which explains why later during the rise in androcentrism, Hatshepsut was reviled and removed from many Egyptian monuments because a woman who held the ultimate power and wielded it well was suddenly superior to men, creating a form of cognitive dissonance for those who cannot view females and males as equal. It is this same androcentric fear of females in power which has created such distrust of women in power by Americans.

Cooney, like so many female scholars still blinded by androcentrism, admits that "when Merneith was a young girl, kingship in Egypt was shiny and new," but Cooney's androcentrism makes her avoid asking an important question that is rarely raised by Egyptologists: were the first rulers really "kings"? What word, exactly, was used by the ancient Egyptians to describe their rulers in Dynasty 1? Cooney admits that hieroglyphic writing from Dynasty 1 was "often indecipherable because Egyptian writing was in its earliest stages of development." So why is she assuming that a male-oriented title was used to describe the earliest rulers of Egypt, especially given the fact that at least one, possibly three, of the "kings" listed on the Abydos King List, written in Seti I's Mortuary Temple (circa 1190 BCE; post rise of androcentrism), were female names? The term used to label a ruler on the Abydos King List, which was written, to emphasize the point, **after** the rise of androcentrism, was *nisu*, which could literally translate into "gods vessel" or "vessel of the gods" (When Women, 30-31). While *nisu* or *nsw* is considered the masculine form of the word, and *nswt* or *nisut* is considered the feminine form (Scribe), the scribes, even if they were female, who wrote the Abydos King List would have been well immersed in an androcentric world by the time they produced the list, so would have used the masculine form, despite listing at least one female on the list, just as many writers use the pronoun "he" today, when they write about people in general. The idea that a leader was a "vessel" would not have been a natural fit for a male, though, since women were not only associated with sailing "vessels," but also associated with cooking "vessels" and storage "vessels," not to mention carried babies like "vessels," so it is more probable that somewhere further back, such spiritual leaders were women.

We know the word used to describe a ruler in Dynasty 1 was not "pharaoh," because that word was introduced by Hatshepsut as she sought to legitimize her power in the 18th Dynasty, roughly 1000 years after humanity could scientifically prove that males had a role in procreation. Yet Egyptian scholars use the title "pharaoh" regularly as though it was always the word used to describe ancient Egyptian leaders, which is highly misleading.

It's important to note two things here:

1. The word "pharaoh" means "great house," which not only provided Hatshepsut with the royal lineage she needed to justify her rule, but also returned Egyptian thinking to the past when "houses" were undoubtedly identified by the women who built and led them—a matrilineal culture. Clear evidence exists around the Mediterranean that the women were not just "housekeepers," but also the house builders (aka homemakers), and many of the homes built and extended by women housed several generations of their children, with many adjacent burial grounds, thanks to increasing DNA studies, demonstrating that many cultures were matrilocal, meaning children were buried near their mothers, and a person's lineage would be matrilineal, meaning the children would have identified themselves by their mother and their grandmother's names, as Bachofen noted.

2. The word "regent," which is typically used by androcentric scholars to describe the women who

ruled Egypt while their children were too young to rule, is a sexist way of describing women rulers. If a woman was powerful enough to make all the important decisions, regardless of whether or not she had a child waiting to officially take the throne, she was the ruler, not just a regent. We never call men who have children waiting to rule "regent," so why do we relegate women to such a minor role? Egyptologists must stop insisting that only male rulers can be called pharaohs and should use the appropriate term used by the culture for their leaders from the time periods being examined.

We have to acknowledge, then, that the androcentric bias carried into the beginnings of Egyptology by the likes of Samuel Birch, who wrote the first Egyptian Dictionary, "in the first quarter of the nineteenth century" (Budge v), has influenced Egyptology since that time, including works by female scholars like Cooney who believe they are feminists reshaping an inaccurate view of history, even though they miss this vital point about the beginning of the Bronze Age—there is no evidence that men were already in a superior social position in any of the cultures around the Mediterranean. Remember, if a woman can be a patriarchal country's leader, men can lead in more gynocentric cultures, too.

Understandably, the original translations of hieroglyphs (and cuneiform) were crude, rudimentary, and lacked some acknowledgement that Egyptian symbols and syllables could have more meanings than what these early scholars put forward, such as the idea that a ruler was viewed as merely a "vessel for the gods." However, we should continue to demand more accuracy and fairness from all scholars studying early cultures.

We now know so much more about both Dynasty 0 and Dynasty 1, thanks to ongoing archeology, so we should be prepared to discover that the primary reason ancient Egyptians had a "penchant for feminine rule" was not merely because they thought having a woman rule would "shore up a weakened regime," but because women make good rulers. A fact that has been reinforced by the COVID-19 pandemic, wherein the countries led by women, such as Iceland and New Zealand, have fared better than countries led by men, largely because the women see their role as akin to being mothers, nurturing their people, while most men see their roles as the "ultimate" power to do whatever they decide (even when they clearly do not have such power).

Once more scholars of any gender accept that androcentrism often blinds them from seeing the actual facts before them because they assume the patriarchy has always existed, we might just be able to see more accurately what life was like for these ancient people as they lived through the world's first industrial revolution, commonly called the Bronze Age.

Chapter Eighteen
How We Got Here:
My Discoveries of the Rise of Androcentrism

I realized, when I began writing this book which I had originally titled, *Pitiless Bronze: How the Iliad Symbolizes the Rise of the Patriarchy*, that I had several obstacles to overcome in order to prove that humanity has not always been androcentric, let alone patriarchal, and that power structures shifted clearly from more cooperative egalitarian societies to hierarchically suppressive androcentric cultures during the Bronze Age around the eastern Mediterranean. I quickly realized, however, that the main obstacle appeared to be a form of willful ignorance on the part of so many scholars, who just assume that males have always been every cultures' leaders, ergo that humans have long been patriarchal, although a few, a very few, acknowledged that patriarchal ideas could not arise until humanity learned males play a biological role in pregnancy.

Therefore, I began researching the many cultures in and around the eastern Mediterranean that existed circa 3000 BCE, which is considered by many scholars as the general beginning of the Bronze Age.

The important obstacle I had to address early on was that continued bias toward androcentrism, exhibited by many scholars unaware of (I hoped) their particular biases toward viewing the world through a males-have-always-been-dominant gender perspective, which seems to have led them to believe that the patriarchy has existed forever. Most archeologists of this period, historians, and translators, especially, seemed to carry this bias with them, but many are having their assumptions rightly questioned. Archeologists, for instance, are having to walk back their assumptions that any human remains buried with weapons had to be those of male warriors because both history and DNA research had demonstrated that this assumption of weapons only belonging to men, as well as the assumption that only men could reach six feet in height, were incorrect.

Since I do not speak or read any form of Greek, I also began to question translator bias when the first four translations of *The Iliad* that I read to get a full understanding of the epic all called Briseis, one of the three main female characters in *The Iliad*, a "girl." Understanding that she had been captured as booty by the Mycenaeans who had murdered her husband and parents and knowing Briseis could have been as young as 12 when she was married so at least 22 at the time of the events in the poem, the poem never mentions her having any children. However, children are clearly only in the poem to represent motherhood. To Homer's credit, even Hector has one touching scene with his young son. Not even Helen, the "woman of discord" who began the tale's melodrama, has children by Paris,[144] so the women who we are supposed to see as only sexual objects—Briseis, Helen, and Chryseis—are not tied by motherhood to any other people. Still, calling Briseis a "girl" is clearly an androcentrically biased view, since she is old enough to be viewed as a sexual object by the men in the story. Continuing to call grown women "girls" is a clear indication of sexism because it diminutizes (keeps small) women as individuals and collectively. Granted, the Greeks sometimes practiced pedophilia, but other translations I later found do not call her a girl, so it is likely that the Greeks had not actually called her one, either.[145]

I became more conscious of this misogynistic bias the further back I went in time. The Inanna priestess who is sent to change Enkidu from a beast into a human being in the Sumerian epic poem, *Gilgamesh*, is often called a "hierodule"[146] in translations, or simply

---

[144] Robert Graves, reportedly, claims Helen and Paris had three children, but they are never mentioned in *The Iliad*.

[145] It is possible that the term "parthenos," which gives so many translators problems, was used in the epic poem, but it does not mean "girl" or "virgin" as it is often translated. Instead, it was the first term used to define girls who passed menarche, so were to be treated as young women. While the Greeks of Homer's time might have sought to continue to diminutize women in importance, that tendency still exists among modern scholars, too.

[146] Hierodule is a Greek word for a sacred prostitute. The scholars who use this term want to give credit to the Sumerian idea that these women are sacred, but they cannot seem to give up the idea that a woman who has sex with strangers is a prostitute.

a "prostitute" or "harlot." The most generous translators call her a "sacred prostitute," but only Judy Grahn acknowledges that the woman's name, Shamhat, actually means "taboo woman," and makes a great argument that "taboo woman" did not mean "harlot" but, instead, refers to a woman only certain men could touch (Grahn 5), a *harîmtu*. The trained and spiritually powerful *harîmtu* performed the sacred sex act—an act so powerful that it could transform a beast into a man; an act so powerful that Inanna (or her representative on earth) could change a mere mortal man into a king with the divine right to rule through the *qursu* sexual rite (Assante 32). My previous research into sexuality in matrilineal cultures helped me see through the androcentric biases presented by so many translators.

In a close reading of the differences in the subsequent versions (e.g. Akkadian and Babylonian) of *Gilgamesh*, I noted that the later scribes or translators also eliminated the fact the Gilgamesh upset the Urukian elders council, not just because he was having sex with all the new brides before their husbands could— which was apparently a kingly right that came later—but, in the Sumerian version, was having sex with whomever he wanted, male or female. A decidedly Western bias against homosexuality, possibly developing during the Bronze Age but fully embraced by the various translators from the Victorian era onward, became clear.

I admit that I cannot read cuneiform, no matter which of nine or ten possible languages it is written in, but I have to wonder, as a scholar relying on the veracity of the translators who are imparting these ancient ideas for us, how much of almost any translation I read now is accurate and how much is drastically altered simply by the translator's particular biases.

As a scholar, then, I have had to verify, verify, verify, as my doctoral professors sometimes chanted, that the evidence I was gleaning was accurate. Given the fact that the gender of the translator, historian, archeologist, etc. has no real effect on their biases, since women raised in androcentric cultures have to overcome the same mental programming that men raised in those cultures have to overcome, this task has been the most time-consuming part of writing this book.

Another obstacle was trying to prove that humans have not always known that men have a biological role in procreation. After all, how can we have a patriarchy without a clear idea that men are actual biological fathers? Given the fact that there are, on average, 40 weeks between the time of heterosexual coitus (thus possible conception) and birth, what had to happen for the ancients to link heterosexual intercourse with pregnancy? I had to think like a biologist to imagine the circumstances involved in such a discovery, and my farm background helped greatly.

The facts of the discovery spread around the Mediterranean, surely, within a few hundred years, but some cultures were more reluctant to accept the idea that sex has anything to do with pregnancy.

We know, for instance, that the Hittites switched from matrilineal descent to patrilineal descent when King Telepinu, circa 1550 BCE, changed those laws. However, Telepinu also, apparently, did not like his first born son, so his third part of the law preserved matrilineal descent, if the king chose to use it (Collins 41)

Egyptians, though, largely remained matrilineal, requiring any male who sought the main position of power in Egypt to be directly related to or married to a female who was of royal blood (thus blessed by the gods or actually related to the gods) for most of their dynastic existence, which is why so many of the male rulers chose to marry their own daughters—to further augment their social status and power by drawing on the natural Female Magic thought to be imbued in the royal women and their blood connections to the gods. Few people admit that it was Hatshepsut, a female ruler, who began using the Egyptian term for "great house," referring back to the matrilineal houses through which royal power descended, as the title for the ruler of Egypt, with way too many scholars claiming the term "pharaoh" should only be used to refer to men. Their bigotry relegates most female rulers to mere regents, proportionally distorting the reality that was Egypt, where women exercised more rights and ruled far more often than many of their neighboring cultures allowed during the same time period.

The Sumerians, however, proudly proclaimed they had a female ruler, Kubaba, a tavern keeper turned ruler, on their Kings List. It's just one of the many signs that the Sumerians loved their beer, but also loved sex, since taverns were notoriously sexual places and the Sumerians prior to the Akkadian take over were egalitarian, if not outright democratic in electing their leaders (Assante 2002).

It was, however, my close examination of the poems or hymns about Inanna by Sumerians that gave me some of the best evidence that women were once not just revered, but that their gender was so powerful that having sex with them provided magical powers—from healing ailments to making human beings from beasts. Inanna, by most respects as conceived of by the early Sumerians, was the most powerful divinity conceived of by humanity, so much so that later cultures reassigned her powers to male gods as androcentrism rose—from her control over the sacred *me*, which included the "kissing of the phallus," that allowed kings to rule their subjects wisely to her roles as both warrior and sacred sexpert to her ability to change a person's gender at will, a magical skill undoubtedly used by male rulers to negate the threat of male rebels or usurpers through castration.

That translators simply call Inanna a "love goddess" is to undervalue her important role in Sumerian society. While some of her hymns sound extremely "loving," they are all highly sexual and nowhere mention any intention on Inanna's part of becoming pregnant nor implies an intent to isolate her sexuality to any one male. Because Inanna is so highly sexual, she is often, by those who assume humans always knew heterosexual intercourse can lead to pregnancy, called a "fertility" goddess, but her only real ties to "fertility" are in making Dumuzi's seminal fluids flow, since semen was often associated with water, especially streams and floods, which make crops grow, like men's penises, so the symbolism should be obvious. I explore these ideas more fully in other chapters.

This lack of differentiation between sexuality and fertility is probably the most egregious error of all—perpetuated by teachers, professors, and scholars alike.

ΩΩΩ

Heterosexual intercourse had nothing to do with human fertility in the ancient mind.

Assante describes it best when she points out that sexual intercourse—whether hetero or homosexual intercourse—is a liminal activity that invites good magic, which is why the ancient Sumerians posted mass produced clay plaques of couples in bed over their doors (32). Sex, like a doorway, was a portal to connections with the divine, a liminal activity that allows mere mortals to touch the sacred space occupied by more powerful spiritual beings.

This divine connection between participants during sexual intercourse, especially *hashadu* or sacred sex, is the sole reason Inanna is able to pass along her divinity to the males (wannabe kings) with whom she has *qursu* or ritual sex via her high priestesses. In the Sumerian world, the only way a man, who has no divine connection to the cosmos, can become a legitimate ruler is to have Inanna's sexual blessings, so many wannabe male leaders vied for her affections, even going so far as to write poems to her. Often dubbed "love poetry," these works were, more often than not, braggadocio on the men's parts, claiming they were the superior lovers, not their foes.

This sexual-sacred connection is also evident in the ancient Egyptian idea that the rulers are "beloved of Ma'at," who personified the necessity to balance life's order against potential chaos. According to some myths, Ma'at stood with Ra as the *ben-ben* (mound of creation) rose out of the primordial waters created by the goddess Nun. Despite the fact that Ma'at had her own temple built for her by Hatshepsut (Mark), many scholars overlook this goddess' importance as mere ritual or ideal. Ma'at had to sanctify a ruler, much like Inanna did, or her/his/xis reign would end in chaos.

Traditionally, the rulers' wives or daughters (much like Sargon's daughter Enheduanna becomes Inanna's High Priestess) also served as "The God's Wife" of Amun, enacting a similar sacred sex act on the male rulers sometimes called "creative masturbation" (Mark, God's Wife) by scholars. Ejaculation with help from another is not masturbation, but it is another form of sexuality.

I have yet to find any scholar who directly examines Akhenaten's change in lead god (***not*** a change to monotheism so often touted by scholars trying to promote their Abrahamic religious views) from Amun to Aten was a result of wanting to remove The God's Wife who carried forth Amun's power, since this woman was the only real power challenger to most pharaohs, but especially to Akhenaten and his father, Amenhotep III.[147]

So, then, if sex in its many forms was still a sacred activity that was not tied to procreation, the ancients were not yet certain that heterosexual intercourse led to pregnancy or had anything to do with fertility.

But they could have suspected there was a connection somewhere.

Indigenous Australian men, for instance, still insist that they are the ones who summon spirit children to their wives' wombs, but that act does not happen through sexual intercourse. The spirit children have to choose the mother from whom they desire to be born, so a woman can become pregnant by merely passing near a sacred place where spirit children congregate. So, while the men still claim to be responsible most of the time for their wives getting pregnant, they have found a way to circumvent the idea that heterosexual intercourse is the cause.

Other people around the world have been just as creative in explaining how pregnancy occurs. The Aztecs firmly believed women bled monthly because they were fighting an ongoing battle with the Spirit Realm to bring human beings into this world. Thus, a woman who died in childbirth was honored as a warrior.

So how then did human beings finally determine that male contributions to procreation were more than spiritual energies?

They had to domesticate an animal that had both a short gestation period and were monogamous sexually in order to make a clear, irrefutable connection between heterosexual intercourse and procreation. Swans are the only animals humans have ever

---

[147] Both Amenhotep III and Akhenaten's original names included "Amun," signifying the importance of Amun to Egypt as a whole, but their political wars with the God's Wife is most likely what led them both toward sanctions against the Amun priests and God's Wife, who served as the High Priestess of not only that temple, but for all of Egypt.

domesticated that are more monogamous than humans, and they have a short time span between coitus and the presentation of offspring.[148]

Around 2388 BCE, Ptah-hotep, vizier to Pharaoh Djedkare, bragged in his mastaba inscription that he owned 1225 swans (Swans in Ancient Egypt 2018), but it is possible that these birds had been being kept on large estates in Egypt for some time. Arguably, humanity probably finally understood the heterosexual coitus connection with pregnancy circa 2400 BCE.

As we might suspect, the rising androcentrism, which led to more brutal forms of hierarchy, impacted women's rights in an effort to affect and to control women's sexuality. Polyandry, marrying more than one man, became illegal in 2375 BCE when King Urukagina of Lagash proclaimed that women having sex with more than one man could be stoned to death (Stone 39). According to Merlin Stone, the edict actually reads, "'The women of former days used to take two husbands but the women of today would be stoned with stones if they did this'" (39).

Marriage became a legal aspect of humanity's social lives sometime **after** Argon adopted the *qursu* ritual into a *heiros gamos* rite with his own daughter, Enheduanna, circa 2350 BCE. Potentially, the idea of sacred marriage was extended to the common people as a way to force the idea of marrying only one man on women, but it could also have been an idea thought up by a priestess to help common people feel they, too, were sanctioned by the gods, when they pledged to marry one partner for life, thus creating the idea of monogamy among humans. As androcentrism rose, however, marriage became a legal contract between a man and the woman's father, with the husband gaining the right to sell his wife and children to wealthier men in order to settle the husband's debts.

This act of relinquishing common men's rights to sexuality by selling their wives is noticeable in DNA traces. Almost exactly around 3000 BCE, "for every 17 women who were reproducing,

---

[148] Aztecs never domesticated swans or quetzal birds that we know of, but they did keep quetzal birds, which are also monogamous, in royal zoos, so the increase in androcentrism in Central America probably began after observing the quetzal birds, for which the raising of the offspring is the male's primary responsibility.

passing on genes that are still around today—only one man did the same" (Diep 2017), according to Melissa Wilson Sayres, a computational biologist with Arizona State University.

Clearly, the rising hierarchical patriarchy not only limited common men's sexual access because they were forced to sell their wives and many were castrated when their cities were conquered, but also curtailed most women's sexual activity by making everything associated with female sexuality negative. How did the ruling "elites" convince common men to relinquish their sexual "property" and rights?

First, Sargon established the first standing army—a point in history often lauded by unsuspecting military-gung-ho male scholars, especially historians who tend to love studying battles more than they do studying everyday life. But Sargon was wise in many ways. He knew that fewer men were needed in most communities as technology in farming improved. Since plows replaced hoeing fields to ready them for planting, and since oxen began pulling those plows instead of men, there were far more younger men who had no real role to play in the communities near the growing cities of Hattuša, Sumeria, Babylonia, and Egypt. So Sargon offered these men an identity, building up the idea that it was noble to die for one's country—a concept the ancient Greeks will turn into the Cult of the Hero two millennia later. It is also possible he encouraged greater communal labor forces, also known as corvée labor, to build irrigation canals and ziggurats, as well as water-sewage systems through some of the cities.

Economic systems were also set in place that kept many farmers and crafters so deep in debt that, in order to pay off their debts, they had to sell their wives and children to richer men, sometimes for years, in order to "pay off" the debt. This form of economic suppression still takes place all over the world, so that, sometimes, desperate families sell their daughters into prostitution.

At what point in his conquest of Sumeria did Sargon discover that he could keep more men alive if he castrated them in the name of Inanna?

According to Lazlo Bartosiewicz Dixon Ford, and Lee Kreutzer, cattle, which had long been domesticated and herded before they were used to pull plows,[149] sledges and war wagons, are perfect beasts of burden when their horns are still intact because the horns act as another element of the vehicle being pulled, from acting as the brakes to enabling the oxen to reverse and push the vehicle backwards (Ford & Kreutzer 10). The best oxen, which can be male or female, are castrated males, aka steers, because, after castration, "they grow larger and stronger" and have more stamina (Ford & Kreutzer 6), so can pull more, but become much more docile, thus are more trainable, than bulls (Bartosiewicz 331).

How long after learning about how few intact males are needed to impregnate a whole herd or flock, and how long after learning that castration "improved" male livestock performance, thus creating steers which became the perfect oxen, did the social elites begin castrating human males to turn them into better servants?

Human males often view "the patriarchy" as a great thing that somehow helped them "rise to power" over all things feminine, but the evidence points to the fact that men who are not rich and powerful were and are as oppressed as women are under such brutal hierarchical social constructs.

Noting this significant fact, who would ever want to live in a patriarchy?

I discovered many patterns that demonstrate telltale signs that there were, indeed, gynocentric cultures that preceded androcentric ones; symbols often overlooked by androcentric-biased scholars who might not have ever considered that there was a definable moment when the concept of "the patriarchy" arose for ancient humanity. The gynocentric cultures were often more egalitarian, even democratic, in structure than most historians who discuss the ancient Bronze Age in Eurasia are willing to admit. The cultures I examine in Section Two practiced matrilineal and matrilocal traditions, remnants of more gynocentric views of the

---

[149] Bartosiewicz argues that "mixed farming was based on cattle husbandry (50% of animal bones found) and cereal cultivation forming a single system, animals providing [both] traction and dung" (331). Ford and Kreutzer emphasize that cattle, unlike horses and mules, can eat "rough forage"—from sagebrush to evergreen needles, things that could kill a horse—an ability that is vital when crossing areas that are not lush with grasses (18).

world, early in their histories, but most went on to become patriarchy-heavy cultures, if not actual outright patriarchies.

What has become clearest is that the human obsession with pyramid-like structures comes from our concepts of birth through women's *mons venus*, the literal mound of creation, so that some cultures built such structures in order to commune with the gods, but also as rebirthing machines for people who were often considered the most sacred, the most important people of their cultures—their spiritual and political leaders, many of whom actually associated themselves with the gods through various rituals that were undoubtedly ancient when they practiced them, rituals that were changed to suit the political needs of the rulers of the time.

Section Two

Postpatriarchal Examinations
of Prepatriarchal Cultures

Because human beings are merely another form of mammal, most of which have gynocentric cultures that rely primarily on female leadership, we know that the patriarchy had to have a beginning, and most scholars would agree that the concept is deeply embedded by the time we get to Judaism's rise in the eastern Mediterranean, so that, while the early Jewish people probably did not invent the idea of a patriarchal-based culture, they clearly influenced how future cultures could take the concept and create a truly oppressive type of hierarchical regime that actively oppresses women, crossgenderers, and poor men. While I could examine the Hebrew culture's changes from polytheistic to more monotheistic, other scholars have already done so. My main reason for not examining Judaism, however, is that they arose fairly late into recorded history. In fact, we are almost as far from their date of origin as they were from the origins of the Sumerian and Egyptian historical cultures. These earlier cultures most clearly demonstrate changes from gynocentric to androcentric social views well before human beings became partially or even fully patriarchal.

This section examines why so many cultures began leaning toward androcentrism, analyzing cultures like the Aztecs, the Sumerians, the Egyptians, the Hittites, the Minoans, the Mycenaeans, finally leading to my original focus for the book, critical interpretations of Homer's *Iliad*, which, to a literary archeologist, leaves numerous clues about the Greek's prepatirarchal views even at the time Homer wrote the book circa 750 BCE.

I lead with my examination of the Aztecs because they left clues in writing, but also because they demonstrate nicely how some patriarchal cultures meet other cultures assuming that all male-led cultures must be patriarchal, since most people raised in patriarchal cultures cannot envision any other way of viewing the world.

Chapter Nineteen
Not the Beginning but a Start:
A Woman of Discord

One Aztec story encapsulates what happened in many places around the world that led increasing numbers of people, first, to enlarge the importance of males and their desire to wage war (e.g. to kill at will; to demonstrate the idea that "might-makes-right"; in other words, to embrace toxic masculinity as a way of life) in order to elevate men above the spiritual importance of women and their seemingly supernatural abilities to create life, and, second, to begin focusing on the sun as their primary deity, moving purposely away from the time and season telling feminine influences of the moon.[150] The tale demonstrates how the Aztecs, also known as the Nahuatl or the Mexica, pushed that boundary to create a war-oriented culture that transformed the idea that death was a "rebirth" in order to use ritual murder as a method of intimidation, on the once-life-giving mound of creation (aka pudendal pyramid).

The tale is often called, "The Woman of Discord," and goes as follows:

> Sometime around 1300 ACE, the Aztecs paused in their wanderings of the world of the Fifth Sun in a barren land "to the south of the great, salt-water Lake Texcoco in the Valley of Mexico" (Dodds Pennock 15).

> Their nearest neighbors, the Culhuacans, remained wary of these "'evil people, of bad habits,'" who had settled near them. As legend tells it, "Huitzilopochtli, god of war" and discord, hated the passivity of the farming life, so he told the Aztecs that Tizaapan, the place where they were living, was not where he

---

[150] As with the position of the sun on the horizon, equinoxes can be predicted by moon placement and phases: "During the autumnal equinox, the moon phase is in the Third Quarter Moon. During the Spring equinox, the moon phase is the First Quarter Moon" (High Touch).

intended them to flourish. He commanded them, undoubtedly through the auspices of a male priest, to find a "woman of discord" to be his wife, so he could properly teach the people the value of aggression (Dodds Pennock 15).

The Aztec priests dispatched emissaries to "Achitometl, the *tlatoani* [or Speaker, usually translated as king] of Culhuacan, requesting that he bestow on them his daughter to become the bride of Huitzilopochtli and a living goddess" (Dodds Pennock 16).

While Achitometl should have been more wary of his neighbors' intentions, he granted his daughter permission to accept this role, which he believed was an honor that recognized his influence and leadership. However, when he and his court traveled to Tizaapan to celebrate his daughter's marriage, "the king entered the darkened temple and began to perform many ceremonies, sacrificing birds and offering incense and flowers. Casting a handful of incense into one of the braziers, Achitometl saw the temple illuminated by flames, revealing a priest dressed in the skin of his daughter" (Dodds Pennock 16).

War, of course, between the two peoples ensued. Exactly as the person(s) in power among the Aztecs wanted to happen.

This story could be a literary account of an actual historical event, but metaphorically it tells a tale similar to ones found around the world: men assuming women's roles, literally, here, slipping into a woman's skin, in order to reorient their cultures' mystical views from centering on women (gynocentric) to centering on men (androcentric), so it is vital that you recognize, here, that it is not the woman creating the discord, but men are the ones imposing their will on the woman in order to create discord.

Through this appropriation process, many cultures changed their religious focus from the moon to the sun, and from the miracle of giving life to the power of making death. In the process, societies that had once been more egalitarian became, some gradually, some rapidly, more oppressive and more brutally hierarchical, giving legitimacy to the male tendency to fight, especially to kill, others for power.

In ancient Sumeria, for instance, the goddess Inanna demonstrates her two most important roles of 1) kingmaking through the sacred sex ritual of *qursu* and of 2) guarding the magical *me* powers the gods gave humanity. Both of these sacred powers are eventually stripped from her and given to male gods. Similarly, and a bit later in history, the Akkadian goddess Tiamat is changed from a great Mother Goddess deity to an evil serpent creating monsters[151] to fight her sons.[152] Inanna was the highly sexual daughter of the dark phase of the moon, Sin, and women in Akkadian were called *sinnestra* or *sinneštu* because of their sacred monthly blood-ties to the moon. For the Akkadians, who came after the Sumerians timewise, the sun god, Shamash, the moon god, Sin, and Inanna, who was believed to be both the morning and evening

---

[151] In order to reduce female gods from The Prominent Creators, androcentric cultures reduced female gods to creating monsters when they create on their own, something many scholars refuse to acknowledge as a clear method to reduce the potency of Female Creative Magic. For instance, Leick, who is examining a poem, "Enki and Ninmah," believes Jacobsen's androcentric interpretation of the poem, which has Ninmah aborting a "foetus [sic] before its time," so that a monstrous creature called Dumul is born, "unable to eat, drink, sit, stand, or even lie down," which gives Enki a way to blame Ninmah for the monster, "reminding her that what the creature lacked was her, i.e. the woman's contribution in successful gestation" (27). Even though Leick traces the origins of the Sumerian gods back to an all-creative female god, Nammu (22), she seems willing to adopt Jacobsen's androcentric reading of the text without asking more important questions. Because incestuous relations occurred so frequently in ancient Sumeria, since the people made no connection between heterosexual intercourse and procreation, they struggled to understand how birth defects occurred, so women and female gods became easy scapegoats for fetal deformities.
[152] Like Coyolxāuhqui, an important Aztec moon goddess, Tiamat was dismembered with her various body parts creating elements of the earth and sky.

stars (aka the planet Venus), had been the holy triad of spiritual powers before the rise of Enlil (god of wind) and Enki (god of water), two often conflated male gods who variously end(s) up raping his(their) daughters and granddaughters in Akkadian mythology. Eventually, Ishtar, who is more violent, less loving than the Sumerian Inanna (Assante 29-30), will fall into disrepute simply as the "Whore of Babylon" in Judeo-Christian texts[153].

In ancient Egypt, the mother-goddess Hathor is changed into the goddess Sekhmet[154] who wages war on human beings at the behest of the male sun god, Ra, who then has to subdue her with beer made by premenarchic young women that is dyed red to look like blood. The fact that a goddess acts while the male god sits back is endemic of typical lion behavior, where the females hunt, and the males wait to feast. Such unmacho behavior might be one reason why a later Egyptian pharaoh sought to elevate one divine aspect of the sun, the Aten, as the king of the gods. Additionally, Isis, although a later addition to the pantheon, is made to steal magical powers[155] from her father, Geb,[156] in order to breathe new life into her dismembered husband/brother Osiris. By the time the Isis legend is created, the Egyptians have possibly determined that heterosexual intercourse creates pregnancy, so the male organ that was worshipped by women and men alike as a sexual pleasure device becomes an important member for Isis to re-attach, so she can have intercourse with Osiris to create their son, Horus, who is,

---

[153] Note that calling someone a "whore" or "harlot" for failing to meet up to Judeo-Christian standards is widespread in Judeo-Christian texts, as though selling or giving away one's sexual services is the worst possible thing a person could do. This bias against women's sexuality underlies far too many interpretations of cuneiform texts.

[154] According to Pheobe Weston, "in an annual festival of intoxication held at the beginning of the year [when the Nile inundation occurs], the ancient Egyptians danced and played music to soothe the wildness of [Sekhmet] and drank great quantities of wine ritually." Sekhmet was known, as a goddess separate from Hathor, as the Lady of War, which makes her very similar to other goddesses around the Mediterranean.

[155] Isis steals sacred knowledge from her "father" the same way Inanna is depicted as doing once androcentrism rises in Sumeria. Coincidence?

[156] Similarly, to facilitate the change from goddesses with power to gods with power, Inanna was also made to steal the sacred *me* from her "father" (a title of respect, not biology), Enki, instead of having them in her own right.

ironically, a god worshipped further back into antiquity than either of his parent gods probably were. The act of Isis finding, replacing, and then using Osiris' penis seems to point directly at the newly acquired knowledge that men play a biological role in procreation, thus this myth was probably used as a re-educational tool for the masses. However, since turgid penises were long associated with the rapid growth of plants, and since Osiris is almost always depicted as green because of his association with plant growth, an older, previous symbolism of the phallus, having to do with the annual flooding of the Nile River, is more likely originally at play.

In ancient Hattuša, home of the Hittites, the great goddess Arinna changes from a mother-goddess known as Mother of the Land, Hepat, to a sun goddess after Hurrians influence the Hittites. Among the ancient cultures of the eastern Mediterranean, the Hittites were one of the last early historical empires to give up remnants of their previous gynocentric world views, including keeping matrilineal descent as a third choice when the laws of inheritance are rewritten. Rather than giving up their great goddess, they change her into a sun god.

For the ancient Mycenaeans, Persephone in various forms preceded Zeus and Apollo, both of whom were female in earlier forms. Zeus, as time progresses, fully assumes the role of mother-creator by giving birth in one way or another to most of his own children, just as his father, Cronus, did, in Greek efforts to rewrite their mythology into an androcentric form. Hestia,[157] arguably one of the oldest known Greek goddesses, is never mentioned in Homer's epic story of *The Iliad*, even though she would have been recognized ritually at every hearth, including when the Mycenaeans offer a hecatomb of cattle to appease Apollo at the beginning of the tale.

---

[157] According to Thomas Palaima in "Mycenaean Religion," "The central hearth reminds us of the importance of fire to Greek communities and of the goddess of the hearth, Hestia. She is unattested in the Linear B tablets for the same reason that she was not anthropomorphized in the historical period. She was a stationary concrete object [presumably the hearth itself], the central focus of the Mycenaean palatial centers and, by extension, of the palatial territories" (353), which still does not explain why Hestia is not honored at the hearths of the warriors, especially when they sacrifice the hecatomb of cattle.

And, among the Aztecs themselves, their earth-goddess/mother-goddess, Coatlicue, gives birth to their war god, Huitzilopochtli, after being magically impregnated by hummingbird feathers, only to face decapitation by her daughter, the moon goddess, Coyolxāuhqui, who is killed and dismembered by Huitzilopochtli as he springs fully armored[158] from his mother's womb.

It is this second Aztec story, which more fully demonstrates the drastic changes that occurred among the Aztecs as they recreated their religious world views, justifies not only placing men above women in importance socially and spiritually, but also revering war as a tool for prosperity and growth as they grew to dominate the Mexica valley. The tale also demonstrates why human sacrifice at the top of a pyramid re-enacts an ancient menstrual rite gone wrong because the Aztecs continued to re-enact Coyolxāuhqui death and dismemberment through human sacrifice atop their main Tenochtitlan temple, with blood and bodies spilling down the pudendum-pyramid to fall on top of a huge, round, moon-like carving depicting her dismembered body at the base of the pyramid. Undoubtedly, this moon-blood was originally meant to bring the moon back into fullness after hiding from humanity during her monthly cycle—a cycle most human beings first used to tell the passage of time and to predict the seasons, or, more drastically, the ritual sacrifice could have begun as a way to drain blood from the blood red moon that appears during a lunar eclipse.

These cultural beliefs and stories I will explore more fully in the following chapters.

But, first, I must address the obvious dichotomies presented by the basic opposition of sun and moon, light and dark, male and female. Such oversimplification of ideas is misleading because the sun is not wholly good, nor the moon bad; light is a symbol for enlightenment and illumination, but it can also blind; the dark of night, like most unknowns, is often feared out of proportion; and, as we will see in many of the upcoming chapters, oversimplifying one type of person as male and another as female dismisses their

---

[158] Athena, Greek goddess of war, also sprang fully armored and fully grown from her father Zeus' forehead. Being fully armored at birth symbolizes these might-as-right gods' readiness to fight and to kill.

many similarities and differences, leading many people in many cultures to assume that there are no common grounds between the genders, when, in fact, gender is and has always been much more fluid than fixed (SciShow).

Why are human beings seemingly attracted to the sun, light, and males? Why do so many cultures believe these three things represent "good"? Why do so many humans no longer value the moon, darkness, and females? Why do so many cultures in the world still view these things as akin to evil?

These questions will be addressed as the book progresses, but here's an image to take away before we get too immersed in details.

Start with the Olympic torch lighting ceremony, which is a solemn ritual held prior to the start of each Olympics competition in Olympia, Greece. To honor the competition's ancient Greek roots, the Olympic torch is lit by the high priestess of the Temple of Hera as Vestal Virgins[159] look on. Using only a parabolic mirror and a special wick on the torch, the priestess lights the flame. Afterwards, the flame is passed from person to person around the world, even traveling into space in 2014, until it reaches the stadium wherein the Olympic competitions will officially begin, where it then lights a huge cauldron that burns until the games are complete. As in ancient times, precautions are taken to ensure that this flame, once lit, never goes out, with contingent flames held as back up in case something were to happen to put out the traveling Olympic torch.

People often turn out in droves to see the Olympic torch carried through their communities because its presence is meant to honor that city, that community, just as the Olympic competitions are meant to unite the world.

Many ancient cultures around the world—from Europe to Africa and even the Americas--had similar fire traditions. In some of those traditions, women played a vital role as keepers of the hearths, which were the centers of every home, with the women themselves carrying the newly lit community fire into their homes

---

159 We know poets love alliteration, but could the poet Virgil be having us on with his nickname, which is clearly based on the Latinate term for "young maiden"?

to set their own family hearths afire. But in some traditions, women were shut away because the ceremonies were often held at night, which is when women's connections to the moon were believed to be strongest. The Aztecs, for instance, believed pregnant women, especially, would be prone to becoming monstrous animals if they were not hidden away during their fire ceremony, which only happened every 52-years when the two calendars—the solar and the religious calendars—met again at their starting points.

Fire often symbolizes the warmth of the home, the hearth or heart of a home in ancient times. That warm, protecting, nurturing environment is what most female symbols we find around the world continue to show us as an important value our ancestors found in women. Through Female Magics—Creative Magic, Protective Magic, and Power Magics—women once led our ancestral communities in full cooperation with the men who also resided there, more often than not. In fact, many ancient cultures were much more democratic than most autocrats or oligarchs want us to know. These people who seek to rule by might and not by nurturing others, continue to project these dichotomies on our present cultures in their attempts to control our modern narratives.

Until we recognize that dichotomies are often projected on the two most common genders, usually linking women with the left hand side (implying being out of sync; gauche), placing men on the right (a word that means "correct" as well as a noun that means exercising one's moral and legal entitlement), humanity will not make earnest progress about becoming truly egalitarian and democratic.

Chapter Twenty
Aztec Sacrifice Equals Menstrual Ritual
Gone Bad

What happens when a patriarchal culture meets and attempts to destroy a nonpatriarchal culture? Furthermore, what happens when a possibly guilt-ridden priest from the destroying patriarchal culture attempts to "save" aspects of the culture his people are destroying? How accurate will the materials he collects be?

Scholars studying the Aztecs, also known as the Mexica or Nahua, have long been confronted with this issue because groups of priests were purposely sent to the New World to record information about the populations in the Mexico Basin after their conquest by Spain. For example, Bernardino de Sahagun, a Franciscan friar sent to New Spain in 1529,[160] collaborated with groups of male indigenous students he himself had trained to create what is now known as the Florentine Codex.[161] His students asked groups of older male Mexica predetermined questions about their culture and religion. Because painted manuscripts with pictures were the common form of writing for the Nahua, the men wrote their answers in this form.

---

[160] For time reference, Hernan Cortes "completed" his conquest of Mexico in 1521, so Sahagun arrived a mere eight years later.

[161] Estimates indicate that there are around 500 colonial-era codices preserved from the Aztec and Mayan cultures. The Florentine Codex is available online through the Library of Congress website. However, none of these codices, neither the Mayan ones nor the Aztec ones, are original documents, which were all destroyed by the Spanish. All of the codices carry forward biases toward both pleasing the Spaniards and of "elevating" the status of their specific cultures with the Spanish. Realize that the Spanish treated the indigenous nations harshly, condemning their religion as satanic and their sacrifices as evil, never realizing the irony of believing in the redemptive value of a horrible, sacrificial death to appease their own god. While every culture understandably wants to be in control of its own cultural narration, we have to recognize that these patriarchal biases are important impediments to understanding the original cultures of the Americas.

Later, a group of Christianized indigenous nobles, also all males, who knew Latin, Spanish, and Nahuatl, interpreted the paintings made by the elderly men, expanding the answers as they saw fit, and transcribing the ideas into an alphabetical form of Nahuatl.

Sahagun and his students went further, creating new illustrations and translating much of the text into Spanish. The twelve volume compilation of the Florentine Codex was then sent to Spain in 1579, fifty years after Sahagun had arrived in New Spain, and a mere 58 years after Cortez conquered the Aztec capital, Tenochtitlan.

In the Florentine Codex, astute observers can see the differences between traditional Nahuatl forms of painting and the European-influenced form that included shading and perspective to indicate three-dimensional forms, something the Mexica had not done before. Astute observers will also note that only half of the Mexica population was truly represented in the stories, despite the fact that the men, ensconced in patriarchal Spanish culture by this time, mention aspects of women's lives in some of the stories.

While willing to admit that the Florentine Codex cannot be completely accurate or unbiased, some scholars defend Sahagun because of his methodology, calling him "'the Father of Anthropology,'" blind to the primary, but then unconsidered, issue of viewing the material from the point of view that "the patriarchy" has always existed even in the New World.

Caroline Dodds Pennock, for instance, finds no reason to believe a Catholic missionary from a patriarchal nation like Spain would "deliberately" distort data he has collected (8). She does admit, however, that "considerable portions of the Nahuatl text either remain completely untranslated or are merely briefly summarized" (Dodds Pennock 8). She, of course, also seems completely unaware, herself, of a bias that both she and Sahagun have in common—viewing the Nahua through androcentrically biased lenses that assume "the patriarchy" existed everywhere in the world. To her credit, Dodds Pennock does repeat a question asked by scholar Betty Ann Brown: "'were Aztec men privy to the knowledge of women's ritual activities?'" (9) However, this logical question, which opens up the very real problem of just how much the men could have known about the Female Magic practiced among Aztec women, is just the start.

Unfortunately, because the codices were stilted presentations that assumed the Mexica were patriarchal, like the Spanish themselves, these post-conquest pieces of literature are not accurate representations of what that traditional pre-contact Mexica people believed or how they truly acted, since the narratives are both self-censored and amanuensis-biased.

But, because the invading Spanish destroyed all of the original Aztec and Maya books, the Spanish-produced codices are the only written histories scholars have to work with, other than the writing left on various monuments, and, unfortunately, most of the Mayan glyphs have yet to be deciphered, and those that have are often viewed from a clearly androcentric point of view.

For instance, Simon Martin, in his interesting, if problematic, examination of "The Old Man of the Maya Universe: A Unitary Dimension to Ancient Maya Religion," attempts to argue the Old Man divinity was actually present in many forms, "posing him as a better candidate [than any previously proposed] for the unitary entity referred to" in the Mayan codices, which present as many problems as the Aztecs codices as far as Christian-influenced biases introduced into the subject matter. What Martin really seems to want is to find Yahweh hidden among the pantheon of Mayan gods. Overlooking or somehow trying to avoid accounting fully for a possible "womb entrance" (aka vagina) on one depiction (194) of the Old Man figure, ignoring the crocodilian-motherhood[162] ties that have a crocodile giving birth to the Old Man through its mouth, or an image of the "Old Man in appropriate 'serpent birth' imagery'" (205, note 29), hoping that claiming he was "dual-sexed" will be sufficient (218) to claim the Old Man has "'multi-presence'" (222), his main argument comes down to the idea that the only thing that "distinguishes monotheism from polytheism is not the number of supernatural agents in the system, but its more exclusive classification of 'deity'" (224). Martin's weak conclusion: "the Old Man...had unusual significance within the Maya religion," so that his presence becomes "the common thread tying [images of other

---

[162] Crocodiles and alligators both carry their newly hatched young about in their mouths until they are large enough to fend for themselves.

deities] to a greater purpose and meaning" (226). The implication seems to be that the Maya worshipped a god just like Yahweh, pervasive and ubiquitous.

While it would not be inaccurate to state the Mayan penchant for human sacrifice was not unlike that of the ancient Judeans, Martin's point is clearly biased toward not only an androcentric view of the Mayan world, but also a patriarchal one.

Many scholars admit that the Mexica were probably related to the indigenous American Indian nations that still live in the American southwest, but, since most of the southwestern cultures practiced matrilineal succession and matrilocal burials, so were clearly gynocentric, there should be clear reasons for these scholars to doubt that any of the cultures from Mexico, including the seemingly blood-thirsty Aztecs, were full-on patriarchal cultures.[163]

Yet debates still rage among Aztec and Mayan scholars about whether or not these nations practiced actual patrilineal descent. Arguably, the most thorough examination of the arguments for each perspective—from arguments for patrilineal descent to arguments for nonunilinear (blood relation not required) or even dual linear descent (from both mother and father)—comes from Susan D. Gillespie's article, "Rethinking Ancient Maya Social Organization Replacing 'Lineage' with 'House," published in 2000. While she makes some great arguments for viewing social organization in Mayan cultures as collective great "houses" wherein everyone living in or associated with an influential house was counted as a member, blood relative or not, overall, Gillespie is still only guessing, since DNA findings from burial practices to establish actual consanguinity (or its absence) have apparently not been conducted very widely, if at all, despite the fact that many cultures in the Mexico Basin buried at least one or more people below their house

---

[163] According to many scholars, the Aztec were more related to the indigenous nations of the American southwest than they were the Maya or the Olmecs, which might explain where the Anasazi went. According to Wally Brown, Navajo historian and elder, the Anasazi were slave traders, which was why the Navajo disliked them, since they often kidnapped Navajo to trade them as slaves. Interestingly, the people who would become known as the Aztecs migrated into Mexico about the same time the Anasazi abandoned their towns in the American southwest.

structures in order, it is believed, to establish inheritance rights (Prindiville & Grove 80).

Were the Aztecs a full-on patriarchy? If so, when and how did they realize males have a biological role in procreation? If not, why are they so often regarded as such?

Most scholars who study the Aztecs, however, seem to assume the culture was strictly patriarchal, even though few go so far as to label it as such, perhaps because women, we are learning, actually held quite a bit of autonomy in the culture. Women not only made most of the household goods traded at the Mexica marketplace, but also organized the marketplace and traded the goods themselves. The Spanish clearly expected to see, so saw, "married" couples living in homes together, and texts like the Florentine Codex seem to demonstrate that the Aztecs practiced monogamous marital rites of some sort. But did the men assisting the completion of such documents write down what they assumed the Spanish wanted to hear? After all, they were subjected to Spanish criticism and outright condemnation for most of their religious ideas and social practices since the time the Spanish set foot on the continent, not to mention witnessed the Spanish destruction of their own historical books because of a rabid Christian bias against other religions. Perhaps they were afraid that the royalty for whom these codices were intended would hate their people and their practices so much that they would finally just wipe all Aztecs from the face of the earth, so they used keen observation skills to note what the Spanish valued most and chose to imitate it in these codified stories.

While we can determine that the Egyptians and Sumerians learned that males have a role in procreation from their having domesticated swans, the only monogamous animal ever domesticated by human beings, we have no evidence that the Mexica, or indeed any ancient culture in the Mexican Basin ever domesticated swans, although we can be certain they encountered swans, mainly trumpeter swans, in the wild.

Most humans derived the idea of seasonal marriages from birds, all of which seasonally get together to have sex, but do not always raise their offspring together once they have hatched. Monogamy, then, really only occurs naturally among some species, like swans, but the bird nearly every culture in the Mexican Basin admired the most was the colorful quetzal bird, which is also monogamous—meaning they choose mates for life, although, like albatrosses, they only live together while incubating and raising their young (Animalia). Native to the "cloud" mountain ranges, where only the Mayan city of Chinkultik was located (Miller & Taube 141), the quetzals were prized for their feathers, which held many iridescent colors, so were often requested as tributes for the leaders of most city-states, including the Mayan cities, in Mesoamerica. The long green-blue tail feathers of the males were especially favored in some of the more elaborate royal headdresses. Most importantly, though, someone at some point acquired live quetzals, which were present in Moctezhuma II's animal menagerie that he kept at his palace at the time of the Spanish conquest (Miller & Taube 141), so it is quite conceivable that many rulers kept such birds, which were both spiritually and socially important to many of the cultures of Central America.

While I believe the ancient Egyptians and Sumerians had to actually domesticate swans before they closely observed their mating habits, thus discovered male participation in procreation, the Mayans and the Aztecs would not have had to domesticate the quetzals in order to observe their mating habits, mostly because the quetzals are so showy about their mating rituals and their incubating duties, so that any observant hunter seeking to gather the male tail feathers would have known exactly when the males shed these feathers in order to gather them. In fact, "hunters were forbidden to kill the birds; rather, they stunned them with a blowgun, removed the feathers, and set them free" (Miller & Tabue 140); most likely, this capture-and-release program happened at the end of the chick-rearing season, which was overseen by the male quetzal birds, which would have been a good demonstration of fatherhood for wannabe monogamous people.

Hunters seeking the feathers would have observed that the quetzal males do not grow these extravagant tail feathers until spring, which is when they use their flourish to attract mates. Not able to fly far, the males get as high as they can in the trees, then swoop toward the earth to impress the females whose attention they have attracted. Seeing these glowing green-blue feathers swooping to earth like raindrops or elegant ribbons, would have seemed like a magical summoning of rain, since spring rains often seemed to accompany the growth of these feathers—both announcing spring's arrival.

The feather hunters would have also observed how the males tend to the nests during the day because their nearly meter-long tail feathers stick prominently out of the holes in the trees wherein quetzal make their nests. Unlike most other birds that build nests, the quetzal couple hollows out trees for nests without adding nesting material. Some descriptions of the feathers protruding from the tree trunks are "like a bunch of ferns growing out of the hole" (Animalia), but, since the feathers can glow with a blue iridescent coloring in certain light, that color would have made them distinguishable from ferns.

However, does the fact that a society imitates one bird's monogamous mating practices, thus inspiring the institution of marriage among humans, mean that that society was patriarchal? We have to ask this question because many scholars argue that the institution of monogamous marriage was created as a social means to limit women's sexuality, especially since so many cultures allowed men to continue to have coitus outside marital bounds or with additional "wives." The natural drive toward pair-bonding helps ensure the survival of offspring, but how long those pair-bonds last in another issue altogether.

If the peoples of the Mexica Basin were imitating the quetzal bird in order to imitate pair-bonding, they were not doing so literally because, after the young are old enough, the female leaves the final part of the rearing to the male (Animalia). In fact, both males and females live solitary lives until spring, when they once again meet up to procreate and to raise more young.

If the Mayans and Mexica derived their ideas about monogamy from quetzals, that suggests the possibility that some of the buildings in many of their communities were meant to house "bachelor" males, or males who only visited their wives' homes seasonally or for sexual bonding, which are both recorded matrilineal social patterns. Indeed, some archeological finds suggest that there were mid-sized buildings—larger than the average home, but smaller than elite homes or palaces, where many men might have lived together. For instance, one building, Structure 5, excavated at Yautepec was "far larger than the other [nonelite] houses," but smaller than most elite houses considered, so that excavators Michael E. Smith, Syntha Heath-Smith, and Lisa Montiel were uncertain how to categorize it (149). What few of the archeologists excavating Mesoamerican ruins seem to look for are buildings that would house men separately from their families, since they assume that each small building structure was sufficient to house "a nuclear family or a joint family," although they do not seem to give much consideration to extended families (Smith, Heath-Smith, and Montiel 147).

Why assume, even if a culture has marital rites, that men and women live together in one house, and that each house only houses a "nuclear" family, which is a fairly recent concept?

Inga Glendinnen postulates that the homes, no matter how small, had "'women's quarters'" and that men could at least meet at "the local warrior house" (169).

However, there is very little evidence to suggest that the Mayans or the Aztecs valued monogamous marriages (the men supposedly took many wives, something often limited to the elite of most hierarchical societies, but little has been said about whether the women could take many husbands) or that men lived in the women's houses or shared actual living quarters with them, prior to the Spanish invasion.

Struggles over how to deal fairly with concepts of sexuality and gender, both of which have impacts on determining whether a culture is matrilineal, patrilineal, or dual linear, pepper the ongoing research about ancient Mesoamerica.

Some scholars call the gender-organization of Aztec culture "gender parallel," meaning that, while there were socially enforced gendered expectations for each sex, both were seen as equal in importance (Horne 12).

However, there is some evidence to suggest that when Tlacaelel I (1397-1487 ACE) acted as *tiachoccalcatl* (second in command, the primary military leader) of Tenochtitlan, he not only elevated Huitzilopochtli, the war god, to the top of the Aztec pantheon of gods, but also increased the number of human sacrifices (Townsend 33). Perhaps more tellingly, Tlacaelel I, who was also named *cihuacoatl* (Nahuatl meaning "female twin"; a role as supreme leader under the Aztec emperor that was created specifically for Tlacaelel I), assumed the sacred role of the goddess Cihuacoatl, the snake goddess, drawing into himself her sacred feminine and earth-bound powers, and calling himself the mother and father of the Aztecs (Horne 12-13).

The role of cihuacoatl, if it existed before Tlacaelel I, was clearly that of "inner chief" to the tlatoani's role as "outer chief." Among many American Indian nations north of Mexico, there were two leader roles for most sedentary communities—one leader was concerned with the internal affairs of the community, the "inner chief," and the other leader was concerned with external affairs, relations between that community and their neighbors, the "outer chief." As Jayme Horne describes it, "the tlatoani oversaw the affairs of the Aztec state," while the cihuacoatl or snake woman "oversaw the internal affairs of the city" (12). When Tlacaelel I becomes cihuacoatl, he is using "cross-dressing...as a tool to appropriate the powers of another gender" (Horne 12). He also pushes the envelope of the boundaries between genders[164] when he names the conflicts he chooses for the Aztecs to pursue the "flowery wars" (Horne 12).

Some scholars even go so far as to claim that Tlacaelel I, who was never an Aztec emperor, was still "'the founder of the Mexica empire'" (Schroeder 5) because of the radical changes he made over the terms of the five emperors he did serve.[165]

These steps on Tlacaelel I's part are similar to steps Sargon the Great took to ensure his rulership over all of Sumeria by

---

[164] The Aztecs apparently enjoyed transgendering quite a bit. Remember the "woman of discord" story in the previous chapter? Aztec priests often flayed female sacrifices, then wore the woman's skin in specific ceremonies, such as during the ceremony of Xochihuitl, which "fell on the day 1Flower each ritual year," wherein gilds of sculptors, weavers, and metal smiths, who were patronized by the goddess Xochihuitl, "offered a female captive, the goddess' living image, to her temple for sacrifice. Once her heart had been taken from her chest, she was flayed and her skin placed upon a priest who, pretending to be [the goddess], went to the steps of the temple and began to weave. Craftspeople, dressed as monkeys, dogs, and felines, danced before [the priest in goddess disguise] and shook their tools of trade" (Flood).

[165] As Susan Schroader points out, some of the mythology about Tlacaelel might be considerably inflated because "some of his descendants [might have been] trying to capitalize on his fame for their personal benefit" (7).

enacting the sacred *qursu* ritual with his daughter Enheduanna who Sargon had installed as the High Priestess of Inanna's temple, so perhaps Tlacaelel I had similar ambitions, and, like Sargon, he had to reduce women's influence by taking on women's prior spiritual roles in wielding Female Power Magic in order to achieve these political and economic goals.

Like many rulers who choose to oppress their people, Tlacaelel I also burned the historical and religious texts that had been written during the pre-empire period in order to rewrite Mexica history (Townsend 33). While many scholars, like Townsend and Horne assume he burned the books to remove social expectations that "his half-brother's sons were the ones destined to rule" (Townsend 33), Tlacaelel's real motivation could have been to erase the more gynocentric history of his people. However, remnants of a matrilineal system still existed even after the Spanish conquered Mexico (Kellogg 326), which we will explore further.

As Susan Kellogg points out in her examination of wills drawn up by Aztecs from the mid to late 1500s, Mexica women still preferred to pass on their materials goods to their daughters and granddaughters, not to their sons (318). Because her examination was limited to a small number of wills available, Kellogg, like Townsend, propagates the idea that a form of gender parallelism (dual lineality)[166] was at work in the form of a cognatic kinship system, which quite often means that males inherited from males and females inherited from females (Kellogg 326); such a system is often a remnant of matrilineal descent. While many scholars now question the idea that was, at first, assumed by most historians and archeologists that the Aztecs were a full-on patriarchal culture, many still miss important clues about the Aztecs' pre-empire gynocentric beliefs and practices.

Probably the most startling gynocentric practice was the use of pyramids, symbols of Female Creative Magic because they are shaped like female pudenda, as platforms for human sacrifice. Tlacaelel I not only burned historical books upon taking command,

---

[166] Dual lineality and cognatic kinship system mean the same thing. Realize that anthropologists and archaeologists love to make up new terms to keep people guessing about what they mean or to attempt, sometimes, to clarify what they really mean.

but also increased the number of ritual sacrifices, especially human sacrifices. How much he increased them and why are still somewhat mysterious, but the pattern of his assumption of power follows a familiar pattern—he needed, as a male, to revise the accepted religious beliefs to make them more male-dominant in order to move his culture away from a more female-centered view of the cosmos in order to garner more Magical Power for himself.

For instance, if the Aztec pyramids represented the female pudenda, like most pyramid-like structures around the world do as either birth mounds or rebirth temples (aka the "mound of creation" or a *jenseits maschine* or *widergebut maschine*, afterlife machine or rebirth machine), the act of spilling human blood on the steps could be viewed as a menstruation or birthing ritual. The Aztecs, themselves, viewed death as merely a necessary step into the next life. Various clues exist that indicate the pyramid structures were, indeed, gynocentrically based, especially since the Aztecs themselves called the pyramid now known as Templo Mayor, Coatepec, or Snake Mountain, because it stood for the home of their earth goddess, Coatlicue, from which she presided over all of creation.

A major clue that scholars seem to misinterpret regarding the Templo Mayor in Tenochtitlan involves the Coyolxāuhqui stone disc discovered at the foot of the temple in the 1970s. Coyolxāuhqui was the Aztec goddess of the moon, so clearly has ties to women and menstrual cycles, although the stone disc depicts her as a warrior. This last point is important because so many Aztec scholars overlook the idea that the Aztecs believed pregnant women were warriors—fighting a spiritual battle to bring life from another realm to this one.

However, just as the Akkadians and Babylonians invented new myths placing male gods in superior positions to the goddess Inanna, the Aztecs created a new myth, possibly during Tlacaelel I's time, that Coyolxāuhqui was murdered and dismembered by her brother, Huitzilopochtli, who later shared the Templo Mayor with the rain god Tlaloc. I have yet to find a scholar who questions why the Aztecs would pair two male gods at the top of a temple meant to signify the primary mountain of creation governed by a female god.

To understand the importance of Coyolxāuhqui myth and the gynocentric views it attempts to conceal, we have to unpack the story.

The story begins with the goddess Coatlicue, the Aztec earth mother goddess. One day while she is surveying the earth from Serpent Mountain (Coatepec), she witnessed some iridescent hummingbird feathers fall to earth. Coatlicue gathered the beautiful feathers together, tucking them into her belt. Magically, she becomes pregnant.[167] Strangely, Coatlicue had 401 children already,[168] but, when her one daughter, Coyolxāuhqui, discovered her mother's pregnancy she knew the child[169] would bring non-stop warfare to the people, so she convinced her 400 brothers to help her kill their mother before the child could be born. Huitzilopochtli, from inside his mother's womb, overheard the plot, and calmed his frightened mother. However, Coyolxāuhqui and her 400 brothers climbed Serpent Mountain and beheaded Coatlicue.[170] Immediately, blood and serpents sprouted from her head.[171] At the same time, Huitzilopochtli, the Aztec war god, emerged from his

---

[167] There are many creation myths wherein goddess or other divine women become pregnant magically, via wind, feathers, or glances, including in Judeo-Christian myths. These myths demonstrate how humanity has struggled to understand how pregnancy actually occurs.

[168] No Aztec story explains who the father of these 401 children was; further indication that the previous iteration of the culture focused solely on mothers, not on fathers, because they were unaware of procreation. For clarity, 400 is the number used most often by the Aztecs to indicate an extremely large, sometime infinite, number. Here, the 400 brothers are stars.

[169] In some renditions of the tale, Coyolxāuhqui is angry because her mother's pregnancy indicates an adulterous relationship, but this version comes, clearly, after the Aztecs learned of procreation.

[170] Note that some versions of this tale have Coatlicue not being beheaded, but other Aztec tales indicate that, instead of two male gods sacrificing themselves at the beginning of the Fifth Sun to create the new world, it was actually two female gods, one of which was Coatlicue, who was beheaded, which is where this idea might have originated. Some monuments of Coatlicue show blood-serpents (i.e. the Aztecs used snakes to symbolize blood flow from human bodies) instead of a human head.

[171] In many Aztec images, blood is often depicted as serpents jetting out of wounds. Undoubtedly, someone among the Aztec had witnessed a nest of snakes emerging from the ground in a similar fashion. Realize that snakes have long been associated with birth because they shed their skins like women give birth.

mother's womb fully grown, bearing a spear and shield. Moving quickly, he killed and decapitated Coyolxāuhqui. Some versions of the myth claim Coyolxāuhqui arms and legs came off from her fall down the mountain, but others claim Huitzilopochtli cut them off and tossed them down the mountain himself before hurling her head into the sky where it became the moon (Luna & Galeana 12-13).

Lunar Surface, NASA, beside photo of Coyolxāuhqui Stone by Dennis Jarvis from Halifax, Canada

We can almost imagine the surface of the moon as impressions from Coyolxāuhqui dismembered body (see images above). While her breasts are clearly exposed, which was undoubtedly shocking to the Spanish, she is dressed as a warrior, ready to do battle. Without the bells on her cheeks (which is what her name means) and her obvious breasts, would scholars have recognized her as female?[172]

While many scholars will admit that Templo Mayor was the equivalent of Serpent Mountain for the Aztecs, few see the pyramid

---

[172] Many figures have been declared male by scholars on very little evidence, such as the Olmec heads, which are usually carved with ball player helmets on, but that does not mean women never played ball, so that the heads could also be of women. Personally, I think the Olmecs were having some fun with those huge heads, since they are mostly shaped like the balls the game would have been played with.

structure (or mountain formations) for what they symbolized for most ancient people: the mounds of creation, aka human female pudenda, that "mound" from which all human life arises. Adding the symbols together—pyramids shaped like female pudenda, blood spilled down the steps like menstrual blood, and a circular stone in which Coyolxāuhqui dismembered image imbedded at the bottom of the temple resembles the moon itself—we can recognize that the sacrifices made on the steps of this temple probably sought to capture the Female Creative Magic of menstruation.

Why was menstruation so magical? Because between menarche and menopause, nearly every woman menstruates—bleeding three to five days without dying—every month, unless they become pregnant or develop diseases of the uterus or ovaries. Remember, we get the terms menarche, menstruation, and menopause from the ancient Germanic word for moon, which is also why we call our menstrual cycles our monthlies, since each month represents another full moon.[173] In many ancient cultures, the menstrual cycle was such a powerful natural blood magic that women either voluntarily or involuntarily secreted themselves away to prevent their magic from overwhelming and possibly negating men's magic. For the Aztecs, specifically, the monthly blood flow women experienced was evidence that they were fighting a cosmic battle to bring human souls from Another Realm to our Earthly Realm.

Menstrual magic, which includes both Power and Creative Magic, has been seen as powerful by most ancient cultures, so much so that men have long attempted to take on this female power for themselves—from circumcision to other forms of penile ritual bloodletting, such as was performed by many Mayan and Aztec male leaders.

---

[173] There are actually 13 moons approximately every two years, but one aspect of making things associated with women negative has been the superstition associated with 13, which is why the zodiac signs, even though there are really 13 of them, one for each moon, has been reduced to 12. The number 13 is a magical number for many cultures around the world, primarily because of its associations with the number of full moons roughly every other year. In 2143, there will be the very rare 14 full moons during the year.

Figure in the fact that Coatlicue was made pregnant by sky/spiritual magic,[174] even after she has given birth to an infinite number of sons,[175] and we have a typical magical creation story.

Besides the obvious Mother Earth symbolism that stems from the goddess Coatlicue who presides over all earthly creations from the top of Serpent Mountain, we also have to acknowledge that serpents were long associated with women and childbirth because of the way snakes shed their skins, which looks a lot like childbirth, and because, like snake mouths which seem to grow in size in order to swallow larger objects, vaginas expand during coitus (Grahn 58). The fact that so many Western cultures see snakes as evil probably arises from the desire by some to see women as evil and sinister, therefore all symbols relating to women, especially those having to deal with Power Magic, are often seen as evil and sinister in patriarchal cultures.[176]

Pregnancy, itself, is a spiritual, if not outright magical, process with most ancient people struggling with determining how, exactly, pregnancy occurs. Here, we have Coatlicue becoming pregnant without heterosexual intercourse, which demonstrates the general human ignorance about how pregnancy occurs that preceded the knowledge of procreation. Here, too, is the persistent idea that women who reproduce without direct copulation with a male, but through "divine" intervention, are powerful women who go on to Mother whole cultures and religions.

However, what seems to have escaped even some feminist scholars' reading of this Aztec tale is the blood connection. Jennie Luna and Martha Galeana argue that Coyolxāuhqui represents not just the moon and women's moon cycles, but also was the symbol of "preparation for the ceremony of menstruation as well" (15). Blood running down a mountain/pyramid that symbolizes human female

---

[174] Similar to one Aztec creation story, numerous American Indian nations have creation stories wherein Mother Earth and Father Sky have to be forcefully separated by their children to make the surface of the earth.

[175] According to some sources, the number 400 encompassed the Aztec concept of infinity.

[176] Remember, the Akkadian word for women, "sinnustra," is what becomes the word "sinister."

pudenda could be nothing less than a menstrual ceremony drawing on that Female Magical Power women have to bleed and not die.

How, then, do we deal with the irony of the required death of human beings as a justification for the continued salvation of whole populations?

It seems ironic, of course, that the Aztecs used the graphic death of a sacrificial human victim to recreate, thus possibly stoke or take on, the power of menstrual magic. Would ancient human beings have known the difference between blood from our hearts and menstrual blood? Of course not.

Few ancient cultures separated life from death, and we must be careful not to assume modern ideas about the separation of death from life. For the ancients, death was simply one part of life—an inevitability that they sought to stave off, at least for a time, through the magic of ceremony.

For nearly every ancient culture around the world, however, blood was sacred. Blood brings us into this world during birth; blood courses through our veins like "a kind of vital force coursing through the universe" (Martin 222 Note 66), similar to the Milky Way's path through the night sky; and blood stops flowing through us when we die, so, for the ancients, blood was one of the most sacred substances on earth and in the cosmos, so they had to keep it flowing. For many of the ancient people around the world, human sacrifices were the ultimate form of offering to the divine energies that kept the cosmos flowing. Bloodletting also turned out to be a great way to demonstrate one group's dominance over other groups.

However, one element stands out among the facts about the myth: the Coyolxāuhqui Stone was embedded at the bottom of the Templo Mayor. Most scholars attribute this placement to the idea that the broken goddess at the bottom of the steps is a symbol of defeat. As Emily Umberger puts it, "all sacrifices on the temple platform were meant to reenact Huitzilopochtli's victory of his sister and, on the cosmic level, the daily triumph of the sun over the moon"[177] (411). For Umberger and many other scholars, this story

---

[177] Realize that the Aztec tales emphasize not so much a "triumph of the sun over the moon," but more of a feeding of Mother Earth—a ravenous monster by this time in Aztec history—in order to free the sun from inside her. Thus, it is the

265

symbolizes how the sun, rebirthed every morning from the eastern edge of the earth, "defeats the moon (Coyolxāuhqui) and stars (four-hundred or innumerable brothers)" (412). At the time Umberger wrote her article in 1983, it was not widely accepted that Coyolxāuhqui was a goddess of the moon, nor had the earliest[178] iterations of the temple been uncovered.

Some scholars admit that topping the pyramid with two temples to male gods with a female god's dismembered likeness at the bottom seems very androcentric—seemingly symbolizing how males are presumed to be better than females. However, most of these scholars also assume Coyolxāuhqui is nude, which they naively believe would have been shameful for a woman living in a hot tropical region of the world, when she is actually dressed like a warrior (Templo Mayor Museum). Note that the Spaniards who discovered one monumental sculpture of Coyolxāuhqui immediately "proceeded to destroy it," probably considering her naked breasts unseemly (Templo Mayor Museum). Discomfort with exposed female breasts is a Western/Abrahamic/androcentric preoccupation.

Modern scholars must address the imbedded androcentric biases in their points of view before they can genuinely understand the Aztecs, since the ritual blood spilled down the pyramid and onto Coyolxāuhqui stone, which symbolized the moon itself, still draws attention to the fact that the original sacrificial rituals were probably meant to return the moon to fulness, not the sun to the sky. In fact, the first such ritual was probably conducted during a lunar eclipse, when the moon itself turns blood red, sometimes for hours.

And, while topping their pyramids with two temples is a common Mesoamerican construct, their symbolism as two exposed breasts seems to escape most scholars, so that few question the idea that they were always dedicated to male gods. Perhaps male gods always need to be suckled?

If Tlacaelel I was wrestling his people away from a more gynocentric view of the cosmos toward a more violent, oppressive

---

frightening female monster who must be appeased to enable the sun to be reborn each day.

[178] The earliest layer of the temple supposedly has not been excavated because of water issues.

androcentric view in order to make his political enemies afraid, he twisted something that was probably a sacrificial ritual drawing on Female Creative Magic that was conducted at the new moon (when the moon goddess was assumed to be "hiding" her face) to ensure that the moon would return, into something that happened every day, moving the focus from the more female connected moon magic to the seemingly more male connected sun orientation in the process.

The ritual would have been a new moon event because the people would have worried that the moon would not show its face again without a blood sacrifice, and the menstrual blood tie comes from the fact that women traditionally "hid" their faces like the moon or sequestered themselves away from men during their own menstrual flows in order to keep their stronger magical powers from affecting the men's magic. Through Tlacaelel I, women were removed from the ritual magic altogether, and he used the ritual sacrifices to create terror among the neighboring communities, which were expected to pay tribute to Tenochtitlan, including sending sacrificial tributes. But giving humans as tributes could go both ways, as Michael Smith points out: "the Mexica tlatoani Ahuitzotl sent 40 captives to sacrifice at the dedication of a new temple in Cuauhnahuac" in 1490 (80, 1986), so we know the Aztecs gave other communities human sacrifices for the other cities' rituals.

Additionally, the *zacatapayolli*, balls of dried grasses used to soak up sacrificial blood, were probably imitations of the woven grasses women used to soak up menstrual blood. In fact, Coatlicue's hummingbird feathers that she tucks into her belt could have also been a form of menstrual pad used by royal women. The fact that women would have burned these early menstrual pads would have been to ritually gift the more potent menstrual blood to the gods. Later, men will also imitate this menstrual ritual by bleeding onto paper, something the Maya originally used for ritual purposes instead of for writing, that they will then ritually burn for the gods (Smith 80, 1997).

In effect, Tlacaelel I changed the Aztec focus from the powers of creation to the power of destruction, changing a monthly human sacrifice and bloodletting into a daily one, twisting the original logic from helping the moon to show its face again toward the idea that the sun required blood to rise each day because it supposedly had to fight its way through the womb of Tlaltecuhtli, the Mother Earth goddess who was so fierce she not only hungered for human blood, but had to be torn in half by the gods Quetzalcoatl and Tezcatlipoca. Her bottom half became the sky, interestingly enough, but her top half became the earth, so that she was imagined to swallow the sun every night and give birth to him every morning.

It is also possible that Tlacaelel I assumed the role of the Serpent Goddess, Coatlicue, in order for men to appropriate all female-creative magic for themselves. Coatlicue, whose personal tale is more complex than most, was supposedly an Aztec priestess, tasked with maintaining the temple at the top of Snake Mountain, Coatepetl, but her magical pregnancy changed her fate. This portion of the tale, not as often told, indicates that Huitzilopochtli, the male war god whose shrine sat atop Temple Mayor above Coyolxāuhqui stone, was not originally supposed to have a shrine at the top of the temple, which would have been Coatlicue's home, Snake Mountain. Instead, his side of the temple was probably originally dedicated to Coatlicue, who, together with the rain god Tlaloc,[179] brought the Aztecs nourishment and prosperity—a truly prosperous couple dedicated to the survival of their people because they were "the agrarian gods of growth" (Soustell 103). The necessity for both women and men to work together to create is also reflected in Aztec religion, wherein "deities appear in either male/female pairs or, in many cases, with both male and female aspects" (Dodds Pennock). Since the Mayan word for world was *kajulew* or "earth-sky" (Tedlock, 29), if they had truly influenced the Aztecs as much as most scholars believe, it would be more believable to have the main temple house these two deities, rather than include the god of war.

Like the earthly lava that often flows from volcanoes,[180] such as the ones in Mexico and Guatemala, Coatlicue's shrine had been painted red, which also coincides with the color of menstrual blood

---

[179] We should consider, however, that the two temples originally probably symbolized the two halves of Tlaltecuhtli, who becomes Coatlicue and Tlaloc.

[180] In the documentary, *Guatemala: Heart of the Mayan World*, images are shown of lava leaking from the sides of the volcanoes, instead of spewing from the top, much like blood escaping human skin.

and with blood spilled in war. Many scholars admit Coatlicue was associated with both childbirth and warfare, so it must not have been too difficult a transition for the Aztecs to choose Huitzilopochtli, a male war god, in place of Coatlicue, a female war god.

If the shrine above Coyolxāuhqui stone was originally meant for Coatlicue and not for her son Huitzilopochtli, the shift would clearly demonstrate the rise of androcentrism over what was once a more gynocentric world view among the Aztecs.

In fact, we do know that there was an Aztec ritual that was performed solely by men that required all pregnant women to wear "maguey leaf masks and all women would be hidden away in granaries"—the New Fire Ceremony (Horne 11). Horne reports that "during this ritual, men would climb to their rooftops and await their fates," but the women had to be hidden away "because it was believed that if the fire were not drawn [the women] would transform into 'fierce beasts' and consume men" (11). According to Julia Flood, however, the pregnant women could only be transformed into monsters if they were "discovered by ghoulish Tzitzimeme,"[181] the angry spirits of women who died during childbirth and of sacrificial victims, all of whom reside in the heavens as stars (Flood).

Since the New Fire Ceremony only came at the end of each 52-year-cycle, the people feared that an unsuccessful relighting of the sacred fire through the ancient way by rubbing sticks together would spell the destruction of their entire world because the Tzitzimeme, who appear only in darkness, would infest Tenochtitlan, destroying everyone and everything (Aztecs; Flood). In preparation for such a possibility, "the people also destroyed their most treasured possessions. They tore their clothes and broke their furniture and utensils. Even gods and idols were hurled into the rivers and lakes" (Aztec).

---

[181] According to mexiolore.co.uk, while Tzitzimeme were believed to come from spirits of specific kinds of dead people—mainly those who were buried, instead of cremated—"they had it in their power to commit both good and bad deeds. They might inflict an illness such as epilepsy or dropsy on one unfortunate Aztec, yet cure his neighbour [sic] of a similar fever."

The maguey[182] masks pregnant women wore were intended to keep them from turning into wild animals, again, possibly symbolic of their closer spiritual ties with nature and the world of the spirits (Aztec). As Camilla Townsend puts it, menstruation and pregnancy were viewed as a kind of battle with women having to "seize a new spirit from the cosmos" and bring it to earth (50). Pregnant women's magic was so powerful that, if a woman died while giving birth, her body would become a totem highly sought after by warriors, so the midwives, who helped in birth and death, secreted the woman's body out of the house through a hole in the back wall (Clindinnen 178). To keep the powerful Female Magic from harming the community, the body was buried at a crossroad— a spiritually liminal place—and guarded by the husband for "four nights as the magic power slowly dispersed" (Clindinnen 178). These women were the ones who were believed, primarily, to become Tzitzimeme—those marauding spirits that had to be kept at bay during the New Fire Ceremony, lest they claim more pregnant women among their number.

After all the fires were extinguished, the men followed the priests to the nearest summits, particularly to the Hill of the Star where the actual relighting ceremony would be conducted, so that men from distant villages could "get a view of the Sacred Hill of the Star to watch for the reappearance of the divine flame, the New Fire" (Aztec). The time for the ceremony was determined celestially, by the position of the Pleiades. After a man was sacrificed via the traditional removal of his heart, a priest would place a piece of soft wood on his open chest, then "twirl the hard stick with great energy…until the softer wood powdered, smoked, took fire" (Aztec).

But the bloodletting was not over. The assembled men, after briefly rejoicing that the flame had returned, "drew blood from their ears with [maguey] thorns and threw [the blood-soaked thorns] in the direction of the blessed New Fire" (Aztec). If the blood kept flowing, they would soak it up with bits of paper that were also

---

[182] The agave plant, called maguey by most scholars, is a native cactus plant to Mexico. Mayahuel was the Aztec goddess associated with the plant who was also called The Woman of 400 Breasts. As the goddess of agave, she was also the goddess of pulque, which is an alcoholic beverage made from agave juices, grandmother of tequila.

thrown into the fire. As the men ran their torches now ablaze with the New Flame to their hometowns, they celebrated that the gods had guaranteed another 52 years of life for the Mexica people.

How soon after the flame was reborn were the women allowed to come out of the granaries?

Unpacking the New Fire Ceremony, we can glimpse its probable gynocentric origins. While the 52-year cycle seems arbitrary, it directly parallels the number of weeks in a year, although the Aztec did not count their time scales the same way we do. Because the Aztecs used two calendars at once—a solar calendar (365 days broken into 18 months of 20 days, plus five "nameless" days) and a ceremonial or sacred calendar[183] (20 periods lasting 13 days each)—the two calendars overlapped approximately every 52 years, prompting the New Fire Ceremony. Coincidentally, 52 years is close to the average age of a woman who reaches menopause. While menopausal women are not considered as spiritually

---

[183] Susan Milbrath suggests that "The 260-day divination count probably originated as a subdivision of the solar calendar, first developed to mark the length of the agricultural cycle around 1000 BC among the Maya in Guatemala and the Olmecs in the Gulf Coast area. The synodical lunar month was certainly observed in relation to this 260-day cycle, but the months did not 'fit' precisely, so a separate **lunar** calendar was developed by the Classic Maya in Guatemala and Mexico, ca. AD 250–900" (emphasis mine). However, the 260-day cycle dates back to circa 3000 BCE, possibly developed to "approximate [the time interval] **between a missed period and childbirth**, how long it takes maize to grow, or the product of 20, the fingers-plus-toes base of Maya math, and 13, another common Maya number...justified by the number of days between a first crescent Moon and full Moon" (emphasis mine, Sokol). At least one depiction of a Mayan astronomer shows a woman, possibly a depiction of the person who corrected the mathematics for calculating the movements of Venus circa 900 ACE (Sokol), which helps us tie women, Female Creative Magic, and the cosmos closely together. Anthony Aveni even concludes that the Dresden Codex's Eclipse Table demonstrates that the Mayans were predicting lunar eclipses (111), with calculations based on that table "used by Maya day-keepers in their prognostications...[in order to] recover the same day of the *tzolkin* with only a slight change in the phase of the moon," demonstrating that the table "must have bene intended to record the dates of actual [lunar] eclipses in the ritual calendar for divinatory purposes" (114). For clarity, the *tzolkin* is the 260-day religious calendar—all based, apparently, on the moon and its eclipses.

powerful[184] as women of childbearing age, their added wisdom, and their ability to pass on cultural knowledge would have made them important contributors to their people's lives.

Requiring pregnant women to wear maguey (agave) masks has intriguing elements because most masks served similar purposes, almost all of which were spiritual in some way. Ian Mursell implies that the maguey fibers were used to make the masks for pregnant women and babies during the New Fire Ceremony because they were precious materials meant to protect important people to the community. Traditionally, death masks were added to mortuary bundles in order to "animate" the dead person being remembered and to grant the bundle, as a whole, the spiritual power necessary for the person's soul to journey to the afterlife (Mursell). Similarly, priests and priestesses wore masks to acquire the spiritual energy, perhaps even to assimilate with the particular spiritual power of a god or animal.

Were the Aztecs, then, trying to transform pregnant women into agave plants?

If so, were they trying to spiritually encapsulate the plant's immobile nature, or did the Aztecs see a similarity between women of childbearing age and the maguey, which tends not to bloom until it is between 20 and 30 years old? Probably not, since the maguey plant dies after it blooms, unless the association comes from the fact that some women died during or after complications in childbirth.

Octli, or pulque, is a sacred alcoholic beverage the Aztecs made from the agave/maguey plant's milk-like sap, so were the masks pregnant women wore during the New Fire Ceremony believed to keep them intoxicated? Probably not, but the agave plant also supplied fiber for weaving, although most Mexica agave fibers are "rigid and course," so must be blended with softer yarns like cotton to make cloth for clothes. Usually, the agave fibers are used to "make baskets, mats, hats, belts, and fillers in mattresses" (Textiles). In fact, "if the leaves are cut and used in the production of fiber, the plant will live for many more years [because] the plant

---

[184] Martin makes the point that, for the Mayans, the elderly were seen as "community sage, seer, and diviner—a character feeble in frame but magical in mind" which "places enchanted knowledge as the foremost of divine powers" (226).

continues to produce new leaves instead of flowering" (Cloth Roads).

Perhaps the agave masks simply provided a protective coating?

The primary myths about the maguey center around the god Quetzalcoatl and/or the goddess Mayahuel. In one, Quetzalcoatl falls in love with Mayahuel and, when they embrace, they turn into a tree, which is then chopped down by her vengeful grandmother, killing Mayahuel. Quetzalcoatl buries her remains, and the first maguey plant grew from them. The second tale involves Quetzalcoatl getting so drunk on *octli* that he has sex with his sister, Quetzalpetlatl. His shame over this action, which we would see as incest,[185] is one reason given about why Quetzalcoatl abandoned the Aztecs. More importantly, Mayahuel was also called "'the woman of 400 breasts'" because of her association with maguey, which contains that milk-like sap which is stored inside the heart of the plant, which is what pulque is made from (Cartwright). Associating a plant that gives milk with women about to give milk for their newborns would be natural symbolism.

Why add the idea that the women will turn into formidable beasts if the fire ceremony does not kindle a New Fire? Why prevent women from participating in what is clearly a moon-oriented ceremony?

The possibility is great that the Aztecs feared pregnant women would die during the New Fire Ceremony, thus become Tzitzimeme, a type of "formidable beast" that terrified many.

But another possibility also exists. Since women's Female Creative Magic was so great, women around the world often hid themselves away, just like the moon does in its new phase, in order to prevent their stronger magic from interfering in other people's personal magic. Therefore, it could have become traditional for pregnant women to stay hidden during the new moon. If they are

---

[185] Since we only have the word of elite males who have been assimilated into Christian biases by the time they wrote down the Mayan and Aztec myths and legends, we might never know what the peoples of the Mexica Basin knew about procreation or felt about incest, since some "incest" taboos arose to keep men from attempting to seduce their mothers-in-law.

hidden, like the moon, they could not run around the hills lighting fires like the men could.

Because all of the flames—community fires and household hearths—were doused before the ritual sacrifice and ceremony to light the new first, the ceremony takes place in darkness illuminated only by starlight. Could it be that the Aztecs associated the dark with sexual intercourse? Were Aztec men afraid of their women's sexual appetites?

Since the Aztecs believed the Tzitzimitl only appeared at night, since they were associated with stars, and they mainly haunted the crossroads, especially where those who were not permitted to be cremated were buried (Flood), this mythology clearly feeds the Aztec community's desire to feed human blood to the sun (or the earth, in this case, who had eaten the sun) in order to ensure the sun would return. Except for the fact that it could still be hours before the sun would rise again.

As with the association of pyramids with women's pudenda or creation mounds, the fact that the men went to the tops of hills and mountains near their home cities in order to participate in the New Fire Ceremony has at least a three-fold purpose. First, so the men had an elevated view to see when the fire was successfully lit; second, to draw on the gynocentric spiritual power of the earth by performing the ceremony at the top of a natural symbolic pudenda, a hill, in order to guarantee success; and, third, to directly feed their self-sacrificial blood, and that of the victim, to the earth mother/monster that swallows the sun every night.

Having the priest ignite the fire on the gaping chest of the sacrificial victim might seem callous, but, if we view the sacrifice as one of creation, a literal birthing of a human soul through his chest and into the next life, we can understand why the Aztecs would believe that the spiritual energy discharged through the sacrifice would lend itself to the successful lighting of the New Fire. After all, the ritual was also supposed to keep the sun rising for another 52 years, and the sun was associated with fire.

Why, then, is the New Fire lit in the middle of a moonless night? Is it possible that the New Fire Ceremony was so old that it was originally meant as reverence for the moon, and not the sun?

After all "the grammatical root of the Nahuatl voice Metztli, [from which we get the word Mexica]...means moon" (Tena-Colunga 2), with some scholars arguing that Mexica means People of the Moon (Tena-Colunga 9), indicating that they, like the early Germanic people, identified themselves as children of or worshippers of the moon.[186]

According to Susan Milbrath, "Judging from the frequency of deity images in the Postclassic period, the lunar goddess was more important than the Sun God" for the more southerly Mayans.

So why would a gynocentric culture founded on a reverence for the moon's magical ties to Female Creative Magic ever begin to consider human sacrifice as an important part of that cosmic life?

We will probably never fully know why the Mayans, possibly even the Olmecs before them, or any of the many other cultures who lived in the Americas originally resorted to human sacrifice. Each instance of sacrifice, whether it was an *en masse* human sacrifice by the Egyptians/Chinese to accompany their ruler into the next realm, a single human sacrifice by the Pawnee at Pawnee Rock in what is now Kansas in the United States or a monthly one by the Aztecs in ancient Mexico, each sacrifice probably has its own catalyst. Most scholars believe such sacrifices are meant to appease certain gods or spirits, who are only happy with extreme sacrifices, such as Yahweh's requirement for someone to be brutally murdered in order to appease him enough to forgive humanity for what he perceives as their violation of his covenant with the people of Judeah. However, many cultures, including people from ancient Judeah, often believed more than one person had to be sacrificed, especially in times of danger—such as when threatened by an enemy state or when threatened by drought and famine.

According to Richard Wrangham, there are two basic types of aggression that lead to violence: reactive aggression and proactive aggression. Reactive aggression occurs when a person or

---

[186] The word "men" in Old Germanic indicates the same idea—that the people worshipped or determined time by the moon, giving us the words menstruation, menarche, and menopause, all relating to women's "moon" cycles. The word "wommen" thus means the moon worshippers with wombs, with "men" being a general word for all humans (aka the people of the moon) and a specific word for those without wombs.

animal reacts immediately to perceived threats with violence, such as through what Wrangham calls "character contests," wherein two men, males being more likely to participate in such violence, confront each other over perceived insults, which Wrangham says sometimes accounts for 35% of all homicides in some cities (26). Proactive aggression is planned violence or intimidation that "involves a purposeful attack with an external or internal reward as a goal, rather than an effort to remove a source of fear or threat" (28).

Human sacrifice often appears to be proactive aggression, especially since some cultures clearly made the same type of sacrifice repeatedly, such as the ancient Celts who practiced a form of triple death sacrifice wherein the sacrificial victim, who was selected in some way or even volunteered, was knocked unconscious, garroted, and then had his throat cut before his body was left in a bog as a sacrificial offering to the spirits or gods of the land, with the remains so well preserved that we learned of this grizzly tradition centuries later. Presumably, this method of sacrifice, while frequent, was possibly only stimulated by outside threats to a community, so the people sought to proactively ward off the threat.

Again, sacrificing a specific person to stop the potential death of hundreds of others might seem ironic to most people.

The same type of proactive logic holds for the human sacrifices made by the Aztecs, except that they were usually the people threatening their neighbors. As they made their way into the Mexican Valley originally, though, they claimed they were often seized upon by people who already lived there (Townsend 28). In fact, they have a legend that explains not only why they fought so often with the people they met along their route to their new homeland, but also possibly why they make human sacrifices. It comes to us from the *Historia Tolteca Chichimeca*, written in 1540. At the time of the story, the Aztecs called themselves the Chichimeca, their self-effacing word for barbarians (Townsend 27). According to the tale, Huemac was a foundling, who the Chichimeca adopted, not realizing he had been left by a trickster god for them to find (Townsend 28). As Huemac grew, he became demanding, insisting the neighboring people, the Nonohualca, bring him women with large buttocks at least "four spans wide," but he was not satisfied with the first women they brought him. He demanded women with even bigger buttocks. Huemac refuses to marry the women, however, instead choosing to sacrifice them on an obsidian table. Enraged that their women were sacrificed by this selfish, arrogant person, the Nonohualca began to fight with the Chichimeca, but their leader pointed out that the violation of social etiquette was Huemac's fault, not theirs, so "working together, the two groups managed to defeat their preposterously evil enemy," Huemac. However, the Nonohualca knew they could not trust the Chichimeca, so packed up their village and moved out of the territory (Townsend 29).

The fact that the Aztecs, as they marched through northern Mexico on their way to their new homeland alienated and killed many of the already established groups living there seems to be a source of pride for them, but they apparently believed, originally, that "the most terrible thing one could do to a conquered people's women" was to sacrifice them (Townsend 29). However, archeological evidence and their own stories makes it clear that the Aztecs were not above sacrificing women.

If the Aztecs had once limited these human sacrifices to once a month in order to feed the moon to help it recover to fullness, they could have made blood sacrifices by pricking various parts of their bodies, bleeding on paper, and burning the paper—traditions they continued to enact for later rituals. Perhaps, though, a lunar eclipse, which lasts up to two hours, had seriously made the Aztecs worry that blood from an ear or a penis was not enough to fully resurrect the moon, so they had resorted to killing a prisoner or slave in order to appease the spirits of the cosmos to free the moon, or even, potentially, to drain the blood from the surface of the moon. Since lunar eclipses can paint the moon red, the ancients who began this menstrual sacrifice ritual might have assumed only making a lot of blood flow would work to set the cosmos right again.

In fact, common methods of controlling excessively violent men, who tend to be more aggressive and violent than women because of higher levels of testosterone, were imprisonment, castration, or execution for many cultures (Wrangham 130). Perhaps this reasoning—controlling men who tend toward violence—was part of the rationale for the first human sacrifices in the Mexico Basin. While the Aztecs enslaved people captured from neighboring cultures, many of whom ended up being used as sacrifices, there is little evidence that most had to be forcefully imprisoned because the social pressures of being chosen as a sacrifice for the gods would have convinced the sacrificee that her/his death would elevate her/him or even their family in the eyes of the divine. Potentially, such victims were also promised that their sacrifice would make them divine, too.

However the sacrifices began, the Mayans, according to the *Popol Vuh*, called the sap of the Croton Draco tree the "blood of sacrifice," which the hero twin brothers, Hunahpu and Xbalanque used to heal a falcon's eye, which they had shot out (Tedlock 114-5). This sap was traditionally used to stem bleeding, but also acted "as an antifungal agent and itch reliever" (Russell). The tree, however, is not native to Mexico, but grows in northern Colombia, demonstrating a connection between the Mayans and the peoples of South America.

But the Mayans, according to the *Popol Vuh*, had already determined that there were two types of basic aggression:
> [The former lords of Xibalba] are makers of enemies...
> They are inciters to wrongs and violence;
> They are masters of hidden intentions as well
> (Tedlock 139).

Violence is wrong, and hidden intentions are wrong. Both cause discord, anguish, and one can stimulate the desire for the other.

Yet it is deception, a form of hidden intentions, that helps the hero twins, Hunahpu and Xbalanque to defeat, to murder, One Death and Seven Death, two of the main leaders of Xibalba, literally the "'Place of Fear'" often described as an underworld that contrasts greatly with the face of the earth where the Mayans dwell (Tedlock 34). This tale demonstrates that the Mayans were conflicted about what is good and what is evil because that determination is largely concerned with *who* is doing the act. If the deception and murder are by the defenders of the Mayans, the actions are good and just. If the deception and murder are committed by someone outside the Mayan world, the actions are evil. Thus, the *Popol Vuh* cautions its listeners (was its contents ever read aloud in ceremonies?) or readers that the masters of Xibalba are also "masters of stupidity, masters of perplexity," who think nothing of using confusion and stupidity to perpetuate their evil (Tedlock 139).

Human sacrifice for the Mayans took mainly two forms: beheading and heart removal. Usually, heart removal occurred after the beheading, though, so the fact that the Aztecs perform the heart removal first, as the primary method of killing their human sacrifices, speaks volumes about how the religious ritual of human sacrifice changed in Aztec hands.

While we know that the assertion that the Aztec god of war, Huitzilopochtli, rose to prominence during Tlacaelel I's influential era in Tenochtitlan, probably replacing the earth goddess, Coatlicue at the top of what will become Templo Mayor might be objected to by some, it is not farfetched to trace other changes that ripple throughout Mexica cultures as a result of males assuming a more prominent role socially, religiously, and, especially, politically in the region.

Like the ancient Greeks' Cult of the Hero, adult men sacrificed by the priests in Mexica cities were expected to die "honorably." Dodds Pennock quotes the Codex Chimalpahin's version of what the priests told the sacrificial victims:

> We welcome you and say to you that you should be consoled that no womanly187 nor infamous deed has brought you here, but manly feats [have been responsible]. You will die here but **your fame will live forever**. (emphasis mine, 14)

By this time in Aztec history, sacrificial death is equal to death on the battlefield for the ancient Greeks: it makes men heroes. Just as the Greek propaganda scheme to get more young men to volunteer for fighting the Greek city's battles for *kleos*, which includes being deified after death, those sacrificed by the Aztecs were promised eternal glory and eternal life, or, as Dodds Pennock puts it, "spiritual survival" (14). To prove to these victims that their deaths were not in vain, the Mexica hung their skulls on skull racks[188]

---

[188] The ritual of hanging skulls on large racks was said by the Mexica to be a method of intimidating their enemies. However, it was such an honor to be sacrificed, having a person's head on such a rack would also help bring the

situated near their primary temples where they could be honored and glorified. But, even though Dodds Pennock prefers to believe that it was primarily only males who were sacrificed, recent discoveries demonstrate that some of the skulls on the Tenochtitlan rack, called the *Huey Tzompantili,* "belonged to women and children" (Gershon).

Quoting Barrera Rodriquez, a National Institute of Anthropology and History archeologist, "'We do know that the [victims] were all made sacred, that is, they were turned into gifts for the gods or even personifications of the deities themselves, for which they were dressed and treated as such'" (Gershon).

While Dodds Pennock cites a source to verify her claim, she has no actual evidence to support the idea that the Mexica were "a warrior culture" from the moment they settled in the valley of Mexico (15) because the source she quotes is clearly a biased one: Bernal Diaz del Castillo, a soldier under Cortes, who wrote about his experiences in a book called *The Conquest of New Spain.* The Penguin Classics version of the book admits that Diaz did not begin writing about his experiences until "he was over seventy, and the last survivor of the Conquerors of Mexico," all the while "fearing his literary abilities were not up to the task" (Editor's Notes, Diaz). The work was published in Spain in 1568, but not translated into English until 1963. What is most notable in this edition of Diaz's work are the repeated claims that Diaz, who had been granted land and "Indians" as slaves by Spain for his part in the conquest, and who claimed to be "governor of the most loyal city of Santiago de Guatemala," was poor, leaving no monetary value for his family except his book, with the Editor noting Diaz had "a graphic memory and a great sense of the dramatic" (Editor's Notes & Introduction,

---

individual, as long as his face was recognizable, that promised fame. It also pays to consider that this ceremony is very similar to their cultural practices wherein the skulls were removed from the dead long after burial for religious purposes. Since our heads/faces are our most defining features as individuals, we should also consider that the act of placing an individual's head among the heads of so many other sacrificial victims was a way of compounding the spiritual energies of all those deceased consciousnesses. Given that the Aztecs also believed the sacrificial victims became spirits that could harm them during the New Fire Ceremony, the act of keeping their skulls on display could have been another way to appease or to control those souls.

Diaz). While having first person accounts can be valuable to a historian, they are also problematic because there are often few methods of verification for the claims they make. What kind of proof could Diaz possibly give to prove that the Mexica were a warrior culture from their start, since he had not been there?

We can trust neither Diaz's version of Mexica history, nor the Aztec versions of their history for the simple reason that Tlacaelel I destroyed all the written histories of his people in order to rewrite that history in the way he saw fit. For all we know, Tlacaelel I alone rewrote the history to be centered fully on his favorite god, Huitzilopochtli, the god of war and discord, because it was Tlacaelel who recognized that discord among neighbors gave the Tenochtitlans a reason to fight them, to conquer them, and to take their land, thus allowing the Aztecs to force the conquered people to send tributes—both people and goods—to the city.

Archeologist Michael Smith also cautions scholars about believing Mexica propaganda meant to highlight their power over other cultures in the Mexica Valley, warning that "the political ideology or propaganda of the Mexica state...permeates the accounts of the postconquest nobility recorded and synthesized by such Spanish chroniclers as Duran and Sahagun" in an attempt to celebrate the "greatness and invincibility of the Mexica state" (84, 1986).

We must admit that the Spanish imposed a staunchly suppressive patriarchal cultural view on the Aztecs. According to various accounts, when the Spanish first encountered a sculpture of the Aztec earth-mother goddess, Coatlicue, they immediately set about destroying it. Coatlicue is important to our examination here because the explanations for her snake-riddled image is wholly gynocentric, despite the fact that Western scholars seeing the snakes as evil manifested because of biblical accounts instead of the association of snakes with birth. For so many nonnative archeologists studying Mesoamerica, snakes are meant to provoke fear, symbolically, not awe. When, in fact, the simplest explanation of Snake Skirt's snakiness comes from the creation process itself, as illustrated through Coatlicue's often undervalued and little examined Aztec creation myth.

One similarity between ancient Akkadian and Babylonian beliefs is the idea that a god had to sacrifice himself either to create humans or to create the sun. In one ancient Mesopotamian creation story, the god Kingu is sacrificed, so that human beings could be created. Kingu, it should be noted, sided with Tiamat, the goddess whose defeated body becomes the earth and its waters, in a great battle against Marduk, a sun god who replaces Tiamat in the pantheon. As Morris Jastrow put it in 1919,

> Marduk's first task after overcoming Tiamat [aka female-created chaos] is to pass across the heavens, assigning fixed positions to the stars—i.e. to the gods—and regulating the calendar through the phases of the moon. With **the sun in control of the universe**, the movements of the heavily bodies are regulated, vegetation springs up below, and the earth [Tiamat's corpse] is thus prepared to support life" (emphasis mine, 278).

As we will see in many burgeoning patriarchies, the sun god rises to the top of the culture's deity pantheon, ostensibly to create "order" out of chaos. That the male sun god's idea of order is ultimate control of everyone and everything comes at the detriment to freedoms once enjoyed by women and poorer men.

According to one Aztec creation myth, Tonatiuh, a god representing one aspect of the sun and fate, "flew into the heavens" after he sacrificed his former self, Nanahuatzin, because, originally, he was "a miserably ill and poverty-stricken god" who "had no reason to continue as he was, so [Nanahuatzin] voluntarily cast himself into the fire" in order to (Barroqueiro). This tale emphasizes that, for an impoverished person to become an exalted deity, s/he merely had to allow her/himself to be sacrificed.

One version of the tale claims that because the god Tonatiuh was so weakened by his sacrifice, he had to be fed human blood to sustain his daily orbit (Roos), but another claims he "was thirsting from the great internal heat" as the sun, so "had to be nourished and cooled by offerings of the red cactus-fruit (which meant human hearts and blood)" (Barroqueiro). Still another version claims that the Earth Goddess, Tlaltecuhtli, was such a terrifying lizard monster (think dragon like Tiamat from Sumeria, but more like a caiman or alligator, both of which are highly protective mothers) that

Quetzalcoatl and Tezcatlipoca had to kill her by squeezing her in half with her top half becoming the earth—her mouth the cave from which the Aztecs emerged[189]—and her bottom half the sky, from which she gave birth to the stars and moon (Maestri). Because of Tlaltecuhtli avaricious hunger, she ate the sun god every night, so that blood sacrifices had to be made to appease her, so she would release the sun.

Either way, the Aztecs created a religion in which they make human sacrifices either to strengthen the sun god, weakened from his battle through Tlaltecuhtli at night, or to appease the earth goddess who hungers for human blood. They built this religion based on human sacrifice and upon the human desire to be valued, convincing countless numbers of victims to willingly sacrifice themselves and, quite often, their children in order to seek prosperity and power in the next realm.

Religious inconsistencies are not uncommon, but they can help us uncover sequences of ideas that take hold or replace previous ideas. Here, it is clear that most nonMexican scholars see Tlaltecuhtli as a monster, some even going so far as to call her that:

> Tlaltecuhtli is depicted in codices and stone monuments as a horrific monster, often in a squatting position and in the act of giving birth. She has several mouths over her body filled with sharp teeth, which are often spurting blood. Her elbows and knees are human skulls and in many images she is portrayed with a human being hanging between her legs. In some images she is portrayed as a caiman or alligator (Maestri).

While male scholars who have obviously never given birth can be allowed some leeway when they believe this earth goddess is all monster, a female scholar who has given birth should not.

Birth is extremely painful and sometimes deadly. When simulations of the pain of childbirth has been applied to men, most males do not have the pain tolerance built up enough to take such strong pain for the hours that the process usually takes. Women who menstruate and go through menstrual cramps, bloating, and mood swings once a month, know they are not experiencing the full

---

[189] One myth tells "that the Mexica came from Chicomoztoc, the Seven Caves, from the northern lands of Aztlan" (Barroqueiro).

force of birthing pains, but all those months of practice in dealing with the pain can be beneficial if they ever actually choose to give birth.

We also have to recognize that the Aztecs did not fear human skulls or blood, even though many nonMexica scholars assume the Aztecs used these symbols to incite fear.

As mentioned earlier, women were believed to wrestle babies they gave birth to from Chichihuacuauhco (not located on the earth, the moon, or the sun, but somewhere in the cosmos), the first mansion of the dead wherein children who had died waited for reincarnation (Xocoyotzin). Women who died in childbirth, which must have been common, went to the Kingdom of the Sun (so called because it was located on the sun), Ilhuicatl-Tonatiuh, in the afterlife, which is where warriors who died in combat resided until reincarnation (Xocoyotzin). These women "became fierce goddesses who carried the setting sun into the netherworld realm of Mictlán," the second realm in the afterlife which was believed to be a parallel existence below the plane of earth that we see (Xocoyotzin). Reincarnation for such warriors, female and male, would be as hummingbirds.[190]

Furthermore, some butterflies, like the Blue Morpho of the Amazon, are gynandromorphic, meaning they are born with both male and female characteristics (Ogilvie). Gynandromorphy occurs in a wide range of animal species, but "affects about one in 10,000 butterflies," where the phenomenon is often stunningly noticeable because one half of the butterfly will be colored like a male and the other half colored like a female (Pavid).

Perhaps Mesoamerican observations of this dual-gendered phenomenon catalyzed the Aztec use of dual-genders in their rituals and political structures.

---

[190] I puzzled for a long time over why the Aztecs would consider hummingbirds and butterflies, seemingly unthreatening and beautiful creatures, as the spirits of warriors. Some scholars believe it is because hummingbirds are territorial, but that claim can be laid to most creatures who have a stake in keeping their nesting areas or food sources safe. What I discovered is that, like butterflies, hummingbirds will sip the juices of corpses or rotting vegetation when no flowers are in bloom. Such a taste for, and a lack of a fear of, death and decay would have been more likely to be the reason why Aztecs saw warrior souls as either type of creature.

Many depictions of humans enjoying the fourth realm, Tlalocán, also called the Mansion of the Moon because it was believed to be located on the moon, show those killed by thunderstorms (aka lightning) and water elements as skeletons enjoying food and fruits in abundance, so that Catholics who saw these depictions often assumed they were seeing depictions of "paradise" (Xocoyotzin), although depictions of skeletons doing things human beings do, like having sex, occasionally freaks out a scholar or two.

Tlaltecuhtli in all her gory detail, then, is not a "monster," but she is a woman in the pains of childbirth; for the Aztecs, **she is a woman at war**. She's fierce because she is battling to extract a whole other human being not only from her own body, but also from another world. She bleeds because birth, and its preparation, menstruation, is a bloody business; for the Aztecs, these bloody moments are moments of intense battle. The jagged knives are symbols of the pains she endures. The snakes, crocodiles, and caimans she is often depicted with are also symbols of birth and motherhood, with crocodiles and caimans being some of the most self-sacrificing mothers of the animal kingdom, carrying their young in their mouths for weeks before eating again, lest they swallow their own children.

While we cannot know exactly what each Aztec believed about childbirth or the importance of menstruation, we can guess at early influences that helped form their cultural beliefs.

Many people note the similarities in how various pyramidal structures are built around the world, failing to recognize the universal symbol for female pudenda, which has also become a universal scientific symbol for transformation. If we take the Aztec ritual and remember that the pyramidal shape is symbolic of the "mound of creation," what we get is a ritualized re-enactment of menstruation—that magical blood spillage that creates all human life—or even of birth. As the priest removes the beating heart, killing the sacrificial victim, he enacts a kind of birth which not only feeds the sun god or the earth goddess the blood s/he requires, but also frees the soul or *yolia* of the person sacrificed. All sacrificial victims' *yolia* were believed to go to the land of the gods or "'the place unknown'" (Townsend 15 & 35), probably Ilhuicatl-Tonatiuh, with some people volunteering or even demanding to be sacrificed.[191]

Just as a woman bleeds her monthlies for many years in preparation for pregnancy, and bleeds during childbirth, birthing a soul into a realm of the afterlife for the Aztecs also required blood. Many Mesoamerican cultures imitated this rebirth blood using cinnabar, just as other cultures use ochre, sprinkled over "high status burials from Teotihuacan and Palenque" (Emery).

Further evidence that the women were more spiritually powerful than the men comes from the fact that the Mexica leaders "sought out princesses of pure Toltec descent as their brides, so that they could inherit the divine right to rule, which belonged to the descendants of Quetzalcoatl" (Barroqueiro). Just like rulers of early Egypt, the Mexica male leaders married females believed to descended from gods in order to become more godlike themselves. These ancient traditions lent themselves to "exchanges" of royal women[192] between male rulers, resulting in a narrowing band of genetics that often resulted in children with birth defects from the physically obvious problems, such as facial asymmetry, cleft pallet, bone or muscle malformation and low birth weight, to the less obvious effects, such as infertility, increased risk of miscarriage, slow growth rate, and increased infant mortality.

While elongated skulls shaped through physically wrapping infants' heads became a fashion statement among Mayan elites, the origin of the practice could have come from genetic defects because of incestuous relationships in order to keep "royal blood pure."

We know the elites used marriage to seal political alliances with the women moving to live in their husband's cities, but how long was this tradition practiced? Did the common people also marry this way? We know that the Mixtecs, a nearby Nahuatl culture, allowed the couple to decide (Jansen 22), but we also know

---

[191] The legend of Shield Flower among the Mexica has her painting herself black and white (the colors of death) and demanding, "'Why do you not sacrifice me?!'" (Townsend 15).

[192] Sometimes these royal exchanges meant men married women who were closely related to themselves. Such an incestuous practice is an indication that the people knew little to nothing of the biological connections between coitus and procreation.

that prominent goddesses, such as Xochiquetzal who "incarnated youth, love and beauty...was amorously pursued by several Aztec gods," marrying one, Tlaloc, and becoming the lover of another, Tezcatlipoca (Flood), had more than one reported lover. Might Aztec women also been able to "marry" more than one man at a time?

While Sahagun's Christian-steeped accounts in the Florentine Codex indicate that monogamous marriage occurred, with dramatic ritualistic words spoken after a meal shared by the husband-to-be's teachers and elderly family members, he adds something of his own opinion to the claim: "'although they were heathens,...the Mexicans were not without good customs'" (Soustelle 174-5).

Instead of just focusing on the elite of a culture, which gives us a skewed image of what the overall culture was like, we need to focus on the experiences of the common people. The Florentine Codex finally relates that the elaborate and expensive wedding festivities described for affluent Aztec couples did not always happen, so that among the "plebeian" population, as Sahagun calls the poor, couples merely "went off together" to establish their own homes, but the codex is not clear where the new couples live, implying that they live—as is customary in a patriarchal culture—in the man's parents' home (Soustelle 178), something Inga Clendinnen corroborates by stating that "married sons usually bring their new wives back to their father's compound for the first years after marriage," but she emphasizes that the Aztecs practiced endogamy—marrying within their own relative group—so that "wives usually remained within easy visiting distance of their kin" (58). In fact, Clendinnen's descriptions of a family's "compound" which comprised several rooms in a square around a central courtyard, demonstrate how freely family members could move into and out of the family zone, around which women's work—performed largely in the family courtyard—was front and center of family life. The fact that the women then sold their produce and crafted goods at the marketplace means Aztec women had a lot more freedom than most Spanish women would have had during the same historical period.

Jacques Soustelle goes into great detail about how Aztec men were allowed as many wives as they liked, but he emphasizes that the first wife was always the primary or legitimate wife, which is unlike marital customs among many American Indian nations of North America where a woman was entitled to divorce her husband at almost any time and without nearly any form of ceremony to take another husband. Soustelle insists, "there is no doubt that the half-barbarous tribes which came from the north were monogamous," reluctantly admitting that polygamy was possibly customary among the Toltecs (179).

Especially informative are their burial practices. Like many other pre-androcentric cultures, the Aztecs sometimes buried dead infants beneath their homes, demonstrating that ties to a family house, thus to mothers, were stronger than ties to culture.

Many people assume women are not violent, not ambitious, and not hierarchical, but these people probably have never seen a soccer game played by only 10-year-old girls. It is important that we recognize that the West and many parts of the East have been socially programming women to be passive victims for nearly five millennia now. So could women have led their people to see ritual blood-letting, even human sacrifice, as an important religious act? Of course.

After all, who could better convince others that dying is a form of rebirth than women? Or that any pain they would feel as death approaches is akin to the pains women feel while giving birth?

Women are inventive, often keenly socially aware, which can lead to manipulative behaviors, and, if convinced they have special powers to see, hear, and commune with the divine aspects of the universe, are just as likely as men to believe they have received those powers for a reason. Otherwise, we would have had no Joan of Arc, Hildegard of Bignen, or Ruth Ward Heflin.[193]

While the Aztecs certainly seemed to be on their way to developing a patriarchal culture, given their separation of men's realms and women's realms, women still held some power in their ancient cities, especially as producers of goods and as traders of

---

[193] Not me. Ruth Ward Heflin was a Christian prophet who predicted Jesus would appear on the Benny Hinn Show, and who could reportedly conjure gold out of people's teeth.

those goods in the city and town marketplaces. The Aztecs, just judging by the willing way the males who worked on the various codices in which they hoped to "preserve" their culture, seemed to eagerly embrace the full-on patriarchy introduced by the Spanish in the Mexican Valley. How much of their prepatriarchal ideas actually got preserved in those codices remains the big question.

Like so many early cultures, the Aztecs still displayed many concepts, especially in their religious beliefs and various stories, of a more gynocentric view of the world, including, especially, the menstrual-based ritual of human blood pouring down the face of their symbolic pudenda pyramids.

Chapter Twenty-One
Inanna: The Most Powerful Goddess
Who Never Conceived

Few people, including most scholars, ever think of goddesses as being powerful because they are not male, therefore are assumed to be inferior to all male gods.

This androcentric bias carries through much scholarship, especially that by scholars who have yet to consider that the patriarchy has not always existed, meaning that there was a discernible historical and pre-historical time period in which women enjoyed an elevated social, religious, and spiritual status before being actively oppressed and suppressed by men.

Some scholars have made inroads toward demonstrating that not only was it probable that early social groups of humans were, at the very least, matrilineal and matrilocal or, at the most, outright matriarchal,[194] but also provable that such societies existed—from ancient and historical groups like the Harappans, the Lycians, and the Lydians because of archeological and DNA data demonstrating that many such ancient people buried the dead near their mothers.[195] However, many well-educated scholars still approach early historical cultures (circa 3000 BCE) as though they

---

[194] Marguerite Rigoglioso lists "Jane Ellen Harrison, Karl Kerengi, and Lewis Farnell" among those who "posit that this older [social] system was matriarchal" (4).

[195] For instance, Timothy Earle notes that the rise of Warrior Societies in Thy created changes from group to single grave burials, which "probably represents a shift from clan [identity, which are matrilineally oriented,] to [male lineage] organization" (163). Mark Kenoyer corroborates similar findings, noting that burials in the Indus Valley were matrilocal with little evidence of hierarchies, and that trade with Mesopotamia began circa 2500 BCE. Further, Stiebing points out that Urukagina's legal "proclamation contains the first recorded reference to 'freedom' or 'liberty,' the Sumerian word *amargi*, (literally 'return to the mother') (46), demonstrating that the Sumerians firmly believed life comes from women, so that upon death we are returned to the bosom of the Mother Creator/Mother Earth, at which time we are "free."

assume the patriarchy was already in full swing by the time humans began recording the goods they produced and traded.[196]

However, we can easily disprove this bigoted notion as a false one by studying one goddess in particular: Inanna. Inanna appears to us in some of the earliest decipherable writings created. Countless prayers, hymns, rituals, and stories have come down to us from ancient Sumeria, giving us a comprehensive picture of a goddess who was probably the most intelligently powerful deity ever conceived of by human beings. What is more is the fact that she was also not known for her personal fertility or as a mother-creator. Instead, she was conceived of as being wholly about sexuality, including determining people's genders, and conflict, especially war, so she was not the typical earth goddess, but was, in fact, a sky-and-earth goddess, very much stylized on human observations of the planet Venus, which was known to the Sumerians as both the Morning Star and the Evening Star.

Many androcentrically biased scholars conflate the Sumerian goddess Inanna with Ishtar, who arises in Babylon some one-and-a-half millennia after Inanna's full bloom in Sumeria. But these two goddesses, while both over simplified as goddesses of "love and war" by most scholars, have important distinctions, with Ishtar demonstrating just how far androcentric bias, which probably did not begin its full rise until after humans could prove males contribute to procreation circa 2400 BCE,[197] had already

---

[196] One such scholar is William H. Stiebing Jr. Stiebing, like many others, believes animal domestication was enough to prove to human beings that males contribute biologically to procreation (28), while acknowledging, however, that the patriarchy did not fully arise in Mesopotamia until later, circa 2500 BCE at the start of the Early Dynastic III period, by his estimations (56).

[197] While some will, undoubtedly, see this approximate date and think of Archbishop Ussher's argument that the date of the "great flood" mentioned in the Judeo-Christian bible was 2349 BCE, this 2400 BCE date is based on human domestication of swans—the only monogamous animal ever domesticated by humans. Humans needed to study a monogamous animal with a short breeding cycle to note that **the act of heterosexual intercourse led to procreation**, so would not have had any verifiable confirmation that males had a biological role in procreation until then. I am aware that many believe knowledge of the biological connections between sexual intercourse and procreation occurred once humans domesticated herd animals. However, the gestation lengths and the propensity for having several sexual partners in a herd would not have been

begun to infiltrate and alter human perceptions of divinity. As Johanna Stuckey points out in her article, "Inanna and the 'Sacred Marriage,'" the sacred sex ritual, *qursu*,[198] performed in Sumeria to legitimize a king's rule becomes conflated with marriage,[199] with Inanna reduced from being a conduit via which the sacred *me*, divine knowledge, are passed from gods to mortals through sacred sexual rituals to becoming "around 2100 BCE...merely the spouse

---

considered verifiable evidence, and many historical cultures, like the Sumerians, who still herded animals were clearly unaware of the sex-procreation connection, clearly calling Dumuzi's semen "milk" in many of the early hymns and poems about the *qursu* ritual; instead, seasonal mating was seen in spiritual terms-- largely female in origin, since most herd animals are matriarchal--as omens for the promise of plenty and pleasure, rather than just as preparation for fertility. Too many scholars conflate "fertility" with a desire to create "plenty."

Jennifer Bell, linguist, who confirms the circa 2400 BCE date, believes that the Sumerian cuneiform for the word seed is proof that "the Sumerians understood procreation" approximately 4500 years ago, presumably because she believes the cuneiform marks depict coitus.

Further evidence that wealthy Sumerian men were asserting their power over others at this time comes from the Standard of Ur, which dates to circa 2400 BCE, and depicts a clear social hierarchy, with the wealthy males on top, livestock herders in the middle, and common laborers at the bottom on one side, and with the king being presented with prisoners at the top, soldiers in the middle, and war wagons carrying supplies and booty at the bottom on the other side (Standard).

As Flannery and Marcus affirm, although these dates are problematic for the Americas, "By 2500 BC virtually every form of inequality known to mankind had been created somewhere in the world, and truly egalitarian societies were gradually being relegated to places no one else wanted" (x).

[198] The *qursu* ritual was originally part of an elaborate New Year festival that began on a designated new moon in what we now call February or March (much like the Christian holiday of Easter). "On the day of the 'sleeping' of the moon, a 'sleeping' place was set up for Inanna…, a bed whose rushes were cleansed with cedar oil, which must have made a satisfying [and fragrant] red wash. The bed, cleaned of [sacred] 'blood,' was then covered" with a sheet dyed to resemble lapis lazuli "to make it safe for the king" (Grahn 201).

[199] After the *qursu* ritual, the divine couple "did not then live together, for this coupling enacted the union of the male and female meta forms, the Bull [of Heaven] with the Morning Star, not the establishment of a new family" (Grahn 202). So what began as a sacred ritual to attune humanity with the gods in the heavens became, as Steibing admits, the view that the Inanna/Dumuzi *qursu* ritual as a form of sacred marriage "was due in large part to the development of kingship and the full attainment of male domination within Sumerian city states" (52).

of the god she served and/or consort of Dumuzi." Many scholars willingly overlook the fact that Inanna never conceives or gives birth[200] in any of the tales about her, despite all the sex she was believed to be having (Assante 39) because they want to overlook the fact that heterosexual coitus and procreation were not yet linked as part of human knowledge at the time. In short, due to the rising androcentrism of the time, Inanna transforms from the most powerful deity envisioned by human imagination to a sidepiece for a god, creating the "institution" of marriage as a legal status as part of the process for subjugating women beneath men.[201]

---

[200] Samuel Noah Kramer's interpretation of "Inanna's Descent" tale, which he titles, "From the Great Above to the Great Below," implies that Inanna has two sons, Shara and Lula (Wolkstein & Kramer 70-1), but, even though the poem may call them her sons, the writer of the poem could simply be referring to them with honorific titles, just as Inanna calls so many male gods "father," even though only one could possibly be her biological (?) father, if the ancient Sumerians understood procreation at that time. I also have to admit that the cuneiform symbol for mother, *ama*, includes several symbols evocative of Inanna—from breasts, which in this case point down, to her so-called standards or gates, which are usually depicted vertically, possibly symbolizing Inanna's vulva and heterosexual coitus, of which she is so proud, and a rosette, which is often used alone as a symbol that Inanna is all-present as both the Morning and Evening Stars. The most simplistic explanation for the *ama* ideogram and its clear associations with the goddess Inanna is that it could symbolize her power to determine the gender of all babies that are born, since the Morning and Evening Stars are "born" from the earth (from an ancient perspective) at different times of the year, spring and summer for the Morning Star, autumn and winter for the Evening Star.

[201] Denise Carmody agrees, arguing that as males' role in procreation became clearer, religious practices changed, leading "to the heterosexual, hierogamic (sacred marriage) practices that were prominent in fertility cults like those of Canaan, against which the Hebrew prophets inveighed in the eighth century BCE. By that time, plowing, sowing seeds, watering and the like were assimilated to human fertilization. Then the farmer (increasingly male?) helped Father Sky fertilize Mother Earth, extending to Mother Earth the solicitude he would offer a pregnant woman" (40-1). Erik Hollager corroborates, pointing out that the goddesses Diwia and Posidaia preceded the gods Zeus and Poseidon, who eventually took the place of those goddesses, in many of Linear B tablets (271). Interestingly, Marguerite Rigoglioso argues that "the ancient Greek title for 'king,' *basileus*...in an earlier form on Mycenaean-age tablets, was, in fact, a derivative of *basile*, an ancient queen's title and the name of a goddess'" (15). However, William Stiebing Jr. exhibits androcentric bias when he reiterates archaeological biased assumptions that all the animals depicted in Çatal Huyuk were "bulls and

Inanna deserves to be examined fully on her own, although this examination can be challenging because, quite often, Victorian androcentric social programming gets in the way of finding fair or accurate interpretations of what the ancient Sumerians meant by many of the words they used in the texts they left behind. Our hope remains with the thousands of untranslated Sumerian texts that earlier, more complete texts referring to Inanna will be discovered.

First, however, we have to address exactly how the ancients view the stars as gods, since Inanna was always associated with both the Morning Star and the Evening Star, which we now call the planet Venus, named after another goddess of sex and war held in high esteem by the Romans, thousands of years after Inanna ruled the skies and earth. Where we merely see stars as distant suns now, the ancients had no idea what those distant vibrant lights were, with most ancient cultures attributing stars to ancestors, or as Anthony Aveni suggests, "luminaries" (13), watching over humanity from the heavens. We undoubtedly owe many different myths about the trysts of different gods from the fact that two of these luminaries became "conjoined" or had a conjunction as a particular notable time, which then becomes either a story about sex or conflict.

The Sumerians, like many ancient cultures, were keenly interested in the stars and fully believed most of them either to be gods or ancestors who were both responsible for determining the fate of humanity on earth. One reason they valued the stars as divinities was because the regularity of the starry skies signaled important changes on earth when those regularities changed. From watching the stars and monitoring their movements, human beings began to predict many things—from time basics, such as the passing of seasons, to when they would be successful at various endeavors, from coitus to war.

As Anthony Aveni describes the importance of astronomy to ancient cultures:

---

rams" (19-20) without question, even though the images are clearly inconclusive and could be all female. His assumption is based on the faulty belief that, once humans began herding animals, they immediately discovered that heterosexual intercourse is required for procreation. This androcentric bias has clearly discolored research in many fields.

the ability to predict the future—to foresee the whereabouts of the luminaries who ruled the sky and the earth—gave prestige, respect, and authority to the adept who bore the gift. Strange but true: Whole cities, kingdoms, and empires were founded based on observations and interpretations of natural events that pass undetected under our noses and above our heads (12).

In his examination of Sumerian, Mayan, and Incan astronomy, Aveni demonstrates how important the planet Venus (Inanna) was in those cultures' predictions and the power those who could "read" the stars engendered from being able to make successful predictions. For instance, he quotes a Babylonian omen text, which states, "'When Venus stands high, pleasure of copulation. When Venus stands in her place, upraising of the hostile forces'" occurs (Aveni 185). This omen, of course, plays fully on the two main attributes of Inanna: sex and war.

Understanding the Venus Cycle, then, is paramount to beginning to understand the powers of Inanna as conceived of by the ancient Sumerians, and possibly by the ancient Ubaidians before them. Observing the planet's movements over extended periods of time, early astronomers noted how this particular luminary moves in predictable cycles. As the Morning Star, this third brightest object in our nightly skies appears in the pre-dawn eastern skies for about 263 days[202], then disappears for 50, before reappearing as the Evening Star in the western skies just after sunset, disappearing once again for eight[203] days before starting the cycle over again. That cycle is Venus' most simple cycle, easily observable over the course of 584 days, her synodic cycle. This cycle also explains the origins of the myth, "Inanna Descends to the Underworld," because her regular disappearance from the night sky

---

[202] About the same length of time for the Mayan *tzolkin* or Sacred Round, which is one important part of their ritual cycle.

[203] Note that Inanna is usually associated with the number 8. Her famous rosette pattern is actually an 8-spoked star. When I was conducting research on the Lakota Star Pattern two decades ago, I traced the origin of that now famous quilt pattern back to Iraq in order to understand how ancient the symbol is, but I was completely unaware that it belonged, originally, to Inanna. The Lakota also call it the Morning Star quilt pattern.

had to be accounted for in some way. More on that important tale soon. Because Inanna, the luminary appears to be "chasing" the sun, myths explaining her seeming attraction to the sun, another god named Utu, are also common, with Inanna, Utu, and Inanna's moon god "father," Sin, forming a sacred trinity for the early Sumerians.

But the planet Venus' cycles do not end with her four-phase synodic cycle. Another, longer cycle also occurs, which I will call her Octo Annos Cycle, since it does not appear to have a formal name among astronomers. Venus' regular movements through the sky allowed humans to predict actual 8-year cycles, *Octo Annos* in Latin. The Octo Annos Cycle allowed ancient astronomers to predict what phase of the moon would occur at any time along that eight-year period. As Aveni states,

> Whatever phase of the moon accompanies the first appearance of Venus, say at the June solstice in 1997, that same phase will repeat again at about the June solstice eight years later, in the year 2005, signaling the return of Venus (43).

Perhaps most interestingly, Aveni notes that cultures that use astronomy to predict future and to explain past events, such as the Asian Indians or the Maya, also date back their creation myths to a very similar date, circa 3100 BCE (43), a date many scholars use to mark the beginning of the Bronze Age around the Mediterranean Sea.

When we recognize the power inherent in being able to predict events on earth from reading the "luminaries" or gods' movements in the sky, we can easily understand how such religious leaders could rise to prominence in any culture. These prophets seemed to understand a power not always available to ordinary human beings, although there are still cultures in the world that use astronomy to determine when to plant, to hunt, and to harvest.

Second, though, we must acknowledge that early Sumerian society—while initially more egalitarian than it will be later because the people elected their leaders to serve on two councils—an elder council and a "free citizens" council (Stiebing 53) composed, Kramer assumes, of elderly "conservative" leaders (History 30)—was already becoming increasingly hierarchical. As with most ancient cultures, the people who claimed the ability to commune with the gods wielded the most power and controlled distribution of many of the goods, especially after communal storehouses became the purview of the temples. As William H. Stiebing puts it, "The good will of deities who control nature was obviously important to early societies, and, in Mesopotamia, by the latter stages of the Ubaid Period, towns seem to have been temple-centered. This suggests to some that priests who controlled the religious rites had become dominant" (27). Stiebing also admits that from "the Early Dynastic III Period (circa 2400 BCE) onward, Mesopotamian society was patriarchal and only a male could become an *en* ('lord,' a priestly office), *ensi* ('steward,' originally a local secular ruler and later a governor), or *lugal* ('great man,' i.e. 'king')" (56). While he is being generous in calling Mesopotamia a full on patriarchy by this point, the time period does coincide with the discovery of procreation by the Egyptians. In other words, Stiebing is correct to note that Sumeria had become purposely androcentric heavy by 2400 BCE. As Laura Culbertson argues effectively, many scholars have a problem with calling even Ur III, which is circa 2100 BCE, a full on patriarchy simply because women are not as visible in the historical record as men are (115), stating that "a clean bifurcation of public and private spheres along male and female gender lines is outdated" (117).

To understand what the Bronze Age push toward androcentrism and full on patriarchy cost humanity, however, we must fully examine Inanna and the sacred nature of Female Magic prior to 2400 BCE.

The best summary of Inanna's peak powers must include the fact that she is all-powerful, even though as androcentric bias rose many of her attributes were parceled off to male gods or to lesser goddesses. Enheduanna (2285-2250 BCE (Mark), the world's earliest identifiable writer and the High Priestess of Inanna's Temple Eanna[204] in the city of Ur, wrote dozens of poems, hymns, and songs glorifying Inanna's greatness, even though the culture was already becoming androcentric heavy.

---

[204] William Stiebing claims "the temple of the goddess Inanna at Nippur goes back to the Uruk period[;] there is no evidence for an Enlil Temple there before the Early Dynastic II period," circa 2600-2500 BCE (45).

One poem, hymn, or performance prayer written by Enheduanna, alternately titled "Goddess of the Fearsome Powers" or "Inana[205] and Ebih," begins with Enheduanna enumerating some of Inanna's most fearsome powers, presumably announcing the actor who will perform the role of Inanna in this drama:

> Goddess of the fearsome divine powers, clad in terror, riding on the great divine powers, Inana, made complete by the strength of the holy *ankar* weapon, drenched in blood, rushing around in great battles, with shield resting on the ground (?), covered in storm and flood, great lady Inana, knowing well how to plan conflicts, you destroy mighty lands with arrow and strength and overpower lands (Faculty, "Inana and Ebih," ll. 1-6).

This poetic drama then allows Inanna voice to describe how she was disrespected by a mountain named Ebih. Inanna details how she will exact her retribution by using battering rams, arrows, fire, and axes to bring the mountain low.[206] Tellingly, Inanna equates the destruction she will wreak with the cities the male gods An and Enlil[207] have already cursed and destroyed because the male gods, of course, had to be given credit for doing first what Inanna is going to do in this performance hymn, even though we see mirrors of powerful, destructive female gods in the cultures surrounding the ancient Sumerians.[208]

---

[205] Scholars are not always consistent with the spelling of Inanna's name, so I will use their spelling when I use their words.

[206] Jeremy Black, in his book *The Literature of Ancient Sumer*, says the mountain Ebih is "tentatively identified with the Jebel Hamrin range in modern Iraq," and that "mountains could also be portrayed as areas of safety" for nomadic tribes threatening invasion (334), so that Inanna is routing these invaders in this poem.

[207] Enlil will become the Sumerian male god who eventually becomes the only god who can confer kingship on humans, taking over Inanna's role without the *qursu* ritual.

[208] Probably the nearest, most fiercesome goddess of war next to Inanna is the Egyptian goddess of war, Sekhmet, whose name is Egyptian for power. She appears in most tales as a lion-headed woman dressed in red, which most

While High Priestess Enheduanna is a woman in a socially, religiously, and spiritually powerful position, she holds her position in Uruk **after** humanity verified that heterosexual intercourse creates babies, and men, like her father Sargon the Great, are rabidly seeking power and social ascendency. Thus, Enheduanna willingly has Inanna show deference and obeisance to the male god, An, who will, for a time, head the Sumerian pantheon saying, "King An, you have indeed given me all this [power], and you have placed me at the right hand of the king" (ll. 29-80), presumably placing Inanna in the obsequious role of favorite god to the king of the gods, a role Enheduanna undoubtedly played in parallel form with her father Sargon. However, the male god An opposes Inanna's desire to destroy the mountain, claiming, "You cannot pass through its terror and fear. The mountain range's radiance is fearsome. Maiden Inana, [sic] you cannot oppose it" (Faculty, ll. 127-29). Perhaps this advice and the accompanying diminution of her status as a mere "maiden" is what Enheduanna heard from Sargon, her father, when faced with foes or when she tried to oppose her father's wishes. Such a diminution of a woman's apparent power by someone who assumes he has greater power acts as a form of pressure meant to keep her from acting under her own power. However, Inanna, and, we assume, Enheduanna, did not always obey her "superiors."

Separating Inanna, the goddess, from the High Priestess Enheduanna becomes tricky because there are many similarities, such as when Enheduanna in her own role as Inanna's High Priestess faces "an attempted coup by a Sumerian rebel named Lugal-An...[which] forces her into exile" (Mark), because both Inanna and Enheduanna refuse to bow to the will of the males in question. Inanna, in the poem "Inana and Ebih," not only kills the mountain against An's wishes, but also brings it "low" (Faculty, ll. 159), possibly explaining why a mountain's physical elevation has been reduced by an earthquake. Enheduanna uses Inanna as a further vehicle for change, however, having Inanna state, "I have

---

scholars assume is the blood of war, but, more likely, is the sacred moon-blood of women, especially when we consider that she often wears a blood red moon (often identified as a sun) disk encircled by a cobra, another female symbol. Like most early Mother Goddess figures, she is associated with lions—which are matriarchal, allowing male lions into their pride for sexual purposes.

built a palace and done much more. I have put a throne in place and made its foundation firm. I have given the *kurjara* cult performers a dagger and prod. I have given the *gala* cult performers *ub* and *lilis* drums.[209] I have changed the headgear of the *pilipili* cult performers" (Faculty, ll. 171-175), possibly giving a religious explanation for why Enheduanna in her new role as High Priestess has made changes in how the Eanna Temple operates, disguising her own desires as those of the goddess.

Similarly, Enheduanna and her father, Sargon of Akkad, might have used Inanna to explain a probable course of action Sargon used to control his newly dominated people. As herders of goats, sheep, and cattle, the Sumerians would have known that castrating a male animal at the right time makes that male grow bigger and stronger than he would have normally been, as well as more manageable (Bartosiewicz 332), so the act of castrating human males captured in battle would have been known to have similar effects—thereby allowing conquerors to keep men and their specific skills, while making certain those men would not foment rebellion. After human beings learned that babies are made by the physical union of both genders,[210] some Sumerians could have seen such castration as a way to control which men could father children, since that biological confirmation was probably made just a century prior to Sargon's rule. In her hymn alternately titled, "The Great-

---

[209] A *lilis* drum is a kettle drum, according to The University of Pennsylvania Museum of Archaeology and Anthropology, but I can only guess that the *ub* is a form of flute carved out of precious stone. The *gala*, literally interpreted as "heart pacification," were special musicians who "were sexually ambiguous" who played music of mourning and spiritual praise (Stewart). According to linguist Jennifer Ball, however, "*gala* meant 'female lamentation singer' **and** 'vulva'" (emphasis hers, Breasts), although James Barrett-Morison interprets *galla* as a vulgar slang term for vagina (Sumerian Language).

[210] Jennifer Ball, an ancient language linguist, believes "we know that circa 4500 years ago, the Sumerian understood procreation, based on this character for

seed" (About): *numun*, which is a compound of *nu*, meaning man or male genitals (and night bird), and *mun* meaning moon and woman. For Ball, the cuneiform sign appears to be a depiction of coitus, so is the first real written sign we have that the Sumerians understood the connection between heterosexual intercourse and procreation circa 2400 BCE.

Hearted Mistress" or "A Hymn to Inanna," Enheduanna "praises Inanna for her gifts of desirability and arousal and notes how she has the power to 'turn a man into a woman and a woman into a man'" (line 121), alerting us to the fact that transgendering—whether through forced castration or actual biological complications which can give people both or no clear sexual organs—was not unknown 4300 years ago.

While many scholars agree that "Inanna's temples and attendant rituals were administered by clergy of both sexes and her devotees were noted for their habit of cross-dressing, blending, blurring, or eliminating the distinction between male and female in pursuit of transcendence through Inanna" (Mark), recent DNA archeology discovered that, roughly in the same time period, 17 women were producing children for every one man who contributed his DNA (Diep). The computational biologist, Melissa Wilson Sayres, who was part of the ancient DNA research group, admitted that "it wasn't like there was a mass death of males. They were [still] there, so what were they doing?" (Diep).

It is highly likely that the reduction of males fathering children is a direct result of the rising elitist androcentrism with one man or a group of wealthy, powerful men being seen as preeminent over every other human being. As Peter Tomkins put it, "Failing households could have been less certain of communal protection and their land, labor, and livelihoods were more open to interference and appropriation by successful households seeking to transform short-term gain into longer-term competitive advantage. ...Some failing households may have opted to stay, perhaps indebting their land or labor"—selling themselves as slaves (40).

Ancient Sumerian males, we know from later written laws, could legally sell their wives and children, not to mention themselves, to pay off their debts, but becoming an indentured servant or an actual slave was not the only way men with influence limited poorer men's access to sex and their chance at progeny.

Sargon of Akkad rose to power and influenced Sumerian life in many ways, but his primary influence, for which he is deemed "the Great" by so many historians, was his formation of the first known standing army with which he conquered most of Mesopotamia. His creation of a standing army was an innovation that helped him accomplish several things, such as:

- control of the young men who were "extraneous"--not needed in the agriculture fields because so much of crop tending had become formulaic enough that workers were only needed for plowing, sowing, reaping, and maintaining irrigation canals; not as many workers would be needed for plowing, once cattle were used to pull plows;
- direct the young men's energy, restlessness, and aggression toward war by promising them war booty;
- remove many young men's sexual pleasure as another way of controlling them through forced castration, probably claiming that Inanna, the goddess of gender, had deemed such actions necessary;
- conquer 65 cities, so create the first verifiably force-united empire in the world.

Sargon probably began castrating, not just his troops in an attempt to make them larger and stronger as well as to have a means of control over them (remember what castration does to male cattle), but also the men he captured from the many cities he conquered to unite Sumeria into what became the Akkadian Empire, probably proclaiming that Inanna willed this gender change[211] for the men, which would have made such mandates doubly effective because the king would be seen as ordering these castrations after having been advised by the reigning goddess. After all, he wanted the men's technical expertise, but he probably wanted to dominate their lives as well as to control their sexual activity, especially since sex meant so much to the ancient Sumerians.[212] The

---

[211] While it is relatively easy to remove an animal's testicles by splitting open the scrotum, most ancient castration of humans would have involved lopping off the penis as well as the testicles, effectively removing each man's "manhood" to transform him into a genderless being.

[212] Sex was not only a social event in the many taverns in ancient Sumerians cities (Assanti 31) with depictions of tavern beer drinkers sipping from straws seen as symbolizing fellatio and sipping from cups as cunnilingus (Assanti 35), but also many forms of sex were used for spiritual transformative purposes, which I will

fact that castration also limited what men could father children had to be an added bonus. Men who are dominated by a larger-than-life-leader often become obsequious,[213] and Sargon used the ancient *qursu* sexual ritual to have himself deemed, first, Inanna's favorite king, and, second, godlike, if not an outright equal to the gods. Sargon promoted himself as the religiously sanctioned king-of-kings in order to create in others a sense of immediate obedience.[214]

Some scholars believe that the poem, "Inanna's Descent to the Nether World" (c. 1900-1600 BCE), is the first clear reference to castrated males. Whatever else we take away from this particular myth, it clearly explains the planet Venus (which was Inanna to the Sumerians) and its synodic cycle, during which it disappears from the heavens for a time, so spends part of the year as the Morning Star and part of the year as the Evening Star, and part of the year absent from the heavens.

In the "Inanna's Descent," the Queen of Heaven and Earth decides to visit her sister, Ereshkigal in the Nether World ostensibly because she wants to attend funeral rites for Ereshkigal's husband, Gugalanna, the Bull of Heaven,[215] but it is far more likely that

---

discuss in more detail. Taverns in early Uruk were so popular that one of the first "kings" of the city was a woman, Kubaba, who ran a popular tavern.

[213] When, exactly, humans discovered that they could control male animals better after castration is unknown, but they might have realized that castration of male cattle at a particular time of their growth actually makes for bigger, stronger, oxen (Bartosiewicz 332), so the earliest castrations of warriors might have happened as experiments to see if the same effect happened in humans. As Kathryn Reusch points out, "the origins of animal and human castration may be linked" since "the first references to human castrates [occur] in the cult of Ishtar [sic] in Uruk" (10). Reush's source was Gary Taylor's book, *Castration: An Abbreviated History of Western Manhood*, in which he claims the genderless beings created in "Inanna's Descent" were castrates (177). Human castration in Sumeria would have undoubtedly had serious spiritual overtones because of the way Inanna is credited, repeatedly, with determining people's gender.

[214] As Earle notes, "By financing the construction of the irrigation canals (and retaining warriors to enforce restricted access), ruling institutions became the owners of the most productive lands" (72), which further cemented those wealthy elites' power.

[215] According to the *Epic of Gilgamesh*, Gilgamesh and Enkidu kill the Bull of Heaven, which Inanna sent to destroy Uruk because of Gilgamesh's rejection of Inanna's sexual attention. The ancient Sumerian tales about bull-gods living

Inanna, who knew she would end up a corpse, despite being a god, had planned ahead better than the earliest known version of this tale indicates, supplying her own means for resurrection.[216] Also likely is the idea that the original tale of Inanna entering the realm of the dead was to defeat (through outwitting) Death Herself, in order to give humanity the hope for resurrection, probably in the form of reincarnation, which becomes a running theme for the Sumerians, since Dumuzi and his sister also "die" and are resurrected annually. Just as Spring brings rebirth to the earth, Inanna's defeat of the Queen of Death brought the promise of an Afterlife to human beings, instead of their just ending up as corpses hung on a wall, like Inanna initially was. Later, Dumuzi's resurrection[217] will do the same thing, and become such a powerful metaphor for life-after-death that Christians will adopt the same springtime (and moon phase determined) resurrection of Jesus for the same purposes.

When Inanna enters the realm of the dead, Ereshkigal, the Queen of the Underworld, first strips Inanna of "the seven divine powers," all associated with specific items of clothing and jewelry

---

underground will not be the last such tale, since the Greeks will revive him as the Minotaur. However, the bull god is probably also the constellation Taurus, which disappears from the night sky from May to July, so this constellation would symbolically die (disappear from the night sky) annually, as well.

[216] As Judy Grahn points out, Inanna is killed in the Nether World, which is ruled by her sister, but she is only dead three days which are the number of days the moon "hides" its face from humanity each month (212), and the same number of days many women would seclude themselves (4) to keep their superior Female Magic from interfering with other people's magic. Therefore, this mythic tale also explains women's menstrual seclusion in symbolic form. Depending on how one counts, however, Inanna could be gone from the Heavens and Earth for at least 8 days, which mimics the actual synodic Venus cycle, with the planet literally disappearing beneath the earth into what the ancients thought of as the Underworld or Nether Region.

[217] Most symbolic rituals now associated with Christian Easter celebrations come directly from Dumuzi's annual spring resurrection rituals (as is reflected in the Old Testament, Ezekiel 8:14, when women weep for Tammuz, having baked cakes for the Queen of Heaven, Jeremiah 7:18), such as baking sweet breads and painting eggs, as well as depictions of fecund rabbits and ducks.

that she is wearing,[218] then hangs her corpse on a wall.[219] Following Inanna's omniscient, prescient instructions, Ninshubur, her minister, goes to several gods pleading for assistance to rescue Inanna. Male gods Enlil and Nanna both refuse to help, but Enki creates two beings from the dirt under his fingernails—the *kurgarra* and the *galatur*[220]—both are genderless creatures (Wolkstein & Kramer). Scholars, like Gary Taylor and Kathryn Reusch believe these genderless beings are the first written evidence of castration as a common social practice. Unfortunately, skeletal remains, since there are no known Sumerian mummies, cannot reveal whether a man was castrated during his life.

After Inanna travels across the earth in order to return to her royal realm in the Heavens, though, she discovers her lover, Dumuzi, has assumed her throne. Since servants[221] of the

---

[218] The fact that clothing and jewelry can be magical should be clear here, since the ancient Sumerians believed in the power of talismans to ward of illness and other bad things.

[219] Neti, Ereshkigal's minister, first takes "the lapis-lazuli measuring rod and measuring line" and a special turban, sometimes called the Crown of Heaven, Inanna was wearing before he allows her through the first gate of the Nether World (Faculty, Inana Descends, ll.129-133), then, in order, removes a set of "small lapis-lazuli beads" from her neck at the second gate (ll. 134-8); another necklace with twin egg-shaped beads at the third gate (139-43), a pectoral decoration "called 'come, man, come'" at the fourth gate (108-13); a gold ring at the fifth gate (149-53), probably mascara called "let a man come" is removed from her eyes at the sixth gate, but the Oxford Faculty have Neti taking the lapis-lazuli measuring rod again; and, finally, Inanna has to remove "the *pala* dress, the garment of ladyship" (ll. 159-63), so that she is naked and powerless before her sister. All of the lapis-lazuli items clearly associate Inanna with the nightly heavens.

[220] Notice the morphemes, "*gala*" and "*kurg*," which also appear in the poem "Goddess of the Fearsome Powers" or "Inana and Ebih" in relation to specific religious cults within Inanna's Eanna Temple as *kurjara* and *gala*. Perhaps, as cis males insisted on ascending in order of importance in ancient Sumeria, they found any males not clearly masculine (whatever their standards for masculinity were at the time) worth less than the dirt under their nails? I predict that, once every tablet from ancient Uruk is found and read, we will find one that gives an earlier version of this important myth wherein Inanna creates these genderless creatures, since she was the goddess of gender, after all, and ordered them to come rescue her before she descended to the Realm of the Dead.

[221] Scholars exhibit their Christian bias when they call these beings "demons."

Netherworld have followed her from their realm, demanding she replace herself with a substitute,[222] Inanna condemns Dumuzi to six months in the Underworld, which brings about a ritual cycle based on the seasons—autumn and spring. Spring[223] is when Dumuzi is resurrected with celebrations of sweet cakes and sex with Inanna in the annual *qursu* ritual. Like Imbolc for northern Europeans, the Spring celebrations undoubtedly included lots of sex for commoners, too, since the ejaculate would magically bring spring rains, thus the rapid growth, like penises, of plants.

The desire to control the sexuality of the population undoubtedly increased as androcentrism grew into a cultural norm, eventually culminating in the social experiment called patriarchy. To act on his god-like power to create and to control, Sargon and his city-king underlings created laws to control their citizen's sexuality. While we cannot prove exactly when the concept of marriage arose, some of the earliest recorded laws regarding matrimony clearly demonstrate that males were given the largest portion of rights, since most marital laws denoted that sanctioned marriages were legal contracts between the woman's father and her prospective husband, with the men having sole rights to the woman's body, a right many men apparently abused. Even in the *Epic of Gilgamesh*, the story begins because Uruk's city elders are unhappy because Gilgamesh is having sex with anyone and everyone he takes a fancy to—either male or female.[224] The gods respond to the city elders'

------

222 The Hittites will emulate the Sumerians in their spring rituals, but their medical magic included the "substitute for death" ritual wherein an appropriate substitute for someone dying from an illness was sacrificed to "the Queen of the Underworld" in order to appease her and allow the ill person to live (Gurney 55).
223 According to Kim Ann Zimmermann, the Sumerians saw Aries, the ram constellation associated with Dumuzi, the shepherd god, rise at the vernal equinox, "but since Earth's rotational orientation gradually changes over time, the vernal equinox has shifted, and the [Aries] constellation now falls within Pisces." D.S. Raevsky points out how Dumuzi was the original Sumerian name for the Aries constellation, but he was renamed by the Greeks (303-4).
224 A significant difference between the Sumerian epic and the Babylonian version is that the Sumerians have Gilgamesh having sex with any young person, male or female, he wants, whereas the Babylonian version has him practicing what is, by then, known as the "right of kings" to have sex with any bride before her groom does.

complaints by creating Enkidu, a beast man who is transformed[225] into a human being through *hashadu* or sacred sex with a priestess, a "taboo" woman only a few were allowed to touch but who is often misinterpreted and derided as a harlot, but who was undoubtedly a woman trained in such transformative forms of sexual intercourse through the Eanna Temple. While Enkidu eventually battles Gilgamesh, thereby subduing Gilgamesh's strong sexual drive, sex in all its versions was important, possibly even vital, to the ancient Sumerians because of its transformative powers. Listed among the sacred *me* that Inanna gifts to the people are the sexual acts of fellatio and cunnilingus (Kramer).

There are few aspects of life that are more powerful than controlling who can have and enjoy sex, especially when sex meant more to individuals in the ancient past, especially to their spirituality, than we modern humans usually give it credit for.

Like the eunuchs who chose castration in order to be among the highest ranks of people close to the Chinese emperors, it is not inconceivable that many Sumerian men also chose castration, if it allowed them similar powers and access to sacred spaces often reserved for women.

We also know that Sargon encouraged the city-kings under his command to enact laws restricting women's sexual activity, such as: allowing men to sell their wives to pay their debts which gives wealthy men more sexual access to more women; encouraging marriage as a legal contract between men, so that a woman' father entered a contract with her husband with little to no say by the woman regarding the relationship; outlawing polyandry for women;[226] and allowing men to kill their wives if they are found having sex with other men. While some scholars have bent and twisted logic to try to argue that these laws are not evidence of the suppression of a previous gynocentric cultural view wherein women were allowed to choose their lovers at will, they have yet to disprove the arguments supporting such a previous cultural norm. Laws, it would seem, were first written down to finalize arguments about

---

[225] Sex in any form was seen as a transformative experience—transforming beasts into men and men into kings, but also used widely as a form of healing.

[226] Urukagina of Lagash made polyandry illegal in 2375 BCE; the punishment for a woman violating this law was death.

who had rights and who did not, with women and poor men clearly on the losing side.

Another early Sumerian poem, "Inanna and Enki" (c. 2112 BCE; 300 years post biological revelation of fatherhood), tells us that Inanna had to steal her powers from Enki. However, we seriously have to consider how a growing preference for males with power would have influenced such a poem to explain why Inanna had so much power at her disposal. Androcentrism dictates that Inanna had to steal the powers she possessed using her feminine wiles and alcohol because she could not have come by them naturally as a female god; at least these circumstances are what the poem seems to want us to believe.

Inanna, though, probably not only controlled the sacred *me* originally, but also probably was originally credited with bringing them to the people of Sumeria from the gods, which was why the *qursu* ritual was the important New Year ritual. As Krystyna Szarzynska points out, symbols "of Inana [in her original three forms] occur frequently in the archaic Uruk texts" (7), but "cannot be found in the later Sumerian religious texts" (9) as androcentrism rises.

Enki's cult arises after humanity learns that men play a role in procreation. He is the male god from whom Inanna reportedly stole the sacred *me*,[227] but he is only credited with this gift because he "already knows the intentions of the gods" (Faculty, Inana's Descent, Segment B, ll. 6-15)—despite the fact that he appears as a god and is rapidly elevated in status among the gods at a time when "the major gods became national gods, identified with narrow national political aspirations" of the men in power (Jakobsen 21).

Enki has that one attribute that Inanna seems to have, but which has yet to be directly or indisputably attributed to Inanna in any of the translated cuneiform documents preceding "Inana and Enki": omniscience. But even the poem "Inana and Enki" calls Enki's supposed omniscience into question, since he repeatedly has to ask his minister, Isimud, where the Boat of Heaven, in which Inanna has escaped with the sacred *me*, is at any given moment

---

[227] Goddesses were refashioned to have to "steal" powers they had probably originally been believed to possess outright. Inanna has to "steal" the magical *me* from Enki, and Isis has to steal Ra's magic to resurrect her husband, Osiris.

(Faculty, Segment H, Inana and Enki). A truly omniscient god would know such things without having to ask. In fact, Enki makes Isimud talk to Inanna as she progresses on the Boat of Heaven down the river toward the sacred city of Unug Kulaba, and Enki pushes his minister to try to capture the Boat of Heaven at least six times, but their attempts, even though they call on various magical creatures to assist them, are thwarted by Inanna's logic and magic at every attempt.[228]

Inanna's logic is impeccable as she confronts Isimud, who insists that Enki's words "cannot be countermanded." Inanna's logical response is that, if Enki's word cannot be countermanded, "How could my father have changed what he said to me? How could he have altered his promise as far as I am concerned? How could he have discredited his important words to me? Was it falsehood that my father said to me, did he speak falsely to me? Has he sworn falsely by the name of his power and by the name of his *abzu*? Has he duplicitously sent you to me as a messenger?" (Faculty, Segment H, ll. 8-19). If Enki's word is so powerful that it cannot be countermanded, then even Enki himself would be powerless to countermand a promise he had previously given. He is, after all, only credited with being omniscient, not omnipotent. Perhaps this poet wants us to understand that Enki is not as powerful as some with androcentric bias try to make him seem? Or perhaps the original explanation for Inanna's powers still holds sway in a culture being asked to believe something new.

I understand that many scholars will assume, because of their androcentric biases, that biological fatherhood was already understood by the time of this poem, which it probably was. However, the term for "father" used in "Inana and Enki" is undoubtedly much older than humanity's knowledge of the physical and biological aspects of procreation because Inanna's biological father (if a god can be said to have such a thing) is not Enki; her

---

[228] According to Segment H of the poem, Enki orders the *enkum* to take the Boat of Heaven first (ll. 1-7), then the fifty giants of Eridug (ll. 34-41), then the fifty *lahama* of the subterranean waters (ll. 68-75), then "all the great fish together" (ll. 110-121), then the guardians of Unug (156-69), then seems to command the Surungal canal itself to stop the Boat (ll. 191-205); lines taken from Faculty, Segment H).

father is the god who is one important phase of the moon, the new moon phase, named Sin.[229] In fact, the use of this appellation demonstrates how androcentrism colors scholarly expectations and translations quite well. The term "father" was an honorary term, much as we still call people "aunt" and "uncle," even when they are not biologically related to us. Many people still call males who are not biologically related to them "father," especially when those men are Catholic priests. We use these terms, which we now consider familial labels, out of respect and/or out of the individuals' places within our extended family or our culture, such as when our Aunt Nelly or Uncle Cleve are our parents' best friends even though they have no biological connections to our family.

So, when Inanna calls Enki "father," she is using a term of respect, not indicating a biological relationship. In fact, Inanna repeatedly has to address the male gods as "father" in other poems to help push the authority of males above that of females: in "Goddess of the Fearsome Powers," she calls An father; in "Inana and Enki," she calls Enki father; and in "Inanna's Descent to the Nether World," she respectfully calls the gods Enlil, Nanna, and Enki all father. Clearly, these writers are forcing the most powerful goddess of their pantheon to pay obeisance to the male gods; or else the translators are insisting that she call these gods by a term of respect in order to elevate the importance of the male gods over this powerful goddess. I have a feeling that the latter is the case, since, not once, in any of the poems I have read, has Inanna been called "mother" or any other title of respect except queen and "princely" given to women in Sumeria.[230]

What is especially revealing about the androcentric bias so many scholars carry into their studies of ancient Sumeria is the fact that most will, sometimes reluctantly, acknowledge that the Boat of

---

[229] The concept of "sin" comes from things done during this dark phase of the moon, with the later Semites drawing heavily on Sumerian concepts, as has been argued elsewhere, including by Kramer.

[230] Krystyna Szarzynska, in her article "Offerings for the Goddess Inana in Archaic Uruk," lists Inanna's archaic forms as: "Inana-nun 'Princely Inana,' Inana-ud/hud 'Morning Inana,' and Inana-sig 'Evening Inana" (87).

Heaven is symbolic of Inanna's genitalia.[231] In fact, humanity undoubtedly gets the idea that boats are female because the early reed boats built in this region of the Mediterranean were, indeed, vulva shaped (see image below).

Sketch of a Sumerian Reed Boat
Author

Undoubtedly, early sailors hoped their vulva-shaped craft were as safe as wombs when they set out in them. Yet those same scholars often refuse to imagine that **Inanna is sailing away in a female sexual symbol**, a symbol that is as liminal as sexual intercourse itself, and a vehicle that is both liminal, transporting a person from one place to another, and protective, like a womb (Miller 543). Inanna is demonstrating her powers as an all-powerful female god, who undoubtedly preceded historical Sumerian culture as a revered deity. It is possible that the poet or scribe (either of which we know could have been a woman) who recorded "Inana and Enki" wanted to remind readers or listeners that Inanna was more powerful than Enki, despite the growing androcentrism of the times in which this poem was recorded.

Androcentrism also leads many scholars, some of whom are perhaps uncomfortable discussing sexual references and sexuality

---

[231] Grahn equates the sacred *me* with "gifts of royal menstrual office" and the Boat of Heaven with the vulva of creation, so that, when Inanna uses the Boat of Heaven to take the sacred *me* to the people, the poem is referring "to the world formation that derives from the vulva" (216). In other words, like Prometheus' theft of fire for humanity's sake, Inanna steals the magic from the gods to give to humanity as part of her role as Goddess Creator of a new world for humanity.

in scholarly terms, to associate sexuality with fertility, so that they can only come as close to discussing sex by linking Inanna to fertility.[232] However, since the ancient Sumerians did not discover the connection between sexual intercourse and procreation until approximately 2400 BCE, these long-lasting images of Inanna as a sexually aware female are **not** the product of male-bias, as some feminist scholars assume, but just the opposite—they are the product of earlier cultures who saw women as powerful, liminally spiritual beings who not only bled for three to five days without dying every month,[233] but also gave birth to whole human beings seemingly out of the magic of their own bodies. For these early people, Female Creative Power, exhibited through sex, was magical, healing, and the best way to reach the divine in all of us. **Sexual intercourse, at that time, had nothing, whatsoever, to do with childbirth or marriage, so little to do with fertility.**[234]

We must acknowledge that Inanna is often misclassified as a goddess of love and fertility by androcentrically biased scholars because of her obsession with her own genitalia and sexuality—

---

[232] Leana Wessels (2013), who attempts to use feminist scholarship to examine Inanna objectively as a likely trickster god, claims that "A feminist reading of the myths as employed in this study aims to reflect awareness that the goddess Inanna/Ishtar is a female divinity that reflects male constructions of the female or feminine, at least to a certain extent" (39). The fact that Wessels willingly conflates the two goddesses, separated by more than a millennia, as the same goddess demonstrates the scholar's lack of awareness of how androcentric bias grew in strength after the biological role of fatherhood was determined. To Wessels' credit, she does admit that "the influence of the researcher's own scholarly bias cannot be underestimated" (39).

[233] Women's moon blood was so powerful that it was believed to disrupt any magical powers men had, eventually leading to cultures assuming it was evil or "unclean." However, many cultures attempted to capture the magic of women's moon blood powers by imitating it in some way, such as through male circumcision, a ritual for the Hebrew people that clearly imitates a girl's menarche ritual. As androcentrism rose, so did male appropriation of female rituals and spiritual powers, until, eventually, men assumed they were more sacred than women (Grahn 44).

[234] The exception is that Inanna caused Dumuzi's semen to flow, and semen was not only associated with milk, but also with water, especially rain and floods. The ancient *qursu* ritual undoubtedly was held in the early spring to create the sympathetic magic necessary to bring rain and the swelling of rivers to fill irrigation canals. Through the rains, the fertility of the fields were assured.

> Her genitals[235] were remarkable...(repet)
> She praised herself, full of delight at her genitals
> (repet)
> (Faculty, Segment A, "Inana and Enki," ll. 1-10)—

not to mention the frequent poems and hymns that refer to her seduction of one male ruler or another. However, Victorian[236] androcentric tact and assumptions dictate that Inanna's full-on sexual character be subdued and reclassified as "love" or misclassified as "fertility." And many scholars confuse the liminal symbols associated with Inanna, such as the temple gateposts, as literal things, rather than viewing them as symbols of Inanna's liminal being—from her sexual organs to her Boat of Heaven. For instance, in 1976, Thorkild Jacobsen made an extended argument in his book, *The Treasures of Darkness: History of Mesopotamian Religion*, that "the gate emblem for Inanna" was to signify that she was the spirit or "numen of the storehouse" (20). Jacobsen claims that

> seeing Inanna as the numen in the date storehouse and Dumuzi-Amaushumgalanna as the power in and behind the date harvest gives us a background against which the underlying ritual of the Uruk Text becomes more understandable...The date clusters that are to adorn the shelves of the storehouse become ritually experienced as traditional feminine adornments and jewelry such as would be suitable for decking out a human bride, and the heap from which they are taken comes close to becoming a jewel shrine of sorts. Its original nature reveals itself, however, later in the text when we hear that Dumuzi-Amaushumgalanna is to be taken to that heap, under its clay plaster covering,

---

[235] I have to wonder, since I cannot read cuneiform, if the real Sumerian word here was vulva, not genitalia.

[236] Feel free to explore the Victorian era's "Cult of True Womanhood" movement, which tried to keep women feminine in an age when women were proving they could do anything, including outshoot men. In fact, Annie Oakley was so conscious of being seen as a threat to men that she purposely wore dresses when she performed.

as soon as he has entered her house, the 'storehouse
of Eanna' (37).

Jakobsen is not only sexist in his regard for who wears jewelry,[237] but also in missing the symbolic implication that the "heap," like the mound of creation associated with women's pregnant bellies, thus with women's magical ability to produce new life, is a source of plenty **and** a source of pleasure, aka Inanna's mons venus. Overall, Jakobsen's androcentrism prevents him from seeing Inanna as anything other than Dumuzi's consort, and his Judeo-Christian bias seems to prevent his seeing most of the sexual symbolism.

Jakobsen also seems obsessed with attributing the gatepost logogram solely with Inanna and her temple's storehouses, but Krystyna Szarzynska points out how often this one symbol appears in so many other contexts, such as being pictured with dragons holding them (11) and on a seal[238] with four nude men holding them (12). She argues that the symbol is a sign of protection[239] and "probably originates from the ancient function of the poles with loops set in pairs beside the doors of reed huts or pens. The loops could be bound together with a string,[240] and in this way shut the entrance to the huts" (note 22, p. 11). Like the red velvet ropes that keep movie goers out of a theater before a movie, the ancient

---

[237] And apparently paid little attention to Sumerian art.

[238] The "Cylinder Seal with Kneeling Nude Heroes, c. 2220-2159 BCE" at the Metropolitan Museum of Art probably depicts four city rulers—whose penises are important enough to show to demonstrate that they are not castrated men. All four men look alike with curly hair and beards, indicating young men in full bloom of health, but each has a different symbol above their heads—one is a bird with a scimitar-like tail, another is the eight-pointed "rosette" pattern often interpreted as one of Inanna's symbols as the Morning or Evening Star, one of which is an amphora with flowing liquid pouring out of it, and the last is a fish. While these signs could be interpreted astronomically, they could also symbolize specific cities or could be a list of Uruk gods—Lugalbanda symbolized by the Anzu bird, Inanna symbolized by the star, Enki symbolized by the water, and Dagon by the fish.

[239] If this symbolic pole is a logogram for spiritual protection and is most often seen with Inanna, could that mean that Inanna, herself, is a symbol for spiritual protection in all things? See my discussion of the Isis Knot elsewhere for comparison.

[240] As a farm girl, I can assure you that string would never be strong enough to keep most animals in or out of a gateway they decided to leave or enter.

Sumerians probably used these poles to prevent people from entering sacred spaces or spaces being prepared for special events. Their link, then, with Inanna could be as simple as indicating "taboo" activities, such as the *qursu* ritual affirming a ruler's reign or the sexual healing ritual, *hashadu*, a ritual that was taboo for common eyes to see which Shamhat performed on Enkidu to transform him from beast to man. They could, however, be functioning in much the same way the *djed* columns or parallel papyrus stems in ancient Egypt functioned (Oriental, Robert Ritner)—to symbolize resurrection and/or protection of life, sacred spaces, or sacred events (compare images below).

Djed Columns

Papyrus Rods

Inanna's Gate Posts
Sketches by Author

What many androcentric-biased scholars have a tough time addressing is the extent of Inanna's sexual powers.

Agreeing with Jerold Cooper, Julia Assante points out that none of the Inanna poems or prayers that have thus far been identified and translated indicate that Inanna was ever concerned with getting pregnant or having babies (39). So why would any scholar classify her as a fertility[241] goddess?

Inanna's sexuality is perhaps her greatest power. We get glimpses of the Sumerian belief in the power of sexual intercourse from other works, including the *Epic of Gilgamesh* and the many different poems surrounding the ancient *qursu* ritual many misguided scholars dismiss as *heiros gamos*[242] or openly seek to corroborate the ritual as a marital act. Derek Collins points out that "the noun *gamos* in the Greek magical papyri often does not mean 'marriage' but 'sexual union'" (110).

Johanna Stuckey, for instance, seems reluctant to admit that the *qursu* ritual "at first conferred authority temporarily on one man, eventually provided religious sanction for male exercise of power" (4). In all probability, the ritual probably originated as a way for the goddess Inanna to transmit the sacred *me* to humanity, with the ritual allowing two human beings to become literal conduits through which these god-granted powers and knowledge were transmitted from the heavens to earth, making heterosexual intercourse a liminal act. As William Stiebing admits, "the rise of this myth and the change in the Sacred Marriage ritual was due in large part to the development of kingship and the full attainment of male domination within Sumerian city states" (52). Originally, the main ruler of a Sumerian city-state was elected to be the "shepherd" of the people, so s/he/xis was not a god but was Inanna's representative for the city (Stiebing 53).

In *Gilgamesh*, Enkidu is transformed from being a beast into a human being through sexual intercourse over seven days with a

---

[241] Realize that I know these scholars use the word "fertility" when they mean sexuality because they are too embarrassed to really talk about sex, but we have to question this choice because it clouds everyone's understanding of ancient ideas.

[242] The idea that sexual intercourse equates with marriage is a very outdated concept. Samuel Noah Kramer was one of the first scholars to conflate the *qursu* ritual with the ancient Greek idea of *heiros gamos*. See his comparison of the *Song of Solomon* with Inanna-Dumuzi hymns where he claims the Songs are "devoid of any religious, theological, moralistic, or didactic motivation" (Kramer 1962). Clearly, he has no idea that sex can be a spiritual or healing act.

priestess[243] from Inanna's temple who is trained in such sacred sexual rites, the *hashadu*. Through the *qursu* ritual, which had nothing to do with marriage, and everything to do with the transformative sacred powers of sexual intercourse, mere mortal men were transformed into kings. In fact, without Inanna's divine sexual blessing[244] in ancient Sumeria, a man could not be sanctified to rule a city, let alone a growing nation, so that we have to assume that Sargon had sexual intercourse with his own daughter[245], Enheduanna, who acted her part as high priestess as the goddess Inanna to his shepherd-god Dumuzi,[246] in order to sanction his role

---

[243] Unfortunately, many androcentric scholars and interpreters feverishly insist that this woman from the temple is a prostitute, perhaps purposely ignoring that the word used means "taboo," which means untouchable by certain people, not polluted, as so many would like to see her--"a harlot, a wanton from the temple of love" (Sanders). Instead, she was undoubtedly a trained healer and priestess from the Eanna Temple who used sex to transform the ill to wellness and, in this case, a beast into a man. These scholars take the concept of "taboo" out of context, refusing to acknowledge its sacred and religious connotations, despite the continued evidence that royalty in some countries are still seen as "taboo" for average people to approach, let alone touch (Grahn 175).

[244] Since male genitalia were associated with semen, presumably the *qursu* ritual required the king to successfully ejaculate, which is similar to how the Egyptians believed the God's Wife had to use her hands to make the ruler ejaculate.

[245] As much as many scholars want to believe that Sargon never had sexual intercourse with his own biological daughter, we must remember that incest was quite common, and, in some respects, required in royal families in order to maintain a legitimate line of descent for the successive rulers even after procreation was discovered. Incest probably became common because humans **did not associate sexual intercourse with procreation**, and, even though Sargon ruled after the time period when humanity could verify that men contributed to procreation, old habits and traditions are hard to break.

[246] In the Sumerian myth, "Inanna Descends to the Nether World," Inanna returns from the Nether World to discover that Dumuzi has not been mourning her death but has been celebrating and even presumed to sit in Inanna's throne. Because of his transgressions and disrespect for her, Inanna condemns him to take her place in the Nether World, thereby creating the seasons. William H. Stiebing Jr. speculates, "possibly the rise of this myth and the change in [the *qursu* ritual] was due in large part to the development of kingship and **the full attainment of male domination within Sumerian city state**" (emphasis mine, 52). Judy Grahn points out that, "By the time the Greeks were telling [the story of Inanna's Descent to the Nether World,] Ereshkigal had been replaced by

as king of kings[247]. By 1820 BCE, however, the kings appear to stop referring to the *qursu* ritual as a means to legitimize their power (Pongratz-Leisten 58). While such incestuous relationships as that between Enheduanna and Sargon were widely common among the elites of the early Bronze Age around the Mediterranean, such communions probably led to the realization that close relatives should not be conceiving children,[248] so that, by the time Hammurabi writes his law codes (c. 1792 BCE) on his famous stelae, incest is illegal.

For the Sumerians, Inanna had power over everything sexual, including what gender a person was. Reportedly, most of the priests in her temple were males who identified as female, castrated males,[249] and males who were homosexual. The priestesses could have included homosexual women or women with vaginal agenesis,[250] as well, but some of the priests could have also been born with both sets of genitalia, so had been dedicated to the temple by their parents, since it was probably well known that Inanna marked her favorite followers in such a way.

While Inanna's temple in the ancient city of Uruk, Eanna, is referred to in several of the poems about Inanna, many scholars' androcentric bias leads them to associate this temple first and foremost with the male god Anu, whose cult rises after humanity discovers males have a biological role in procreation. While Inanna

---

an active male principle, Hades (Pluto), and from the male point of view, Persephone's descent was an abduction" (213).

[247] It is possible that it was the Akkadians who first coined this idea, the "king of kings," also known as the "first among equals," but the Hittites used the term quite often in correspondence between their rulers and Egyptian rulers (Stiebing 45).

[248] It is around this time that the female gods in myths begin conceiving "monsters" when they try to create children on their own. This phenomenon demonstrates that the ancients were still struggling with how, exactly, babies were made.

[249] A undoubtedly common practice that is difficult to prove because skeletal remains do not show if the male was castrated, and there are few direct indications of castration in the literature that has been translated so far.

[250] Vaginal agenesis is a rare condition resulting in an undeveloped vagina and/or uterus, which can go undetected until the woman never menstruates (Mayo).

was undoubtedly worshipped by the Ubaid people who preceded the Sumerians, she continued to be worshipped through the Akkadian period, an era ushered in by Sargon the Great, and saw a significant revival of interest in her favors when the Eanna District was restored with a new temple for Inanna that included a ziggurat called the House of the Universe built by King Ur-Nammu (c. 2030 BCE).

Others have discussed, to great effect, how the ziggurat influenced the basic temple structure for many cultures that followed, including the Hebrews' Temple in Jerusalem. What few discuss, however, is the symbolic structure of the ziggurat, which is extremely sexual in nature with a long ramp symbolizing the vagina or birth canal, an arched entry symbolizing the cervix, and the holy of holies temple at the top symbolizing the uterus or womb.[251] While many scholars understand that both the ancient Sumerians and the ancient Egyptians believed life arose from a sacred mound, few associate the ziggurat with women's "mons venus," even though the word "mound" is right there in French. I have seen no other scholars associate women's magical reproductive organs with the structure of such temples, even though the symbolism is pretty obvious, once you know it is there.

Similarly, many future goddesses who are weak imitations of Inanna, will continue to be associated with fertility, even when they bear no children, because for most androcentric-biased scholars' basic creation is an act of fertility because they fail to realize that, **for the ancients, fertility was divine, not biological**. The gods decided who would bear children, how many dates would appear, how much grain would ripen, how many calves would survive, etc. Prior to the knowledge that men played a biological role in pregnancy, everyone knew women were the creators of new life. While a pregnant woman's belly seems like a mound, from it and her *mons venus* will emerge new human life, accompanied by many fluids and some blood.

---

[251] While Judy Grahn connected many of the symbols regarding the sacred nature of menstruation and early religion, and while she admits that "the menstrual hut...was the axis around which the village grew" (243), she did not connect the structure of the ziggurat to women's reproductive organs.

It is the female association with blood that ties everything together, as Judy Grahn might say. It is blood, specifically the sacred nature of monthly blood loss[252] and bloodletting during battle that creates the ties for goddesses who are labeled as "love and war" goddesses. Many men, even today, are terrified of women, have no idea how to approach us or even talk to us like human beings, so imagine how frightening many women must have seemed for ancient men who saw women as these all-powerful beings who could bleed and not die on a monthly basis that made them seem forever tied to the cycles of the moon,[253] who could seemingly bring forth magical new life from their own bodies, and who could leave men limp and vulnerable after a good session of rousing intercourse. As the Babylonian version of the *Epic of Gilgamesh* puts it, "let her woman's power overpower this man" (Sanders).[254] These things are true power, powers that still intimidate many men, some of whom seek to control women's bodies to such a degree that they deny women their individual rights as human beings.

The kind of power women wield is what men have sought to control since 2400 BCE, fueling a gender conflict that never needed to arise, and there are many indications that some men did not want this gendered conflict to arise. Circa 2375 BCE, Urukagina proclaimed himself king of the city of Lagash in Sumeria, but he undertook a series of reforms in an attempt to end conflicts between

---

[252] As Charlene Spretnak points out in the Foreword to Judy Grahn's *Blood, Bread, and Roses: How Menstruation Created the World*, "it was women's moonblood, not babies, that was considered numinous...[which] might explain the absence of infants from the upper paleolithic and neolithic goddess statues" (xii).

[253] The Sumerian word for woman was *munus*, and *mun* comes down to us through IndoEuropean changes as moon. The **men** in menarche, menstruation, and menopause all refer to women's moon cycles, not to men, the gender, although it is probably the precursor to the term "men" meaning human beings. Similarly, the Akkadian word for woman was *sinnistu*. Since the Sumerian mood god was known as Sin, *sinnistu* identifies women as those people tied to the moon and its cycles. As Androcentrism rose, however, *sinnistu* will also give us the word **sinister**, things that threaten us in the light of the moon, including women out to seduce us.

[254] Later, the Greeks will credit Aphrodite for subduing men through sex, too (Strolonga). See the many articles about why men fear, even while they desire, sexual intercourse with women.

Lagash and Umma. While his proclamation[255] "contains the first recorded references to 'freedom' or 'liberty,' the [most important] Sumerian word" in the treaty is "*amargi* (literally, 'return to the mother')" (Steibing 46). Like Egyptian ruler Hatshepsut will 1000 years later, Urukagina seemed to hope that returning to earlier Sumerian Female Magic myths could help him increase cooperation and reduce conflict. Unfortunately, Lagash would be soon taken by the leader of Umma, but Umma would almost as quickly become part of Sargon of Akkad's empire.

Ishtar, a Babylonian goddess who is often spoken of as though she was Inanna, however, is a weakened transformation of Inanna that begins to demonstrate just how negative male views of women will become.

In the Babylonian version of the *Epic of Gilgamesh*, Ishtar not only sends the Bull of Heaven to lay waste to the city of Uruk after Gilgamesh rejects her sexual advances, but also reappears in one of her many forms as the snake who eats the sea herb Gilgamesh gathered from the bottom of the sea that would have made him immortal like the gods. While this portion of the Babylonian tale might have originally been meant as a "just so" story to explain why snakes shed their skin[256] and appear to live forever, later, the snake aspects of this goddess will be associated with deception and cunning of Ishtar, and will echo back to us in the Iron Age in the Books of Moses when the snake visits Eve and Adam in the Garden of Eden in the Judeo-Christian myth of Genesis—by then, the snake, probably a symbolic representation of Ishtar or Astarte, the Canaanite goddess, is associated with evil incarnate, but also with

---

[255] Urukagina's proclamation has yet to be recovered intact, so scholars often only work from what other sources say about the proclamation, but some also believe the word "edin" used to describe open country or uncultivated fields between the Tigris and Euphrates Rivers mentioned in the edict became the source for the biblical idea of Eden (Stiebing, Note, 46).

[256] Snakes have long been associated with women's magical creative powers of giving birth because they shed their skins the same way women give birth—head first, and because their jaws can widen to take objects that might otherwise not fit, just like vaginas.

everlasting life—something Adam and Eve lose when they disobey Yahweh.[257] Another just-so story explaining why human beings die.

The Babylonian version of the *Epic of Gilgamesh* changes many details from the original Sumerian version. For instance, in the Sumerian version it is Gilgamesh who is frightened as they approach the lair of Humbaba, the giant,[258] but the Babylonian version makes Enkidu the frightened one, so that Gilgamesh, the hero, must chivvy him along, saying, "Dear friend, do not speak like a coward" (Sanders).

Ishtar becomes the fearsome goddess, like Inanna, but where Inanna was also quite seductive, Ishtar is frightening. When Gilgamesh refuses Ishtar's proposal in the Babylonian version of the Epic, he lists all her former lovers in their animal forms—bull, bird, lion, wolf, stallion, and frog—clearly indicating that, unlike these other lovers, he has no intention of returning to an animal state (Grahn 253). While Gilgamesh loved Enkidu like a brother and deeply mourned his death, Enkidu had taught him that there was more to being human than obsessing over sex.

Perhaps most telling, however, is the story of "Inanna and the Huluppu Tree."[259] As Johanna Stuckey succinctly puts it in her

---

[257] In fact, most of the symbols in the Genesis tale of creation are purposely designed to promote the idea that women are evil and duplicitous, not to be trusted, and, of course, the sole cause of "the fall" of men. See Chapter 10, "Unraveling the Myth of Adam and Eve" pp 198-223 in Merlin Stone's *When God Was a Woman*.

[258] Several times in the Sumerian version of the tale, Gilgamesh has terrifying dreams that Enkidu must interpret for him. One night, terrible beasts circle their camp, but Gilgamesh sits as though frozen with terror, with Enkidu admonishing him, "why do you remain there unmoving?" (Stephany).

[259] J. Alexander MacGillivray makes some interesting connections between goddess names and uses of tree symbolism:

> The Minoan divinity in the most common libation formula is A-SA-SA-Ra, according to most readings. This closely resembles the names of the Luvian goddess Ashassarasmes, the Hittite goddess Ishassaramis, meaning 'my Lady' or 'my Queen,' Ishara, the very ancient Hurrian goddess of oaths, Vedic Ishwara, and the great Canaanite mother-of-gods and mistress of plants and animals Ashera, later adopted by the Hebrews, and their counterpart to Egyptian Aset/Hether (Isis/Hathor). Thus the double axe's living haft might be comparable to the Ashera tree, akin to the tree in which Aset (Isis) in the form of a swallow found her dead brother/husband Asir (Osiris) in Egyptian myth (126).

article, "'Inanna and the Huluppu Tree': One Way of Demoting a Great Goddess," "Inanna kept her high standing among the Sumerian Deities even as society increased in male-dominance. The poem 'Inanna and the Huluppu Tree' gives a mythic explanation of how the throne and the bed used in the 'Sacred Marriage' came into existence and, in the process, records a drastic demotion in Inanna's status" (1). In the Huluppu Tree story, the earth's cycles had just begun with the earth being separated from the sky. Inanna rescues a floating tree that had been uprooted by a storm and plants it in her sacred grove in Uruk, which lends the story a bit of Cretan flavor because they, too, loved groves of trees. Inanna's goal was to grow this sacred tree large enough to make both a bed and throne[260] from it. However, like the biblical tale that comes much later, a snake moves into the tree, specifically settling among the tree's roots—both the tree's roots and the snake symbolic of sacred ties to the netherworld. Another being, this time a female spirit called a *lilitu*, which lends itself later to the Hebrew idea of Lilith, moves into the tree's trunk, and a bird, the Anzu bird, moves into the tree's branches. In this way, all three realms—the Underworld, the Earth, and the Sky or Heavens—are represented by the tree and its new inhabitants. Like many scholars, Stuckey sees this tree as symbolic of "the World Tree" (2).

Stuckey also indicates that Inanna's ability to plant the tree and make it fruitful is part of her whole-unto-herself powers; she needs no gardener, which was a title usually held by male kings to describe how they weeded, pruned, and grew their people like pliant plants, but also demonstrated their sexual prowess, since many metaphors for male sex come from gardening and plowing terms (2). However, Inanna is not a good gardener because "the tree acquired what, in a male-dominated world (garden, womb), would have been considered parasites" (2).

Consider what follows above regarding "Inanna and the Huluppu Tree" in this same context.

[260] As Ian Chambers describes it, the furniture Inanna wants to make "is to be the symbolic seat of Her power, the place from which she will hold council and from which authority and sacred law, the *me*, resides with its owner." From the throne, Inanna will rule Uruk and her other chosen cities; from the bed, she will imbue chosen rulers with the sacred *me* through the *qursu* sexual ritual.

While Stuckey associated the Anzu bird with other winged monsters from Mesopotamian lore, the Anzu bird is possibly a predecessor of the Simurgh or other benevolent birdlike creature associated with weather (5), but the important factor is that the bird flies off with its young, which demonstrates not only motherhood, but also the power of mothers to lead their children. Like Stuckey, Chambers believes the *lilitu* spirit is a mirror entity to Inanna's independence, since "the dark maid Lilith who would come to preside over the liminal moments of womanhood, from first sexual encounters to childbirth, [who was] demonized in Jewish superstition" is similar to the "adolescent Inanna." The snake, of course, has long been associated with women's ability to create life from their own bodies because of the way snakes shed their skin to become something new and because snake jaws enlarge to accommodate prey just like vaginas grow to accommodate different sized penises.

In order to rid the tree of its inhabitants, Inanna implores her brother gods to help, but none volunteer. However, her erstwhile lover, Gilgamesh, does. Using his bronze axe, which is both a symbol of female reproductive power and a new symbol of male oppression, Gilgamesh strikes the snake, causing the other two female spirits in the tree to flee—violence subdues the feminine. Further symbolism is the act of Gilgamesh chopping down the tree, keeping the branches—the symbol of reaching for the heavens—for himself, giving the trunk to Inanna, which she then shapes into her throne and bed.

Stuckey laments that "the destroying of the huluppu tree meant that human beings could no longer count on Inanna and the World Tree to maintain the cycle of life and death. Instead, they were now facing a terrifying, linear world" (6), one where males dominate with pitiless bronze weapons those who were once seen as magical creators of everything.

These literary changes as new cultures adopt and adapt previous stories to fit their own needs demonstrate a reshaping of what were considered masculine and feminine ideals for each successive culture, so that, as androcentrism attaches itself more deeply to the androcentric ideas of the patriarchy, mortal women and goddesses are seen as weaker and less significant than they had been to previous generations.

Inanna was all woman, all fearsome, and all-powerful...at least until humanity verified that males have a biological role in procreation, so jumped to the conclusion that men were planting seeds in women, just like they planted seeds in the soil, relegating women to the status of fields to be plowed, and chattel to be sold. Women gradually lost their rights as human beings, as did poor, powerless men, all to gratify and uphold the concept of the king of kings, the male ruler on high, the "first among equals" (Carmody 45).

Now, after 4300 years of androcentric bias, it is time to set the story straight and learning more about one of the most powerful gods conceived of in the human imagination, Inanna, is a great way to start.

Chapter Twenty-Two
Egypt's Shift Toward Androcentrism

Since Howard Carter discovered the tomb of Tutankhamun in 1922, revealing to the world the many golden treasures buried with the 18th Dynasty boy ruler, the world has been enthralled with ancient Egypt. Reportedly, after Napoleon Bonaparte invaded Egypt in 1799, allowing "one of his foot soldiers [to stumble] upon the Rosetta Stone" (Watson), European cultures dominated the exploration of Egypt in the hunt for such treasures in the belief that true "Western" culture[261] began there, so these hunters were searching for the glorious "roots" of their cultures until the late 20th and early 21st centuries, when Egyptology became, finally, the purview of modern Egyptians.

Probably because these so-called Western scholars dominated the narrative of ancient Egypt for so long, and also because the first serious excavations happened in the more prudish Victorina Era of the 19th century, much of ancient Egypt's fascination with the pleasures and mysteries of sex remain submerged beneath what is believed to be "more serious" scholarship. Also submerged, subdued, and oppressed are the facts that support the idea that, prior to the rise of androcentrism in Egypt, circa 2400 BCE, the culture was more gynocentric than androcentric.[262] There were even women rulers, such as Merneith, possibly accidentally admitted to on the earliest of King's Rolls, which was first created post-procreation discovery.

One problem with studying ancient Egypt is the fact that this culture spanned over three millennia, so scholars who study the 18th

---

[261] What exactly is Western culture? By most accounts, it only began when men began to dominate cultures, and, some continue to firmly believe, began to make more "complex" hierarchies a way of life, around the Mediterranean, but many colorless people who make such claims today are challenged by the fact that their racist views lead them to ignore the fact that people of color were the first instigators of these civilizations. The racism must be addressed.

[262] Some scholars like to call these cultures "heterarchies" because they hate to admit that women ruled, but they are essentially acknowledging that both males and females could rule a community.

Dynasty (roughly 1550-1290 BCE), for instance, might not be as familiar as they should be with the Dynasty Zero or Dynasty One (circa 3100 BCE) cultural evidence. For instance, some might not know that Hathor and Min are two gods who are among the most ancient Egyptian gods, earlier than Isis and Osiris, and possibly older than Horus. Like most cultures, the ancient Egyptians adopted new ideas as they arose, while still clinging to old, familiar ideas.[263]

The result of not knowing everything they should know about the breadth of Egyptian pre-literate and literate history is spotty scholarship about the importance of some gods and some rulers, which is further complicated by the fact that almost every Egyptologist assumes the Egyptians were always a patriarchal culture, when they obviously were not.

For instance, David Wengrow, who is a highly respected ancient cultures scholar, demonstrates his lack of knowledge when he discusses the cult of Hathor, stating that "women enjoyed unusually high social status [in the cult], which extended to ***the reversal of ordinary gender roles*** in presenting offerings to the deity" (emphasis mine, Voyages, 155). Wengrow will repeat his belief that gynocentric cultures arose after androcentric ones in his co-authored book, *The Dawn of Everything*, as well, displaying what seems to be willful ignorance of the fact that women, quite naturally, once led their families and communities as our primary and only known blood ancestors. The inability by so many scholars to acknowledge that women actively contributed to leading cultures because of their sacred connections to the cosmos—menstruation tying them to the moon cycles and demonstrating that they can bleed for three to five days and not die, because of their ability to conjure children from their own bodies, as well as because of their ability to use sex to heal and to transform, which helped maintain Ma'at, the balance of the universe—means these scholars are **willfully ignoring** what must have been normal for ancient people.

For Egyptians, Ma'at was also a goddess who had her own temples in various cities, but most scholars only pay attention to the

---

[263] According to Denise Carmody, the Egyptian hieroglyph for town or city meant "mother" (56), another indication for a gynocentric cultural view.

fact that the rulers almost always gave themselves the subtitle of "beloved of Ma'at," to indicate that they participated in maintaining the sacred balance of the universe, completely ignoring the fact that early male rulers only got to rule because they were related to sacred royal women.

Also surprising to Wengrow is the fact that a figurine uncovered at Byblos "exhibits a unique amalgam of male and female attributes" (Voyages 154). He notes that Helga Seeden described the figurine like this--

'Both arms are peacefully crossed under the breasts. This and the wide hips are characteristic for female figurines... . Nevertheless the male sex is indicated"-- causing Wengrow to comment that the discovery is a "striking anomaly [thus] significant...because it indicates some ***conscious manipulation of gender categories***" (emphasis mine, Voyages 155). Apparently, the fact that nearly 2% of the world's population is born with partial or whole genitals for both genders is another fact Wengrow is unaware of. Instead of the Egyptians "manipulating" gender "categories" held by androcentric cultures, they were undoubtedly depicting what they witnessed—someone born with both sets of genitalia. Such "two-spirited" people were often considered doubly sacred in most ancient cultures.

Most disappointing to me was Kara Cooney's book, *When Women Ruled the World*. Cooney, who by most definitions is a feminist scholar, admitted during a lecture on the book at the Nelson-Atkins Museum of Art in Kansas City on March 5, 2020 that she was an expert on the 18th Dynasty (circa 1550-1292 BCE) of Egypt, but not an expert on Dynasty Zero or One, but her book, like so many Egyptologists before her, demonstrates that she **assumes** the patriarchy was full on at the beginning of the Egyptian culture when she states that women "were utterly powerless, mere pawns of a patriarchal system over which they had no control and could never hope to alter in the long term" (12). For her, the reason why so many women ended up ruling in Egypt was because Egyptians had "a penchant for feminine rule as a necessary aid to shore up a weakened regime" (15). Later, in Chapter One when she discusses

Pharaoh[264] Merneith, Cooney claims that Egypt required female rule "from the very beginning," a fact she believes "differentiated Egypt from other parts of the ancient world" (23); Cooney makes this sweeping statement despite the fact that the Sumerian King's List[265] names at least one female ruler who had been elected city leader by the people of Uruk, and the fact that the Minoans were a gynocentric culture that probably had many female leaders.

While Cooney notes how strange it was that the royal Egyptian harems were never guarded by eunuchs (77), she claims it was probably because "Egypt was such a closed society territorially and culturally" (77), ignoring the vast trade system and written communication networks the Egyptians eagerly participated in. For some reason, it never occurs to her that the Egyptians might have been slow to believe that males truly had a biological/physical role in procreation, best demonstrated by the continued incestual practices among the royalty. It is quite possible that the Egyptians never saw a reason to castrate men,[266] unlike the Sumerians and the Chinese, because they still believed men's most important contribution to conception was spiritual, not physical, and that they believed men's semen needed to flow freely to keep the Nile River flowing, as well. Nor does Cooney make the connection with the

---

[264] The term pharaoh was not used by Egyptians as a title for their rulers until after Hatshepsut began using the title to remind Egyptians that they descend from a royal female House, which is what the word literally means. Despite this fact, many Egyptologists use the term for male leaders, relegating the female rulers, usually, to mere "regent," implying she is a poor substitute for the male ruler who is too young to manage the office.

[265] According to William H. Stiebing, Jr., "the author of the [Sumerian] King List assumed that the gods always intended that there be only one ruler on earth at any given time" (44), an idea the Hebrews will adopt once they "discover" the so-called lost "Book of the Law," Deuteronomy (Nowak).

[266] In Egyptian religious stories, the god of chaos, Set, dismembers Osiris, which included cutting off his penis and testicles (they are usually depicted together), which was, in fact, a form of castration. The miracle of the tale is that Osiris disparate parts are gathered and bound together (supposedly to demonstrate mummification) without his penis, yet images of Isis, as kite (bird), hovering over the resurrected but still prone Osiris explains how their son Horus was conceived. It is important to note that the oldest fragment mentioning Osiris in ancient Egypt dates back to 2436 BCE, just after they learned men played a physical role in procreation (Oriental, Robert Ritner).

important must-marry-a-princess theme that continued throughout most of Egypt's history as a holdover from their ancient matrilineal practices.

While Cooney thought she did her best to cover all the women who ruled at one time or another in Egypt, she overlooked Sobekneferu, who ruled Egypt from 1790-1797 BCE and was buried at Abydos (Penn Museum). To be fair, Egyptologists will likely find many more women who ruled Egypt than they ever believed possible as they continue excavating and analyzing their archeological finds.

Yet far too many archeologists, especially, approach ancient remains expecting to find the patriarchy in full swing. An article published in Psychology Today, claimed that the authors of a study of Neanderthals asserted that women "traveled upon marriage to their husband's family group," based solely on the DNA analysis of two people—a man and a woman (Derr). Further, the scholars assumed the woman was the daughter of the man, when she could have been his sister, if she is, indeed, younger than him, and they died at the same time (Derr). The author of the article expresses skepticism throughout his article, which is to his credit. That archeologists have believed Neanderthals, who clearly can shape rock, which requires striking the rocks together, did not ever learn to create fire is beyond preposterous and has led to an elitist view that Neanderthals were inferior by so many scholars. However, far too many scholars are willing to give even Neanderthals credit for figuring out procreation, when there are still so many cultures today that do not believe sex has anything to do with pregnancy.

Unfortunately, too many scholars approach the study of gender in ancient cultures with such a simple dichotomous view of gender—men rule, women follow--but their androcentrism clearly shows through their biased assumption that having men ruling a group of people automatically makes that group patriarchal, just like the culture the scholars were raised in, presumably. Opponents to the idea that matriarchal or gynocentric cultures existed point out that women have been leaders in patriarchal cultures, but they neglect to add that men could have been leaders in more matriarchal cultures, too. Merely having male leaders, then, is not proof that a culture was patriarchal.

Some opponents of gynocentric cultures use less familiar words to disguise earlier gynocentric practices. For instance, Kristian Kristiansen and Thomas B. Larsson admit that early patrilineal systems were "agnatic—roles descended through the king's brothers first, after which the king's children could rule" (1), meaning, in other words, that matrilineal descent was used before patrilineal descent. They even go on to admit that "'residence at marriage is strictly virilocal'" (237), meaning that sons return to their mother's villages where their maternal uncles raise them. Therefore, all readers must be wary when reading such scholarship.

The transition for Egypt toward androcentrism and, possibly, eventual patriarchy mirrors many other such transitions around the world. First, men had to make new myths and gods to explain why men were now superior to women biologically, spiritually, and politically, which was necessarily a gradual process because many humans can remember things that happened more than a generation ago. This process might not have been slow enough for many people around the Mediterranean, however, since the Aegean Dark Age (circa 1100-750 BCE) could have occurred because groups of people, tired of the ceaseless violence engendered by the move to androcentric spiritual beliefs that had thrown over more nurturing gynocentric ones, finally rebelled. Egypt could have been spared from that overthrow because of their increased reluctance to completely cast off their powerful goddesses or women's rights. Second, old myths had to be modified to explain the new views, so goddesses like Hathor and Inanna, in Sumeria, who were powerful deities in their own right, had to be shown obtaining those powers by deceit, carrying the idea that females are deceitful, duplicitous, and manipulative[267] by nature through changes to popular literature. Egyptians made these modifications to myths in many ways, but especially in depictions on stone[268] and in hieroglyphs, so that the common people would see and embrace

---

[267] I explore how words change in meaning to demonstrate the demotion of women in the chapter on Inanna.

[268] Even David Wengrow admits in his book, *The Archaeology of Early Egypt*, that men used depictions on artifacts to create social norms wherein men dominate, "omitting any direct visual reference to female creativity" (187), in order to make males seem like natural leaders and creators.

them. The goddesses were not obliterated from cultural memory, merely toned down considerably in power.

Like many of the Sumerian creation myths, Egyptian stories also tend to try to explain why things are the way they are—why Inanna had to steal the sacred *me*[269] and why Hathor, one of the most beloved Egyptian goddesses, tried to kill human beings at Ra's insistence in her "former" role as the Eye of Ra.

Before we explore that particular tale, however, we need more background information on who Hathor was to the ancient Egyptians.

Like Inanna for the Sumerians, Hathor was a sky goddess who dictated human sexual interactions, but who was also associated with cattle, presumably because cows are very nurturing to their offspring and because cows also "nurture" human beings through milk, a product that could be transformed (as if by magic) into cheeses and yogurts. In fact, also like Inanna, Hathor wore horns.

Horns, remember, were not originally symbols of male power, but are, in fact, symbols of Female Creative Magic because the cow skull with horns symbolizes the vagina, uterus, and fallopian tubes—the female seat of creative and spiritual power.

Among the Sumerians, the more horns a god had depicted on her head, the more powerful she was, and most images of Inanna have four or more pairs of horns.

Hathor is usually just depicted with one pair of horns because Female Power Magic did not stack for the Egyptians the way it did for Sumerians. In all probability, since the moon only bears one pair of horns as it both waxes and wanes, they assumed one pair of horns was enough to show this goddess' connections to the moon, thus to the larger cosmos. The fact that the moon waxed and waned--just as the flow of the Nile did, just as the growth of crops and fruits on the vine do, just as pregnant women's bodies do, and just as the lifetime of a human being does--was merely one demonstration of this goddess' numinous powers.

---

[269] Inanna's theft of sacred powers from a "more superior" male god mirrors Isis' theft of magic from her father in order to resurrect her husband Osiris. These "thefts" emphasize the idea that the later male gods were the "rightful" owners of Magic Powers.

One of the oldest Egyptian artifacts, the Narmer Palette, has a cow-woman hybrid goddess, most likely Hathor, with curved horns[270] looking down on Narmer from the heavens carved at the top of the tablet (Cooney 25), but there is even older evidence for Hathor's worship.

Bedouin Eide Mariff discovered large megaliths in 1973 at Nabta Playa, believed to be at least 5000 years old. Archeologists excavating the megaliths expected to find human remains in what was clearly a burial at the site, but, "instead, they dug up cattle bones and an enormous rock seemingly carved into the shape of a cow" (Betz). Some scholars believe this cow-shaped rock was an early version of Hathor, the benevolent cow goddess who helped the people who erected this megalithic site survive times of little rain and famine. According to Fred Wendorf, the archeologist who excavated the site, the rock "'could explain the religious significance that Egyptians attached to the cow and cattle in the Old Kingdom'" (Betz).

An androcentric version of Hathor's story, often called the "Book of the Celestial Cow" or the "Book of the Cow of Heaven," was possibly "composed during the First Intermediate Period (2181-2040 BCE)" (Mark, Five Gifts), but first recorded in the 19th Dynasty, circa 1344 BCE, appearing on a shrine in Tutankhamun's crypt (Pinch 74). Tutankhamun's tomb also contained a golden headed cow figure on a pedestal and double golden cows on his couch. The double headed cow couch carvings were possibly that of

---

[270] The severely curved horns of the goddess on the Narmer palette help demonstrate the cow-head's symbolism of Female Creative Magic because they appear very similar to fallopian tubes above the uterus-head of the cow-woman hybrid face. The lips on the face are also emphasized, symbolizing the vaginal lips. While the ruler, who wears a false beard on both sides, and the crowns of both upper and lower Egypt separately on each side, is ruling over his opponents, he is clearly doing so under the observant, possibly nurturing gaze of Hathor, who is sanctioning his deeds. Below the mythical long-necked beasts that form the actual useable part of the palette, is a bull, also with severely curving horns, standing over another figure, but next to what appears to be a curved enclosure surrounding a towered temple. Instead of symbolizing the Apis bull, son of Hathor, whose worship as a god did not begin until the Second Dynasty, this bull undoubtedly symbolizes the constellation Taurus, with the three squares, the human figure, and the temple with enclosure possibly giving details about the date and location of the events depicted on the palette.

Hathor[271] holding the full blood red moon in her horns, but is just as likely to be Mehet-Weret, a cow goddess associated with the sky and primordial waters, who helps the sun be reborn every morning by allowing him to "sail" on her back through the Duat, or underworld—the invisible part of the world below the visible earth. As a goddess who assists with the rebirth of the sun, she was also believed to assist in the rebirth of the dead, which explains why she would be supporting Tutankhamun's funeral bed. If the stone from Nabta Playa was an early image of Hathor, or even Mehet-Weret, belief in a female cow-god as an important deity had lasted at least 3656 years by the time Tut was buried.

"The Celestial Cow" story is more vague than "Inanna's Descent," but it involves having Ra create humanity, then having humanity rebel against Ra, in particular, (it is unclear how they rebelled), because he has grown old and they have ungratefully forgotten all he has provided for them (Mark, Five Gifts), so that Ra sends Hathor to earth as the Eye of Ra (heat-intense drought?) to destroy the humans. After Hathor/Eye of Ra has killed many, Ra repents for the destruction he has caused, but Hathor, deep in her blood frenzy, will no longer listen to his commands. In order to stop the bloodshed, Ra has thousands of barrels of beer dyed red to look like blood by premenarchal girls. Hathor/Eye of Ra/Sekhmet sees the pools of blood, drinks her fill, falls asleep, and wakes up renewed to being the nurturing mother goddess, Hathor, she really is. Because this story is late in Egyptian history, we have to wonder whether this twist on the mother goddess imagery usually associated with Hathor was on purpose to denigrate the love the common people felt for her. The ploy does not seem to have worked, since Tut or whoever decorated his tomb still apparently loved Hathor.

Note that there are no overtly sexual references in the "Book of the Celestial Cow," but there is plenty of blood, and it was blood that was most sacred because, once it stops flowing in our veins, we die—a physical transformation from one state to another. But blood also flows from most adult cis women on a monthly basis, until it

---

[271] Marguerite Rigoglioso insists that the cow goddess in question is actually Neith, who she believes originated in prehistoric Libya (Virgin 28).

"dries up" during menopause, yet another transformation.[272] Blood, as we know, occurs at birth, so most ancient people also associated death and rebirth with blood, sprinkling the dead with red or orangish-yellow pastes and powders to simulate birth in order to help the dead be reborn into the Next World, or, as the Egyptians called it, the Hidden Place. So, while the story of Hathor's temporary destructive nature can explain why so many human beings die during famine caused by droughts or pestilence, it possibly also explains Egyptian attitudes toward older, menopausal women, who are still useful as nurturers, but who can also go on angry rampages, if necessary.

Some versions of the "Book of the Celestial Cow" also have Ra requesting that only maidens, young pre-menarche[273] girls, make the beer needed to pacify Hathor, so that beer, one of the most common forms of payment in the ancient world, is clearly associated with women, both as the makers of beer and as the consumers of beer. This version of the story also nicely frames pre-menarche women with post-menstrual women, both of whom were still valued members of Egyptian culture.

According to some scholars, Hathor, "as the female creative principle[,]...could be the most seductive and alluring of deities" (Pinch 138), just as Inanna was for the Sumerians. Just as the sacred *me* transferred from Inanna to a Sumerian ruler through the ancient *qursu* ritual where Inanna invokes Dumuzi to make his "milk thick and sweet," "the power to rule entered Horus with the milk of Hathor" (Pinch 138), and "each new king absorbed the power to rule from the milk of his mother Mut" (Pinch 168), who

---

[272] Women, unlike men, tend to have at least three drastic transformations in their lives: menarche, when menstruation begins, pregnancy, and menopause. Realize that the "men" part of these words does not come from any association with things male, but come from old wording for the "moon," whose tides and turnings marked most culture's days and nights activities, but with which women have long been associated because of our similar 28 day menses cycles. That root word, "men," is where we get the word month and the idea of "monthlies."

[273] Note that androcentric scholars call them virgins, but, since having sex before marriage was not forbidden, the ancient Egyptians would not have thought about virginity as an important demarcation. Instead, the women were probably pre-menarche because menarche is a moment for most cultures when girls become women.

becomes, after the New Kingdom, a convoluted form of Hathor, making milk the common theme here, since the ancient Sumerians saw semen as a form of milk.

Milky substances also intrigued the Egyptians. As K. Annabelle Smith puts it, "in Ancient Egypt around 2000 BC, lettuce was not a popular appetizer[;] it was an aphrodisiac, a phallic symbol that represented the celebrated food of the Egyptian god of fertility, Min.... The plant was believed to help the god 'perform the sexual act untiringly.'" In fact, the type of lettuce associated with the Egyptian god, Min, who is always depicted with an erect phallus, secretes a milky substance when pressed. Smith admits that "instead of being part of a meal, the seeds from the bud of the flowers [from the lettuce] were harvested and pressed for their natural oils which were used for cooking, medication—even mummification."

Min, as a god, presents my argument for 2400 BCE as the era in which heterosexual procreation was discovered with something of a challenge, however, since he was one of the oldest Egyptian gods ever identified and is usually misclassified as a "fertility" god because of his erect penis. However, I believe the androcentric bias of the scholars who classify him, early on, had a fertility bias (or sex aversion) which is compromised by the fact that, later in Egyptian history, passages of text actually identify him as a god of abundance and increase, which is often conflated with fertility. However, the poetry, such as "The Hymn to Min," clearly describes him as a sexual god:

> ... Min, Bull of the Great Phallus,
> ...
> You are the Great Male, the owner of all females.
> The Bull who is united with those of the sweet love, of beautiful face and of painted eyes,
> Victorious sovereign among the Gods who inspires fear in the Ennead.[274]
> The goddesses are glad, seeing your perfection.
> (Seawright)

---

[274] While "Ennead" refers to the Nine Gods (the Ogoads plus Ra) in Egyptian myth, I cannot help drawing connections between similar sounding words, such as "The Aeneid," an epic poem describing how Aeneas, or "the god inhabiting a mortal body," founded Rome, and, older still, the Eanna also known as the Temple of Inanna in the ancient city of Uruk.

Min and Hathor are probably two of the most ancient Egyptian gods with one (or both; Min symbolized by the erect standing stones) of them being honored in the earliest known carved cow statue from the western deserts. In human form, and, most notably, after 2400 BCE, Min's statues were "carried out to protect and bless the fields" with "the first fruits of the harvest" sacrificed to him (Pinch 164).

Indeed, Min is "sometimes given the title, 'Protector of the Moon,'" because "lunar gods tended also to be gods relating to moisture and thus of fertility" (Seawright), indicating that his penis represents semen as that life giving milk associated with the Sumerian god Dumuzi, not as the fertile "seeds" themselves, at least not until later in Egyptian history when men came to believe they were planting seeds in women's wombs. Moon phases, of course, were essential tools for knowing when to sow and when to reap most crops, as well as for predicting the annual flooding of the Nile River and both Min and Dumuzi can be seen, through the fluids ejaculated through their prominent phalluses, as river gods.[275]

The flail Min is also usually depicted with "might be symbolic of sexual intercourse [because] the flail forms the V while [the ruler's] upraised forearm seems to thrust inside the V" (Seawright), so that the V symbolizes a vulva and the forearm symbolizes a penis.

Therefore, Min, like Inanna, was a god of sex and sexuality, whose moisture from his constantly erect penis represented the life sustaining waters of the Nile and whose flesh, "painted black to represent the fertile land along the Nile," was the soil that brought forth the many grains and other vegetables grown by Egyptians (Seawright). Earth and water are necessary for crops to grow, which are necessary to feed the people.

Min's symbolism, in effect, is also depicted in the famous Egyptian ankh with the loop at the top, supposedly representing Isis for Female Magical Power, and the cross at the bottom symbolizing Osiris, or the Male Magical Power found in two testicles and a penis. However, we must not overlook the Isis Knot (Egyptian *tiet*), which

---

[275] J.G. Macqueen argues in a footnote that, while many gods like Dumuzi and the Hittite god Telepinu are associated with vegetation and/or animal propagation, but might be more correctly associated with the plenty brought by rivers (174, N. 12).

is often called the Blood of Isis (MacGillivray 123), clearly referring to her menstrual qualities and a fertile woman, because the symbol of the "Sacral Knot" is nearly ubiquitous around the Mediterranean, usually depicted as the tie on a priestess' or goddess' cloak

It is easy, when reading general books about Egyptian history and mythology, to assume that the Egyptians were rather prudish and did not appreciate the pleasures of sexuality the same way the Sumerians did. However, the problem might lie more with the prudish nature of early and many current Egyptologists, who were and are clearly fully immersed in the patriarchal cultures from which they grew, so we have to acknowledge that they looked for violence and male supremacy in almost everything they recover(ed) from early Egypt, avidly ignoring most references or indication of powerful females, sexuality, and women's magic.

Take, for instance, W.M. Flinders Petrie's interpretation of a prehistoric Egyptian vase he included in his book, *Prehistoric Egypt* in 1920. Plate number 74 (below) depicts two human figures amidst lines of vegetation and alternating triangles that go up and down the vase.

Sketch of Petrie's Plate 74
Author

In 1992, Béatrix Midant-Reynes reviewed Petrie's interpretation of the actions depicted on the vase, deciding, without any justifiable reason, that Petrie had been correct to label the depiction as combat, despite two earlier scholars, J. Vandier and Elise J. Baumgartel, who believed the smaller figure was female with a protruding vulva and the taller figure as male with his semi-turgid penis (174).

Yet, if we look at this image more objectively with more knowledge of the probably widespread belief in the importance of the spring *qursu* ritual that was still practiced by the Sumerians historically, the images point, clearly, to exactly that sort of sexual ritual. First, the female figure is showing the male figure her genitalia, just as Inanna does in her seduction of Dumuzi when she invites him to ride her "Boat of Heaven" and to "plow her field." The woman's hair is wild and curly, which could possibly indicate her magical abilities, but it could also emphasize her sexuality, since women's long loose hair has long been associated with their sexual desire and availability, which is why so many androcentric cultures seek to make women cover their heads. The male, too, is depicted to resemble Dumuzi, the shepherd god, with his symbolic cow or sheep horns protruding from his head. The triangle, long the symbol of creation because of its association with human female pudenda, climbs the length of the vase (not around its circumference), which mirrors the male's erection, but, just as importantly, that one bottom triangle with its eight steps (eight is also the symbolic number for Inanna) demonstrates the importance of the future pyramid symbolism and its relation to the triangle-pudenda as a generator of life or rebirth. Finally, the rapidly growing tall vegetation indicates the time of year, spring, when the Sumerians, and presumably the Badarians, who preceded the Egyptian culture, could view rampant acts of plant growth and sex happening in among the birds and small animals around them, so believed they needed to join in in order to ensure that Spring would bring everyone plenty of food (generalized as "fertility") <u>and pleasure</u>, thus good health, for the coming year.

In fact, if we read this drawing the same way we read Lakota ledger art, which uses lines drawn between people and objects to indicate names or actions, the line drawn between the woman's throat and the man's chest is indicating that she is using her voice, her words, to seduce the man's heart.

So, while the Egyptians might not have been quite as open about sex and sexuality as the Sumerians were prior to 2400 BCE, we have to recognize that they shared many elements of culture, as noted by scholars like Toby Wilkinson, who described "a facade composed of alternating recesses and buttresses—which create a highly effective pattern of light and shade in Egypt's sunny climate—had first been developed in Mesopotamia, in the middle of the fourth millennium BC" (31).

Another example of Egyptian interest in sex comes from the office of the God's Wife. One important function for the holder of the Egyptian religious and social role of God's Wife was to enact a ritual in which the god, or his kingly substitute on earth, physically ejaculates semen. Many of the scholarly texts that describe this ritual or the associated one in which Hathor personifies "the hand that made Atum ejaculate and the divine 'seed' itself" (Pinch 138), describe the ritual as more like fellatio than sexual intercourse. But the divine climax is important, not so much for producing "seeds," as it was to drive away evil (Pinch 138), like famine, by ensuring that the Nile River would continue to predictably flood every year. For the Egyptians, semen was closely associated with the waters of the Nile.

The Egyptian attitude toward sex, while perhaps more modest, was not dissimilar to the way the Sumerians would hang clay plaques depicting couples having coitus in bed over their doors and windows (liminal spaces) in order to invite good spirits into their homes and to drive evil away (Assante 48).

For both the early Egyptians and Sumerians, women were important as both goddesses and as creators, with some scholars arguing that Egyptian women enjoyed more individual freedoms than Sumerian women did, primarily because Egyptian law never made women the property of their husbands. However, many scholars do recognize that female Egyptian gods, like their Sumerian counterparts, were often reduced in importance later. Geraldine Pinch points out that "it was only after Isis took over many of her attributes that Hathor lost her place as the most important Egyptian goddess" (139).

Another important religious similarity between Egyptians and Sumerians comes from their beliefs regarding death. For

Egyptians, death was not an end, but a moment of transformation, even, potential rejuvenation. As Eric Hornung explains, "The blessed dead and the gods are rejuvenated in death and regenerate themselves at the wellsprings of their existence," citing the Pyramid Texts[276] to argue that "rejuvenation and regeneration [are] the true meaning[s] of death" (160). The ancient Egyptians called their realm of the dead the Hidden Place, where "the fate of death, which is referred to as 'decay' and 'disappearance,' is claimed to await 'every god, every goddess, all animals, and all insects,' in Chapter 154 of the Book of the Dead (Hornung 157). Thus, the Egyptians believed everyone died, including the gods,[277] but that, like the gods, humans and other animals would also be reborn.

Reincarnation is implied in "Inna's Descent into the Underworld," too. She goes, deliberately, to what she knows will be her death, but is able, with help, to return to her rightful place as the Queen of Heaven. One possible connection between death and sexual intercourse could be what the French still call "the little death," that feeling of intense release that comes with a strong orgasm. Such orgasms are clearly transformative moments for both women and men, after all, and it probably comforted our ancestors to believe that death was no more painful than a good orgasm.

For the Egyptians, though, death was not an impediment to enjoying sex. Even Cooney tacitly admits that the disembodied penis was valued, not just for reproduction as androcentrism increased across the land, but also for sexual pleasure with women directing that pleasure. Cooney reveals those sexual-transformative moments as she recounts Isis' resurrection of Osiris' erection. Cooney states,

> In the **last** temple dedicated to the divine feminine, in holy places like Dendara's Temple of Hathor, we see carefully crafted and detailed imagery of Isis helping Osiris to be reborn, in all its sexual glory. We see the god reaching out his hand to his erect member ready to re-create himself

---

[276] From the Pyramid Texts, "You sleep that you may wake; you die that you may live" (Hornung 161).

[277] Ra, the sun god, after all, dies every day, but is reborn, rejuvenated, the next morning.

through sex with himself[278]—but it is Isis who manufactures the circumstances for the miracle of self-recreation to happen at all, facilitating and receiving **the god's seed** to make the next generation of kings, to conceive Horus, the god of kingship himself. It was Isis who bred and bore Egyptian kingship. It was Isis who acted as the magician who pieced her broken husband Osiris back together after [their] brother Seth had cruelly cut him up into dozens of pieces. She is the alpha and omega of Egyptian kingship (emphasis mine, 25).

Here, Cooney reveals more than she probably realizes.

Dendera's Temple of Hathor was built very late in Egyptian ancient history by Ptolemy XII and was nearly completed by Queen Cleopatra VII around 54-20 BCE (atlas). Even after the Greeks, then the Romans, commanded Egypt, the people of Egypt still worshipped Hathor, one of the longest known goddesses to be worshipped, but androcentrism was riding high among most of the cultures surrounding the eastern end of the Mediterranean Sea by this time, with the Romans more patriarchal than any of their predecessors since they were among the first cultures to actually forbid unaccompanied women in public spaces, so intent was their need to control female bodies and procreation. Because of this Roman lean toward outright patriarchy affecting Egypt, the depiction of Osiris shows him about to "re-create himself," as Cooney points out. Yet it is the act of coitus with Isis that not only revives Osiris—a transformational moment like the Sumerian *hashadu* sexual healing ritual—but also explains the creation (a long time after Horus had already been created as an Egyptian god, mind you) of Horus,[279] the king-making god. In fact, the most

---

[278] This same manipulation of the pharaoh's penis is traditionally performed by the God's [Amun's] Wife, who was the most powerful woman in Egypt, so much so that Akhenaten waged a war against the Temple of Amun, over which the God's Wife presided, attempting to force Egyptians to worship a different aspect of Ra, the sun god, by closing the Temple of Amun and erecting a new temple to the Aten. The primary political reason seems to be control over Egyptian wealth, but we clearly have males trying to dominate all women in this moment in history, too.

[279] There were at least two different Horuses worshipped in the 3000+ years of ancient Egyptian culture. "Horus the Elder (or Horus the Great), the last born of

important of the three to five names given to each ruler was her/his "Horus" name.

Cooney clearly demonstrates in the passage above that she was aware of the importance of the "divine feminine," as she calls it, in earlier Egyptian eras. But she also points out that males are assuming the roles of creators in Egypt when she interprets Osiris' gesture as an attempt to re-create himself through masturbation—an Egyptian ritual to verify the king's ability to continue to ejaculate as he aged performed by the God's Wife[280] because the king's ability to issue forth milky ejaculate meant the waters of the Nile would continue to flow and to replenish the land. This ejaculation *per manus* ritual led by the God's Wife and performed on the reigning male ruler made the office of God's Wife, which was the highest ranking priestess position in Egypt, as well as the Amun temple this important woman served, extremely powerful in the 18th Dynasty. Strangely enough, to my knowledge, no single God's Wife is ever named by scholars, despite the power this position held.

However, the myth behind the High Priestess position refers to the fact that the kings are supposed to be fathered, not by living human males, but by the god Amun, thereby causing the royal descent line to be matrilineal, not patrilineal (Wilkinson). Part of Amenhotep II's reason for reducing the influence of the Amun cult during his reign (1427-1401 BCE) was because he despised the power women had gained through this office in Egypt. His son, Akhenaten, would carry on this political battle with the Temple of

---

the first five original gods, and Horus the Younger, the son of Osiris and Isis," according to World History Encyclopedia. Both were sky gods (hence the hawk association). As the Elder Horus, Horus is actually the brother of Isis and Osiris, who fought perpetually with his other brother, Set, in a good versus evil kind of dichotomy. Because so many goddesses were being replaced with male gods, Horus the Younger became associated with both the sun and the moon, which was an attempt by the Egyptians to co-opt previously female territory, the moon, which always appears between Hathor's raised horns and combine its power with those of the more (by this time) masculine-associated sun.

[280] This ejaculation *per manus* ritual led by the God's Wife and performed on the reigning male ruler made the office of God's Wife the highest ranking priestess position in Egypt, and the Amun temple this important woman served extremely powerful in the 18th Dynasty.

Amun and the God's Wife by upending the religious system to elevate another solar god, Aten,[281] as the primary god of Egypt.[282]

Even though there is ample evidence that Egypt struggled with the expectations put forward by the growing androcentrism in their world, there is much to suggest that Egypt and Sumer heavily influenced each other—from the easily comparable pyramids and ziggurats to the reduction of female gods as important in their pantheons as androcentrism, triggered by the biological discovery that males have a physical, biological role to play in procreation, began its rise toward patriarchal domination. Just as they shared cultural ideas, they clearly shared scientific advances, as well, so that we can say that biological sciences easily began somewhere in the eastern Mediterranean circa 2400 BCE.

In the same way Egypt and Sumeria have clear parallels, we can also see through the evidence that Egypt was more gynocentric earlier, more sex-as-magic oriented, prior to the gradual rise of androcentrism.

The real question remains: was ancient Egypt ever a full on patriarchy?

---

[281] Unlike the other aspects of the sun, like Amun, Aten was only represented as the solar disk—never as a humanoid god. In most of Akhenaten's depictions of the Aten, he and his family are present and the Aten is merely symbolized by hands reaching down from the sun disk. In effect, he changed the worship of the god/divine being into a worship of himself.

[282] There are many scholars who try to make Akhenaten's move to suppress the worship of other gods, which he did not outright forbid, as a move toward monotheism. However, even when the Hebrews later followed the Sumerian example and attempted to assert "one god, one king," as their new theology, they met with resistance just like Akhenaten did with most of the common Judean people continuing to worship their god's wife, Ashera, at home (Dever 101). Frankly, there is no such thing as a monotheistic religion when, of necessity, there is another powerful divine being opposing the so-called good one to create a dichotomy of good versus evil.

Chapter Twenty-Three
The Hittites Change Inheritance Laws, Sort of

Compared to Sumerians and Egyptians, Hittites were fairly late arrivals to the eastern Mediterranean scene, with most scholars agreeing that they moved into what is now Turkey between 2300-2000 BCE. Assyrians had, apparently, established trade relations within the Anatolian peninsula before the arrival of the Hittites (Collins 25), but the Hittites were quick to learn cuneiform writing from them, recording one of the earliest versions of an Indo-European language that was fairly easy for scholars to decipher, once they learned the language and basic grammar rules. Archeologists from the German Archeological Institute have found thousands of clay tablets preserved in the ruins of the Hittite capital city, Hattuša, and correspondence, including at least one important peace treaty,[283] exists between the Hittites and rulers in Sumeria and Egypt, as well.

The Hittite language, closely related to Luwian,[284] is considered one of the oldest Indo-European languages (Peaks). For a growing number of scholars, the Luwians were actually a more dominant culture on the Anatolian peninsula and for a longer period of time than the Hittites (Hatti), but less is currently known about the Luwians than is known about the Hittites, although both were matrilineal and matrilocal at the beginning of their histories in Anatolia, what is now Turkey. By many accounts, the Luwians outlasted the Hittites by nearly a thousand years (Who).

Even though the Hittites rose to power after the Egyptians and the Sumerians had learned that heterosexual intercourse is required for pregnancy circa 2400 BCE, most indications suggest the Hittites were reluctant to give up their more gynocentric cultural views in favor of androcentric ones, although they still leaned

---

[283] The Treaty of Kadesh was written in Akkadian circa 1269-74 BCE, but the Hittite account of the Battle of Kadesh presented is markedly different than Ramses II's account, which paints him as the glorious hero of the battle.

[284] Luwian hieroglyphics, which was used from 2000-600 BCE, "ranks, therefore, as the first script in which an Indo-European language is transcribed" (Luwian).

heavily in the direction of becoming patriarchal by the end of their empire.

Even by 1990, scholar Harry Hoffner, Jr. had to admit that the goddess Inara mentioned in an early version of a Hittite story, which he simply calls "The Illuyanka Tales," "was not considered subordinate by the Hittites themselves" to the Storm God (10). Instead, Inara, like Sumeria's Inanna, had the power to make kings by having sex with men through her priestess representative on earth. For some scholars, the male character in the tale, Hupasiya, was "a *Jahreskönig*, a king who with his priestess queen guarantees the flourishing of livestock and vegetation," which would make such a sexual ritual performed at the Hittite Purulli Festival very similar to the Sumerian New Year *qursu* ritual discussed in the Inanna chapter. While this similarity could mean that the Sumerians and the Hittites originated from a similar area of the world, so held similar religious ceremonies, it could also mean that the Hittites were extremely influenced by the Assyrians from whom they took writing lessons. Since the Hittites' love for their primary goddess was so deep, the former possibility is stronger than the latter, so we could then assume that sex was, for the Hittites, a spiritual act that, when enacted between a king and a high priestess, ensured continued prosperity for the land.

Like me, Hoffner dares to argue against the idea that such sexual acts were seen as *hieros gamos* or sacred marriage by the ancient Hittites. Hoffner points out that "the nature (and even existence!) of [*heiros gamos*] in other ancient Near Eastern cultures is still seriously questioned" (11). Indeed, scholars are imposing a much later concept on earlier cultural activities that had nothing whatsoever to do with long-lasting paired bonds in any legal sense.

In discussing the second version of the same Hittite tale, Hoffner notes that "if a young suitor was too poor to pay a bride-price for a wife, he could offer himself as a 'live-in' husband [wife?] to a wealthy father-in-law in exchange for a 'bride-price' paid to himself" (question mine, 13). In other words, just as in the earliest versions of the *Epic of Gilgamesh* wherein Gilgamesh is abusing his rights as Uruk leader by having sex with anyone and everyone he wants, male or female, the Hittites also allowed homosexual relations with one man agreeing to act as the 'bride' to another man, so that the poorer man could earn a 'bride-price' which was required before he could bind himself (become a husband, literally, bound to a house) to a woman in marriage, in essence proving he could provide for a wife. An example from the myth "Telipinu and the Daughter of the Sea God," has Telipinu giving the Sea God a "thousand cattle and a thousand sheep" (Hoffner 27) to prove he was worthy of marrying the Sea God's daughter, but we have to remember that this exchange was made between gods, so the number of animals is undoubtedly a god-scaled exaggeration. At the end of the tale, "The Disappearance of the Sun God," for instance, the narrator asks listeners to repay the male Sun God for reappearing to drive off the cold and death by paying him a tribute of nine sacrificial animals, "and may the poor man give [the god] one sheep" (Hoffner 28). Typically, the bride-price or bride-gift was like a dowry, ensuring that, should the man leave the woman, she had the financial means to survive without him, as long as her family, who protected her bride-gift for her, had not spent it. A bride-gift was paid whether the man moved in with the woman's family (matrilocal practice) or the bride moved in with the groom's family (patrilocal practice) because it was physical proof, usually in the form of something of value like livestock or goods, that he could provide for a woman and any children she might have, not to mention proof that he could care for her elderly parents later in life.

What is most questionable about this particular interpretation of events is the idea that the poor man in the tale is moving in with the rich man he is "marrying" in order to earn the bride-price. It is much more likely that the poor man paid for the funds for his bride-price without having to move into the actual rich man's house in the same manner a female bride would, but, instead, the poor man would have been "attached" to the rich man's household in some agreed way, such as providing labor in the agricultural fields or in building irrigation canals.

But this situation begs us to ask: Who moved in with whom, if anyone moved anywhere, in early Hittite culture when a couple married?

The Hittites had thousands of gods, but they called their designated Mother Goddess, Hannahanna,[285] and she was as wise as she was nurturing and careful. Her personality contrasted greatly with the often less controlled and more violent male gods, like the Storm God and his son Telepinu. In the tale, "The Disappearance of Telepinu," Telepinu, an agriculture god, has gotten angry, thrown a fit, and disappeared in a huff. The tale lists all the negative results of Telipinu's tantrum—besides eliminating grains growing in the fields and causing animal fecundity to stop, he also sends a drought into the mountains where the trees are drying up, which is a fire hazard. The Storm God and many lesser gods search for Telepinu, but cannot find him, so the Storm God sends sharp-eyed eagle aloft to search for Telepinu, but it still cannot find the angry god. Finally, the gods turn to Hannahanna, imploring her for advice. She sends a bee to search for the errant god, but the Storm God berates her, claiming such a small insignificant being could not possibly find a god (Hoffner 15).

However, the bee does find Telepinu, and stings him awake, after which the bee has to comfort and mollify the angry god with a litany of items—from beer to fruit—to get him to remain calm enough to return home, but the god becomes angry again, striking everything with lightning (Hoffner 16).

Kamrušepa, the Hittite and Luwian goddess of healing magic, sees Telipinu's new tantrum and asks the gods for 12 rams to sacrifice over the wayward god, proceeding to pour *karas*-grains through a straining basket over Telepinu, and smudging him with smoke in order to remove "his evil" from his body (Hoffner 16). Repeating a litany of what are probably ritual words to banish negative conditions from the body, Kamrušepa sends the wayward god's negative energy "down in the Dark Earth" to burn up in bronze vats of lead (Hoffner 17). Once Telepinu comes to his senses, he releases all the elements of nurturing and sustenance that had been hidden away during his rampage, renewing, once again, "Plenty, Abundance, and Satiety" (Hoffner 18).

---

[285] Some scholars interpret Hannahanna to mean "grandmother," but it probably literally means "mother's mother," which means it is matrilineal and does not include a man's mother.

Hannahanna, however, was not always a kind and happy goddess (what mother is?). In the tale, "The Disappearance of Hannahanna," the Hittites believe that the failure of their crops to sprout, for the failure of their cattle and sheep to bring forth young, and because "mothers took no account of their children" (Hoffner 29), means that Hannahanna had disappeared in anger. Hittite gods seem to get angry a lot.

After someone pours "holy water of the Queen of the Spring" on top of a mountain or an altar (the tale is on a clay tablet which is fragmented), "a cloud arose. The vapor (?) came up.... It went into Hannahanna's body. It drove from her body [Hannahanna's] anger, wrath, [sin, and sullenness]" (Hoffner 30).

First, realize that interpreting any Hittite word as "sin" is absolutely sinful because that Christian term did not mean the same thing it does now 3500 years ago. Sin was the name of a Sumerian moon god, and Sumerian and Akkadian women were called *sinnestra*—people of the moon and stars. As androcentrism rose, *sinnestra* became sinister, and Sin came to mean "things done in the dark of the moon," like sex, and, later, things done that anger gods. But I have yet to see any proof that this particular concept was embraced by these cultures by 1500-1200 BCE. Therefore, when interpreters use Christian-based ideas to interpret other religion's concepts, I become super wary, as should you.

This particular Hittite tale could have provided various insights or ritual instructions, such as pouring libations of holy water on particular altars or mountains to invite this particular goddess back to nurture her people. However, one thing we learn from the many Hittite myths about their gods is that they anger easily and often disappear in a huff until proper sacrifices are made to appease them. In this way, they explain why their people are suffering from all kinds of negative events—from plagues to drought.

More problematic for some scholars has been getting a handle on how the Hittites viewed their primary Sun Goddess, Arinna[286]. By most indications, the Hittites believed the sun was

---

[286] Arinna as Sun Goddess was also the "Mistress of the Netherworld because of her nocturnal travel" (Kristiansen and Larsson 283) beneath the known surface

female[287] before they reassigned that role to a male god. Part of the confusion for scholars was the fact that prayer tablets to Arinna (named after the city in which she primarily resided) in cuneiform that had been interpreted in the late 19th or early 20th centuries ACE used a term for "lord" while addressing Arinna that "did not agree with the female sex of the Sun Goddess" (Güterock 237). Failing to consider that the term might actually be genderless, interpreters like Hans Ehelolf instead exchanged Arinna's name and title for that of Telepinu, the primary god of agriculture, in Ehelolf's published versions of other prayers, since other copies of similar hymns were clearly to Telepinu. Hans G. Güterbock noted that when Oliver Gurney wrote about the prayers in his dissertation, "he correctly remark[ed] that the solar characteristics do not fit the nature of Telepinu," and that the Telepinu prayer left no space for the obvious hymn to the Sun Goddess, which also "contains the phrases characteristic of a solar deity" (237). However, not even Güterbock, who argues that the Hittites borrowed many elements of their prayers from Akkadian and Babylonian hymns (241), considered that Arinna was being replaced by the Hittites in preference for male gods as androcentrism increased in their culture. Instead, he reverses the probable order, claiming that "the version addressed to the Sun Goddess of Arinna omitted some of the sections contained in the former" (Güterbock 244), even though he admits that "the Sun Goddess of Arinna [was] the highest deity of the Hittite pantheon" (Güterbock 245).

When J.G. MacQueen discusses Arinna, he admits that

> as far back as our records reach, they stress the supreme importance in Asia Minor of a female deity. In Hattian myth, although the Sun-God may be the titular head of the divine assembly, the prime mover seems always to be a goddess—Hannahanna or Inara or some other,

giving evidence from the idols found at archeological sites in Anatolia because "any male figure associated [with a goddess] seems to have an essentially subordinate position" (MacQueen 177).

---

of the earth. Her alternate names often included the title Sun Goddess of the Earth because she could rule even when hidden, seemingly, within the earth.

[287] The Japanese and Cherokee have also long believed the sun to be female.

He finds this organization of power among the gods natural because every goddess represents some aspect of the "Mother Earth, and all being and all fertility proceed from her" (177). Further, MacQueen points out "that the association of divinities with springs and [water] sources reaches back into the pre-Hittite period, to a time of Hattian water-gods and mother-goddesses" (178).

MacQueen goes on to discuss the Indo-European invaders, some of whom were the Hittites, who

> found in Anatolia, as they did in Greece, a goddess heading the native pantheon. Rather than depose her, the conquerors followed the usual polytheistic principle of [their land, their religion] and married their god to the native goddess. In Greece, Hera had for the most part to be content with a secondary place; in Anatolia the influence of the mother-goddess was so great that she gradually gained authority over her Indo-European husband, and by the time of the [Hittite] Empire was head of the pantheon once more (MacQueen 180).

That a female god should be ascendent above the male gods is one indicator that the Hittites were originally a more gynocentric culture.

Like many, if not most of the ruling elites in the eastern Mediterranean, the Hittites referred to their royalty as the "'great family,'" or, specifically, "*šalli haššatar*" (Beckman 442). The female ruler, usually called the queen, "among the Hittites is unusual both for the important part she played in matters of cult and state, and for the fact...that she continued to reign after the death of her husband" (MacQueen 181). Many scholars assume the queen's importance is a remnant of their "earlier system of matrilineal succession" (MacQueen 181) because the queen "was in origin the true ruler of the land, the true representative of the mother-goddess" (MacQueen 181).

Many Hittite scholars struggle with the lines of succession in Hattuša because the matrilineal tradition of selecting a male to rule from the king's sister's sons confuses them. This tradition was used in both early Egypt and in Sumeria, but it fell by the wayside after androcentrism began to encourage people of the eastern Mediterranean to view males as the dominant spiritual and political rulers, replacing the traditional roles of women as political and spiritual guides, if not outright rulers.

It is important to note, at this point, that these same scholars who present various arguments about the rules of succession in Hattuša also struggle with the idea that the Hittites originally were more democratic than they become later. For instance, Gary Beckman spends most of his article, "The Hittite Assembly," trying to argue that the assembly, called the *panku*, which is a sweeping universal term like the word "catholic" to mean everyone, was actually composed of the *šalli haššatar* and "the upper echelons of the state bureaucracy," but held no power to appoint rulers (442), even though the Hittites had separate words for these two groups.

The main flaw in the points of view of all of these scholars is that they cannot imagine a culture wherein women ruled. They especially have trouble envisioning a culture where women rule democratic governments (who would want to elect a woman!?!) even though many admit or come close to admitting that many of the elements of the texts they examine indicate such a thing could happen. Trevor R. Bryce, for instance, admits that the Telipinus Proclamation, which is the oldest document known that explains royal succession, allows "a king [who] has no eligible son to succeed him" to appoint "a husband of the 'daughter of the first rank'" or his son-in-law as his legal successor (12). While often mislabeled as "levirate" marriage, this type of marriage is also indicative of remnant matrilineal inheritance.

As strange as it may seem from our point of view that women could govern and did rule many cultures prior to the discovery of procreation and the increase in androcentrism among these cultures, many of these scholars begin their assumptions from the point of view that males have **always** succeeded males as rulers, despite the lack of proof staring them in the face.

One important Hittite story, for instance, never mentions a male ruler at all. "The Legend of Zalpa" (circa 1650 BCE, after humanity has learned of procreation) focuses on the actions of the "Queen of Kanesh," according to most translations, leaving us to wonder if her title was actually the top royal personage for her city-state. The unnamed queen gives birth to 30 sons, and she orders a servant to put them into a reed basket and give their fates to the river. The servant, reluctant to kill so many babies, instead puts them on the bank of the river, hoping a kind person will take them in, but the river swells, as a divine spirit would, and carries the basket and the boys down to the "Sea of Zalpa" (Gurney 7). Instead of a human being finding them, according to some sources, the gods find the boys and raise them. Meanwhile the Queen of Kanesh gives birth to 30 daughters, who she decides to raise herself. This little detail demonstrates a mother's preference for her daughters—through whom she will pass on her title and power, indicating a matrilineal inheritance system.

"The Legend of Zalpa," however, is not just a story about a possible matriarchal Hittite culture. It is also a cautionary story about incest, since the 30 boys grow up and set out to look for their mother; the gods who raised them apparently neglecting to tell them who she was. At Tamarmara, they are told that the Queen of Kanesh had given birth to 30 sons who had mysteriously disappeared, so they travel to Kanesh. As grown men, they are unrecognizable to the Queen, who welcomes them gladly and allows them to marry her 30 daughters. Apparently, the sons know these are their sisters, and the youngest brother warns the others against marrying their sisters, since such an act would be incestuous, thus violate the will of the gods.

In one version of the tale, the tablet is broken, so readers only know that the tale ends with the destruction of Zalpa (Gurney 8). In other versions, the young women have also figured out that the young men are their brothers, so plot to kill them on their wedding night for daring to marry their own sisters, thus the slaughter brings about the destruction of Zalpa. In another version, the youngest brother and sister agree to leave the others to their fate, but it is uncertain what happens to the youngest couple.

As with the ancient Sumerians, the Hittites valued their war-sun goddess above their other gods, although they later allow the Indo-European sun god, Istanus, nearly equal weight circa 1400 BCE because of a mix of "Luwian and Hurrian cultures," which gave the Hittites "a great mass of rituals, mainly magical, and typically performed by the so-called 'Old Woman'" (Gurney 10 & 16). Many scholars credit this shift in perspective, from a sun goddess focus to a bull/weather god, Teshub, with his consort Hebat, on the androcentric influences from the north (Gurney 16). The fact that the Hittites had many problems with royal inheritance systems because of their struggles in deciding which gender was most important becomes abundantly clear when we study their history.

Bryce struggles, for instance, with why Hattuššili "appointed his sister's son, 'the young Labarna' as his son and successor, but subsequently repudiated him in favour [sic] of his grandson Muršili" (10). Perhaps unfamiliar with matrilineal succession, Bryce did not realize that one common aspect is that succession must come through the women, not the men, since we are not always certain who the father of a child is, but we always know who the mothers are (hence the likelihood that early succession was always through the mother). Emersed in Judeo-Christian biblical tales, many simply dismiss this matrilineal tradition as a "levirate" marriage, failing to recognize the origins of such marital traditions, too.

By most accounts, the Hittite royalty were a blood-thirsty ambitious group, which is understandable when we acknowledge that blood sacrifices were meant "to absolve a house from a blood-guilt"[288] (Gurney 29). After Hattuššili successfully traveled to and sacked Babylon, circa 1595 BCE, making allies as he went (Collins 40), upon returning to Hattuša, he was assassinated by a brother-in-law, Hantili, who assumed the Hittite throne. During Hantili's reign, "his wife and sons were captured and murdered by the Hurrians" and "his remaining heirs," which would have included all of his sisters' children, were murdered by a previous ally, Zidanta, who assumed the throne (Collins 41). Zidanta was murdered by his own son, Ammuna, whose own sons were assassinated by Ammuna's brother-in-law, Huzziya (Collins 41).

---

[288] Blood guilt is an ancient belief that people who murder others—on purpose or by accident—must absolve themselves of the spilling of blood through ritual and sacrifice in order to return balance to their own lives. Blood guilt gets mingled with menstrual blood, later, so that menstrual blood also become "unclean," and women who once secreted themselves away to keep their stronger magic from interfering in male magics begin having to segregate themselves (not even being allowed near a temple or church while menstruating) because the negativity pushed onto all things feminine becomes so strong that women's natural monthly bleeding becomes associated with evil.

Later, clever King Telipinus, turned the family tragedies to his advantage, claiming that the usurpers were being punished by the gods threefold: 1) by drought and famine, 2) by invasion from neighboring enemies, and 3) by the loss of their families to murder (Collins 41).

In order to try to end the extended family bickering over the throne, Telipinus established new rules for succession in order to make it clear who was heir to the throne. First in line was any son from his primary (and legitimate) wife (apparently the order of the sons' births did not matter). Second was any son from a secondary wife or wives (as though that would solve the assassinations!). And third was a retention of the matrilineal version wherein any of his sisters' sons could become ruler (Collins 41).

What fascinates me most about the Telipinus Edict (aka Proclamation) is how many knots scholars will tie themselves into to avoid admitting that the third version is a common matrilineal form of succession. Siim Mõttus calls this form of succession by its Hittite word "antiyant" or uxorilocal, both of which simply mean matrilocal, but would more correctly be called matrilineal (44). Either way, the scholars tend to hide behind big words instead of being more clear and direct in admitting that women were not only vital to maintain royal Hittite lineages, but also important leaders in their own right.

In essence, while Telipinus desired an end to all the family bickering over power for the "king of kings" position among the Hittites, his Edict failed to end the violence. It failed because, as egotistical androcentrism replaced cooperative gynocentrism in these cultures, many began to feel it was not only right to kill for power, but also to believe that "might makes right." Hence, it is not terribly surprising that such a violent culture that worshipped not only violent, angry gods, but also grew increasingly androcentric, unleashing a male tendency toward violence as a means of controlling others that has grown into what we now call toxic masculinity, created a culture that imploded.

Chapter Twenty-Four
The Minoans:
More Gynocentric Than You Know

If you were raised at all with the mention of the Minoan[289] civilization on the island of Crete, you probably heard about King Minos, the mythical ruler who forced Athenians to send a tribute of seven young females and seven young males for sacrifice to the Minotaur, half-man, half bull, who lived in Minos' labyrinth beneath his capitol city, Knossos.

Unfortunately, most early archeologists began their careers trying to prove their favorite myths to be verifiable history, so Minos was foremost on the minds of the first Europeans to dig through the ruins found on Crete. Of course, it did not help much that the first European to excavate on the island was a Greek from Iraklion named Minos Kalokairinos in 1878 (Lerner). Kalokairinos was forced to stop excavating, but, 22 years later, British archeologist Sir Arthur Evans, who had visited Crete in 1894, began excavating the site in 1900 and spent more than 30 years excavating and, unfortunately, recreating the sites he discovered, which he felt was the only way to preserve the palace site.

While many scholars cringe at the recreated parts of the Evans' excavations, and while many scholars are willing to disparage some of Evans' findings, especially about how he dated pottery, scholars like Jan Driessen agree with some of Evans' sexist views, relegating rhytons, drinking vessels with "female attributes," to mere "fertility" devices for specific rituals, but also seeming to agree with Evans' assessment that the stone "throne," that Evans insisted was the throne of Minos himself, was too small "to fit a female bottom" (Chercher 148). A sexist view, indeed.

Driessen, like many androcentric scholars studying ancient prepatriarchal cultures, wants to be fair, so admits that, in Crete, "the iconographical evidence suggests that women, and perhaps

---

[289] What did the people of ancient Crete call themselves? Not Minoans and not Cretans, we know that for certain. Still, I will use these terms because they are familiar.

especially senior women, possessed agency and authority in public ceremonies" (Chercher 147), but he clearly assumes he knows exactly how a matrilineal and matrilocal culture would organize space for living, based solely on the Ashanti culture, because he believes "related women (sisters especially) **should** live together with different non-related males (the husbands of the sisters)" (emphasis mine, Chercher 11), and even asserts that "matrilineal societies tend to be matrilocal: a man goes to live with his wife and her mother's kin" (Matrilocal 270), ignoring the many other matrilineal cultures that practice solitary living, each woman building her own home, with males visiting them, when invited, at night. Otherwise, the males live in bachelor quarters, or, as Driessen himself finally admits should be called the "Men's House" (Chercher 155). Driessen is clearly not convinced that the culture was gynocentric, despite evidence touted by other scholars who note that "throughout the Neopalatial period, elite women in art (and so presumably in life) wear more elaborate costumes than men as a sign of their importance" (Younger & Rehak 161).

Driessen admits, in a different essay, that the primary social unit in most cultures was a "house, in the Levi-Straussian sense" (Matrilocal 364). He attempts to differentiate his interpretation of a house by calling the Cretan ones "Established Houses" in order to "distinguish the term from the mere architectural structure" (Matrilocal 364). Just as Hatshepsut will remind all of Egypt that their royalty descends from a sacred line of women by adopting the term "pharaoh," which literally means "house," a term that scholars will, ironically, use for all Egyptian rulers except the women, Driessen begins to see sense in admitting that Minoan matrilineal descent lines were clearly associated with the women who started them.

Finally, he admits that continually rebuilding on the same site localizes the family group, allowing them to develop a sense of place in association with the land and to define their place in a larger community, "as well as creating a House genealogy" (Matrilocal 366).

Like many scholars, Driessen has trouble seeing women tending livestock like cattle, though. He believes "the massive investment in livestock in the later Mycenaean period [on Crete] fits better with a patrilineal society" (Matrilineal 371) because he believes all "patrilineal societies are…pastoral…whereas matrilineal groups [are] associated with horticulture and trade" (Matrilineal 370). Here, he repeats a common androcentric viewpoint from scholars studying Africa and Eurasia where men often went on raids to steal other people's cattle. He almost immediately contradicts himself, however, when he claims "matrilocality and matrilineage are good solutions in societies where men tend to be away for long periods," using the fact that many men went out on large construction projects as an example (Matrilineal 371), ignoring the fact that many traders were men, so would have been gone for long periods, too.

However, we have ample evidence that just because women ran the communities did not preclude them from sending men (and probably women) out of the communities on various tasks, such as trade expeditions, something Hatshepsut is quite famous for, actually. As one American Indian woman admitted to me, her grandmothers used to joke about having to send the men away from the village during the winter in order to allow the women to get their winter work, such as tanning hides, creating clothing, making pots, or healing those who caught seasonal illnesses, done because the men mostly just wanted to have sex, tell stories, and often sat around arguing, so fell to plotting tricks, and sometimes violence, against others. She firmly believed regular trade expeditions were planned among her people, one of the matrilineal Iroquois nations, simply to get the more idle men away for extended periods.

So Driessen's idea that having women in control only makes sense when the men are away flies in the face of the real logic that women needed more space to make the goods they produce for a better lifestyle by sending the men away to trade, to raid, or to work on large community building projects.

It would be highly amusing if we could verify that huge building projects—from Göbekli Tepe to the Egyptian pyramids—were created by women in order to keep men busy. Of course, androcentrists like Driessen prefer to believe women can only lead when men are absent.

Driessen draws a female scholar, Charlotte Langorh, into his assertions about the primacy of "nuclear families," a modern social ideal, in another article (Recent Developments 91). While they initially admit that "social theory and anthropological parallels" have influenced more recent scholarly examination of Crete (89), they readily admit that earlier outdoor feasting and ritual activity dominated prior to the development of "palaces," which indicates gynocentric rituals, wherein women, who are tied to the moon via menstruation, ground themselves by being barefoot on the earth outside paved buildings. Driessen and Langorh also admit the fact that the growth of palace-like structures took more time over more decades than originally believed (89). They even go so far as to admit that "basic relationships," rather than

> a traditional hierarchical organization with an almost superimposed, esoteric priest-king, a faceless elite and even a more discrete lower class, a more varied and dynamic social structure is now suggested, in which both hierarchical and heterarchical elements were interwoven in a more hybrid way (Driessen and Langorh 91).

Basically, they are admitting that communities on Crete were much more complex than the rather simplistic super-stratified hierarchical structures Driessen had proposed in earlier articles. They also naively assume that "nuclear families certainly existed at every level and time of Minoan history" (Driessen and Langorh 91), despite the lack of evidence for such a claim, and the fact that Driessen reluctantly admitted in another article that there were structures that probably housed men called the Men's House (Chercher 155). Such a space is typical of matrilocal-matrilineal societies where women own the houses and live with their offspring, while the men in their lives only share a roof with them overnight for sex.

Reluctantly, it seems, Driessen admits that "spatial segregation would eventually have led to gender stratification and a different social situation altogether" (155), assuming that such spatial segregation was not present from the beginning when women were in control, but admitting that such segregation becomes obvious when men elevate themselves above women in importance, which presumably occurred post-procreation discovery. The idea that one man and one woman is the ideal basic social unit, aka "nuclear family," often keeps many scholars of ancient cultures from seeing any other social structures, even when they apply anthropology and sociology to their findings, except for men with harems of wives or "concubines." Having a society in which women are free to choose their sexual partners seems completely alien to such androcentric scholars.

To give him credit, however, Driessen comes close to admitting that a gynocentric culture would be focused on menstrual rites as their primary religious focus when he credits Katarina Kopaka for "identifying blood on the horns [of consecration] depicted in the sunken room of Xeste 3[290] at Akrotiri." It was Kopaka who "proposes to see the lustral basin as...a menstruation environment, linked to specific rites of fertility" (Driessen, Chercher, 11). While Driessen wonders whether "all lustral basins [were]... specifically reserved for women and their" menstrual rituals, his reluctance to accept the known connections between menstrual blood and Female Magic keeps him from understanding that the "horns of consecration" were not bull horns, but the horns of the moon when it is in its crescent shape, either waxing or waning. Also seemingly avoided by most scholars is the idea that the smaller horns of consecration could have been dildos, with ritual genital penetration for young women as a part of maturation ceremonies.[291]

Driessen claims that "ritual actions especially took place outside...suggested by specific natural elements (trees, rocks), or with architectural facades and portals as backdrops," claiming that "scenes involving women taking place **inside** are, however, more difficult to find" (emphasis his, Chercher 147), as though women have been relegated outside the buildings, instead of choosing to work their Female Magic where their feet would be more grounded, actually touching the earth. More likely, however, is the fact that matrilineal cultures sought unity, instead of discord, so kept religious rituals outside in order to allow everyone to participate.

---

[290] Ute Günkel-Maschek warns that Akrotiri was not Crete, so it is advisable, instead, to speak of Theran aspects" when talking about Akrotiri, not Minoan, although she admits there were many parallels (117).
[291] See Ute Günkel-Maschek's analysis of the Xeste 3 wall murals depicting girls and women of various ages enacting goddess-centered rituals.

Later, in the Mycenaean era, such rituals will be moved inside, only allowing select individuals to participate.

Reluctantly, Driessen agrees that, "despite the presence of male-oriented ritual space within elaborate residences, a consequence of the matrilineal and matrilocal tradition may have been that men were accustomed to unite outside of residential structures which were, especially where ceremonial functions were concerned, primarily the domain of women" (Driessen, Chercher, 155), since most lustral basins are located within buildings. According to Dario Puglisi, "some monumental buildings [on Crete] dating to the Neopalatial period" correspond "to the historical *andreia*, the men's halls mentioned by ancient Greek literary sources" (66), admitting that it is challenging to identify that the Cretans practiced marriage or organized themselves as nuclear families. Driessen did admit, earlier in 2012, that the length of time—hundreds to thousands of years—in which some places are occupied and rebuilt over time makes it difficult to argue for the idea of "nuclear families" (Matrilocal 265), but he has clearly changed his mind by 2016.

Strangely, Driessen cannot admit that women were the literal "homemakers," not to mention home organizers, nor is he able to even speculate that men lived separately from women in matrilineal cultures, unless invited inside. So tied up in the ideals of patrilineal descent is he that, while he acknowledges that "humans die, the House does not and in this way the House offers its inhabitants a sense of immortality," he fails to acknowledge that a matrilineal and matrilocal (aka gynocentric) culture would probably have given that ascendency and near-god-like quality only to the women.

Sometimes it is quite painful to watch androcentric scholars do everything they can to avoid acknowledging that the women of Crete built and guided the so-called Minoan culture.

Puglisi, on the other hand, considers ancient Crete to have been a type of "acephalous society," which was egalitarian and prevented centralized leadership, even during the palatial development of the island because of "the high level of ritualization, especially developed in the palatial period and focused on initiatory practices; the social stratification and high level of social differentiation; [and] the absence of rulers" (70). He even considers one aspect of the Minoan "system of rites of passage [was] collective marriage," which was historically documented in literary sources (73), believing that, since significant rites marked individuals' progression through Minoan society at specific ages, even marriage would involve all members of the same age group. However, he argues, with no evidence, that the Minoans understood the connection between coitus and procreation, claiming that such a marriage, even as a group, was "aimed to procreate and raise legitimate progeny" (73), failing to explain what these ancient people who were more egalitarian than most cultures around the world would have considered "legitimate progeny," especially if they still believed women were the sole creators of other humans.

As Stephanie L. Budin puts it, "females do appear in prominence in the arts of both [Cretan and Mycenaean] societies, showing them as goddesses, functionaries, persons of authority, watchers, travelers, dancers, and flower-gatherers. Just not [as] moms" (109), as she takes issue with so many depictions of females with younger humans, especially since so many scholars opt for the "Madonna and child" syndrome of interpretation. While Budin doubts that the Cretans worshipped a goddess, her analysis of the Mavrospelio Kourotrophos, which she believes "represents an adolescent girl holding aloft an idol of an adolescent male deity," falls short of explicitly illustrating what she seems to believe the girl is doing with this "adolescent male deity" (110). Her argument for defining the female as adolescent because of her hairstyle, "a proto-Mohawk—a line of somewhat longer hair running along the center of the scalp" (110) is believable, but she never bothers to explain why the object held by the girl is male or a deity. And her logic takes her very close to claiming that the girl is using the doll as a sex toy when she says, "rather than suggesting that the Mavrospelio figural group shows a mother and child/lover, we might instead consider the object to represent an adolescent girl in the midst of ritual action" (111). It would be fascinating, indeed, if young girls were encouraged to imitate coitus, which was possibly a more public activity than many androcentric scholars are willing to admit, with dolls, and even more interesting if they did so believing the doll was a physical representation of a god. The girl, of course, could also just be playing with a doll, a common practice in preparing girls to become mothers.

Budin, in fact, is so obsessed with proving that the people of Crete did not worship a Mother Goddess figure, she leaps full into her oversimplified claim in another article that, because so few images of mothers and children exist among the Minoans, "all this leads to is the ultimate fact that the Minoans actively rejected the image of the *kourotrophos*—the female who intimately nurtures and supports an infant" (, 31), clearly forgetting that the absence of something is not proof that it does not exist. What she so eagerly overlooks are the images of menstruation and sexual attraction that fill Minoan art—from the bleeding lustral basins to the leaping young males (?) tumbling over the backs of bulls.

Why must a "Mother Goddess" be an actual mother? The goddess Artemis, after all, whose incarnation as Kourotrophos gives us the name for the women who are pictured with infants that Budin uses, was always considered a "virgin" goddess, meaning that she always appeared as a youthful young woman forever on the verge of womanhood. See the chapter on Virginity for more about the first definition of *parthenos*.

Susan Lupack argues that "it is clear from the numerous representations of female goddesses in various media that a female deity was the major religious focus of Minoan religion. She appears as the Mother Goddess, the Mistress of Animals, and the Guardian of Cities" (258). In fact, Lupack asserts that "the act of summoning the goddess and her resulting epiphany appears to have been a major part of Minoan religion" (258).

While Lupack's view of the gynocentric nature of the communities on Crete is more accurate, she still falls into androcentric views when discussing some symbols, such as the cattle horns. When describing a set of rhytons, she calls them "bulls head rhytons" (Lupack 257), and links the double headed axes, which are also opposing crescent moons, therefore symbols for Female Magic, to bulls (258). I have to assume that most of these scholars[292] have never actually seen cattle with horns, which is the

---

[292] Helen Whittaker is one scholar who admits that the many cattle skulls and art-rendered cattle horns cannot be all bulls, although she admits that many scholars make such assumptions (109). Eleni Konsolaki-Yannopoulou, on the other hand,

most natural state for both male and female cattle but have only seen polled cattle on farms. Until selective breeding created cattle without horns, even cows' horns would have been large, with some even rivaling bulls' horns in size. As J. Alexander MacGillivray notes, "In Egypt, we find that certain pre-dynastic cults studied by Percy Newberry shows that the X, *djew* (horizon sign) and bovine [horns] were grouped together already in c. 5000 BC, and there is the record of 'priest of the double-axe' in the fifth dynasty" (125), which is arguably the earliest known use of bucrania symbolism.[293] However, since the symbolism is clearly associated with what is called the "Hathor Bowl," which depicts horns with five stars "that make up what archaeo-astronomers call the Egyptian X, which Evans knew as Hathor's Cross" (MacGillivray 125), the horns were more likely to be from a female cow, rather than a bull.

A natural problem that scholars studying artifacts from ancient cultures encounter is the fact that they only see one small piece of the puzzle at a time, so, unless they are heavily immersed in similar cultures from other places in the world, they often make conclusions based on very little actual evidence.

Exactly how Mycenaean, aka early Greek, culture became enmeshed with Minoan culture is still heavily debated, but Linear B tablets found in Knossos on Crete as well as on the Greek mainland have offered us a peek into the similarities between the late Minoan and Mycenaean cultures.

It was only through translating Linear B, for instance, that scholars learned there were earlier female versions of some of the most popular and powerful male Greek gods, "such as Diwia and Posidaia, female versions of Zeus and Poseidon" (Lupack 271), who had "similar powers and effects that the later male gods had (Palaima 480). Linear B tablets also introduced scholars to other

---

claims "the bovine should be viewed as a bull, although the sex is not indicated, since its overall appearance suggests a very strong animal" (50). Clearly, she believes the myth that females are naturally weaker than males.

[293] We really should count the Paleolithic Y symbols painted in many caves throughout Eurasia (Lawson-Tancred) as bucrania, however, since they clearly symbolize the reproductive organ system common to all mammals. While the scholars working with Bennett Bacon are close in assuming the Y symbolizes "two parted legs," the actual symbol stands for female reproductive organs, which are all the same shape in mammals.

female gods they had not heard of through merely studying the Mycenaeans or Greeks. According to Lupack, "the most prominent female deity in the Mycenaean corpus is Potnia. The word on its own denotes a female who wields power and can be translated as 'Mistress." However, the word is associated with several epithets and locations, such as the Horses, the Grains, the Labyrinth, Athens, and Asia" (Lupack 271). In other words, Potnia, the most powerful early Mycenaean god, was also known as the Mistress of Horses, the Mistress of Grains, the Mistress of the Labyrinth, the Mistress of Athens, and the Mistress of Asia.

Because of this evidence proving that many later Greek gods had originally been female Minoan or Mycenaean gods, we have to wonder why so many scholars resist admitting that religions based on Female Magic preceded religions based on Male Magic, especially before humanity learned that males have a biological role to play in pregnancy circa 2400 BCE.

If it was believed that women were magically producing children from their own spiritual connections with the cosmos, it only makes sense that women were seen as both more spiritually powerful than men and more creative than men, which could be one reason why so many of the early tasks that made life easier for early cultures—from making pottery to cook and to store foods to making homes out of available materials—fell to the more magical gender, women.

Much of the physical evidence gleaned from archeological digs on Crete demonstrate a clear gynocentric orientation so strong that we might even be able to call early Minoans matriarchal. Once scholars learn to distinguish between symbols designating Female Magic--many of which, like the double-axe and horns, will be co-opted by rising androcentric cultures later as symbols of male power, including the power inherent in violence— and fertility symbols, they will be properly empowered to overcome their tendencies to be influenced by their own androcentric cultures, resulting in a more accurate picture of what life was truly like for our ancient ancestors.

Chapter Twenty-Five
Hero Schmero:
The Iliad as a Cautionary Tale

When I gave birth to my son, I was in labor over 36 hours; hard labor lasted more than 24 hours. If you have never given birth, you might not understand what any of that means, so imagine having a white hot poker jabbed into your lower abdomen every few minutes for 24 hours while you are trying to push an eight-pound baby out of your vagina. Yet I have met men and women (especially those who have not given birth or who scheduled c-sections instead of enduring vaginal delivery) who believe that childbirth is child's play. Anyone can do it because, of course, it has been done for millions of years by gazillions of women. What's heroic about childbirth, they ask.

Interestingly, nearly 40 years ago, the United States Army determined that it was cheaper and faster to strength train women to the point where they could carry full battle gear into combat than it was to teach men who were enlisting how to read and to write as well as they needed to be able to in order to be effective in modern combat situations.[294]

Ask an Army officer about this research today, and most female officers will smile, shrug, and say, "of course." Most of the men, all officers or former officers, I have tried to discuss this research with hedged, claiming either not to know about the research or tried to argue that most women are still too fragile to carry full combat gear. Really? I ask. And how many men can do it?

It's hard to get a straight answer on that question, too.

Some scholars, especially those who do not recognize that they are reacting to the saga through androcentric views, do similar hedging when they write about the bloody battle scenes in *The Iliad*. Like Hector, the Trojan "hero" in the epic poem, these scholars

---

[294] See the 1980 article by Joseph Knapik, James Wright, Dennis Kowal, and James Vogel, "The Influence of US Army Basic Initial Training on the Muscular Strength of Men and Women," for a start.

assume war is "men's business,"[295] and most find the graphic battle scenes beautiful, even if tragic or outright horrifying.

Unfortunately, it is not just the colorless male supremacists who believe *The Iliad* demonstrates their superiority over everyone not European and not male. As scholar Roy Hack worded it in his 1929 article, "Homer and the Cult of Heroes," the Mycenaean Greeks were the Greeks of the fabled Heroic Age,[296] elevating these earliest Greeks "sufficiently above the level of ordinary men to entitle them to receive **worship** from their **inferiors**"[297] (emphasis mine 60). The goal, then, was to worship certain men as though they were gods, elevated above common human beings. Note that colorless male supremacists and many Hellenic scholars love *The Iliad* first and foremost because it seems to make the

---

[295] For instance, Hans van Wees (1996) claims that "the *Iliad* confirms that a positive liking for war is generally disapproved of" (5), and admits "that 'glory-bringing' is the only positive epithet of war" in the entire poem (23), but he claims "the poet himself approved of the hard line taken by these heroes" (27), wherein pejoratives, such as the "exhortations to 'be men' and taunts of being 'women' or 'little boys' ... indicate that a prime cause of shame is to fall short of standards of masculinity" (22); yet van Wees never describes what those "standards of masculinity" were for the ancient Greeks, who not only cry a lot, but also do not hesitate to call on their mothers (e.g. Achilles) for help. Overall, van Wees believes the combat descriptions Homer provides are "unrealistic," so that "the gory detail is not inspired by acute observation of the horrors of war. It is there to illustrate the superior force of **the ideal warrior**" (emphasis mine, 39).

[296] Concepts I will examine in more detail later.

[297] These concepts will be examined in detail as this article progresses, but a great example of this adoration for the Cult of the Hero follows. This idea that male warriors deserve worship from others who do not hold a similar social status carries over as an assumption by many scholars, including Carla Antonaccio, who believes that the remains of a "cremated male individual in an antique bronze vessel accompanied with weapons" (therefore assumed to be a warrior), has "obvious echoes of both the poetic tradition of 'heroic' burial, and the tradition of 'warrior graves' in the Bronze and Iron Ages" (398). Antonaccio is a bit reticent to assume the woman buried nearby "the Toumba basileus," as she calls the male, was a possible sacrifice to honor this male, even though the woman was buried with "a stunning antique, gold necklace of Babylonian origin that may have been as much as 1000 years old when it was buried with her" (390). Antonaccio does admit that there could be many other reasons why the woman was buried so near the man but likens her burial to the sacrificed 12 Trojan youths at Patroclus' funeral in *The Iliad* (390).

Greeks, often identified as the "earliest" sophisticated European civilization (read "white" for this instance, although I prefer the word "colorless"), especially the men, appear **god-like**. For Hack and others, such a view "converts [the belief in heroism] into a **natural** act of faith" (emphasis mine, 61).

While we know that many of the so-named "heroes" in the epic poem end up having cities or monuments named after them with myths claiming they had personally built those monuments or founded those cities,[298] so were, in effect, worshipped by the Greeks of Homer's time[299] and beyond, do the men in *The Iliad* really appear to be god-like? Perhaps more importantly, how did concepts of **divinity** help define who a hero was to the ancient Greeks?

For centuries now, scholars[300] have debated the heroism claimed in the poem, and some have even argued about whether or not Homer believed what he said about each character in the story. We must explore some of these ideas in order to understand what reactions Homer probably really intended in performing the poem, then by preserving it in writing for the ages.

The Iliad is not so much a brofest as it is a cautionary tale warning us what can happen when we allow hot-headed men who often see themselves as god-like to rule because such an elevation in status encourages them to believe it is necessary and right to

---

[298] Roy Hack notes "there is conclusive evidence of the cult of heroes of the Mycenaean Age in Clazomenae, Erythrae, Samos, Miletus, Panticapaeum, Clarus, Chios, and Teos; and one of the chief centres [sic] of Achilles worship was the Black Sea, in exactly those sites which were colonized by the Milesians during the eight and seventh centuries" (64).

[299] Roy Hack states that "as the Iron Age progressed, the old tribal groupings were altered: new city-states were formed in large numbers during the prolonged era of colonization, and internal political and social changes brought about the constitution of innumerable new groups, the tribes, demes, phratries, and guilds that [become] so conspicuous in Greek life. Each of these groups worshipped either a god or a hero as its divine or semidivine ancestor" (61). I discuss who the original "divine ancestors" must have been in the other chapter on *The Iliad*.

[300] Christopher P. Jones notes that "Plato connected the word [hero] with *erós* or the verb *eirein*, 'ask,' and others such as Augustine derived it from the goddess Hera, a view that has found support in modern scholarship" (3).

suppress other people in order to be the "big men"[301] who lead not just cities, but whole cultures.

To prove this point to you, I need to unpack many ideas that have been discussed about *The Iliad* but **have never fully examined from a nonandrocentric perspective**.[302] Along the way, we have to address the important questions of what a hero is to us, so we can better understand what a hero was to the ancient Greeks, and what social changes, both before and during Homer's era, gave rise to the worship of individual males as semi-divine or even as gods outright.

Note that I expect you, because you have been raised in an androcentric culture, to disagree with almost everything I say, but, because you are intelligent and will verify all my claims yourselves, you will come to grudgingly agree that my perspective on this important epic poem is not only possible, but also probable, largely because so many tales, myths, and legends come down to us, not to

---

[301] Does "wanax" mean "big man," or are its roots actually from "big belly," as in all those big bellied women depicted in rock, clay, and ivory from various parts of the world? Considering that scholars like Thomas Palaima associate the term with "'*maître souverain*'" (2006, 54), which has its own roots in the Latin term "*mater*," meaning "mother," we have to wonder whether the origins of such hierarchical concepts come from applying terms to men that used to be associated with divinely-guided women who led more egalitarian cultures. Palaima concludes that Greek words associated with *wanax* "are meaningful names in connection with blood-line fertility, birth and progeny" (62), going so far as to tie Priam and Paris' names in *The Iliad* with "birth and generation" (58). However, Tobias Kienlin clearly associates both *wanax* and *basileis* with "'big men'…caught up in activities such as raiding and piracy that would not seem entirely appropriate in a system of orderly taxation, palatial control and economy" (25).

[302] While some men might have been considered leaders in various cultures before humanity discovered that men have a biological role in procreation, we have written evidence from every major culture in the eastern Mediterranean that it was not until after this verifiable discovery was made circa 2400 BCE that those cultures began to switch from matrilineal descent laws to patrilineal descent laws. In fact, when the Edict of Telipinus was written, it preserved matrilineal descent as the third choice of inheritance of the throne for the Hittites. As a result of learning males have an actual, verifiable role in procreation, androcentrism began to rise, leading to the patriarchies most scholars assume have existed for at least 10,000 years. This blindness toward the facts of this change have led many scholars to not ever consider that the cultures they are dealing with pre-2400 BCE might not be patriarchal.

teach how to behave, but to teach us how **not** to behave. And *The Iliad* is one of those tales designed to teach us that just because we have the power, the equipment, and the opportunity to oppress others does not mean we should oppress others.

What Is a Hero?

The first thing we have to address is the concept of "heroism" because one of the points of contention about *The Iliad* is whether or not the characters we view in it were truly heroes despite the fact that the poet repeatedly labels them thus because one or two moments in the poem allude to a time **before the time in the story**,[303] which has been called an Age of Heroes.

Therefore, an important question becomes: did Homer really want us to view all the men labeled as heroes in the poem as actual heroes, or could he possibly be using the idea ironically?

Confusion abounds about whether the time being described in *The Iliad* is the Age of Heroes, or if that age occurred at a previous time,[304] since Homer implies the men fighting were weak compared to men of old. For instance, in Book 12, Hector "grappled a boulder, bore it up and on" (Fagles, trans. l. 516), but the boulder was so large that "no two men, the best in the whole realm,/could easily prize it up from the earth and onto a wagon,/**weak as men are now**" (emphasis mine, ll. 519-21). Readers tend to take from this moment that Hector is superhuman, thus semidivine,[305] because he is

---

[303] When Diomedes graphically kills Tydides in Book 5, Aeneas jumps from their chariot to rescue Tydides' corpse, which prompts Diomedes to heft "a boulder in his hands, a tremendous feat—no two men could hoist it, weak as men are now" (Fagles, ll. 337-8) is an example of this type of reference that confuses many readers. Is Homer referring to his time, his contemporary listeners, chiding them for being weak? Or is he referring to a time before the siege of Troy to a previous Heroic Age?

[304] Hans van Wees (2006) clearly believes "that both Homer and Hesiod did think of the distant past as an age when the world was inhabited by a semi-divine race" (363).

[305] Farron argues that Hector is not the hero anyone in *The Iliad* believes he is because he exhibits cowardly actions more often than he does heroic ones, arguing that "the desire to see Hector as a great warrior leads many scholars to make obviously erroneous judgments about individual episodes" (55). Similarly,

stronger than all the other warriors, a kind of superman, but is Homer actually shaming the men of his time for not being as strong? Or could he be using the concept of heroism, which was a growing form of actual religion in his day, ironically, attempting to warn us that such behavior is not heroic at all?

This phrasing, indicating a contrast between either men of Homer's time and the men in his saga or between the so-called heroes in the poem and the super-human males who exhibit great feats of strength,[306] might never be fully resolved, but we have to wonder if super human strength was a/the main quality that made a man a hero to the ancient Greeks.

For Sabrina Hardy, *The Iliad* is a warning, claiming "Homer seems to be signifying to his audience that once the war has ended and the survivors have returned home, the time of the heroes will be over" (2). She examines Achilles' new shield made by Hephaestus, which has no heroes depicted on it, just two cities—one in peace and one devastated by war, and argues that Nestor's observation that he "struck up with better men than" Achilles and Agamemnon (Fagles, Book 1, l. 304), "such men [I have never seen] and never will again" (ll. 306-7) because they were "a match for the

---

so many readers and scholars want to view the men called "hero" within the poem at face value, rather than questioning the possibility that Homer is using the word ironically or euphemistically.

[306] Even Agamemnon, the king everyone hates, exhibits super human strength in Book 11, when he "kept ranging, battling ranks on ranks/ and thrusting his spear and sword and hurling heavy rocks/so long as the blood came flowing warm from his wound" (Fagles, ll. 310-12). Interestingly, once his blood clots, his wound begins to hurt, "spear-sharp as the labor-pangs that pierce a woman" (l. 315), which sends him back to the safety of the ships. First, the equating of blood clotting with women's labor pains potentially points to gynocentric origins. Is the poet acknowledging that women, too, are heroes for enduring the pain of childbirth, a play on, perhaps, an older, gynocentric idea much like the Aztecs held? Or is Homer, perhaps, using a back-handed compliment to reduce the moments of valor Agamemnon exhibits in this book to this point? As a scholar of pre-literature cultures, I cannot help but draw a comparison to Charles A. Eastman's story, "Blood-clot Boy," wherein a magical, heroic boy is born from a single clot of blood, which was a common reproduction theory among many post-procreation cultures that assumed semen was meant to clot women's monthly menses in order to produce children or to feed the infant in the womb, or, possibly, both.

immortals./ They were the **strongest** mortals ever bred on earth" (emph mine, l. 309-10), alludes to Homer's caution for his readers that the Heroic Age was ending. However, all the references to the time of heroes in *The Iliad* can be read as though that age had already ended, which is why, perhaps, this siege of Troy takes 10 years to accomplish.[307] Super human strength, it appears, is very important, but Achilles also demonstrates such strength during battle, despite what Nestor implied about him, so there has to be more to being a hero.

The Iliad presents us with at least two distinct views on the rising Cult of the Hero of Homer's day. On one hand, Achilles makes a distinction between a "hero" and a "coward," while, at the same time, proving that their lives' outcomes differ very little:

> One and the same lot for the man who hangs back
> and the man who battles hard. The same honor waits
> for the coward and the brave. They both go down to Death,
> the fighter who shirks, the one who works to exhaustion

(Fagles, Book 9, ll. 385-88).

The poet could be having Achilles make this anti-hero speech to explain why he continues to refuse to fight, even though Agamemnon's embassy has offered an impressive list of "gifts" to entice Achilles back to battle, or he could be attempting to mitigate the violent rampage Achilles will wreak later in the poem, killing several men by stabbing them in the back as they flee, since, by choosing to come to the siege of Troy, Achilles has chosen *kleos* in favor of any other form of fulfilling life, a kind of life wherein having lots of gifts would be made more pleasant. Achilles, just like the young Greek men enticed by the Cult of Heroes to join battle, has,

---

[307] According to J.M. Scammell, it took Hercules less than a day to storm Troy, depose King Laomedon (who preceded Priam by a few generations), slay him, and give Hesione, Laomedon's daughter, to his friend Telamon, who had assisted him in assaulting "the citadel of Troy at the strongest part of its walls" (422). That kind of feat sounds superhuman or heroic, but we should also remember that he had superhuman strength and was named, if ironically, after the goddess Hera as Heracles. Since Hera was a god worshipped much longer than Zeus was, perhaps there is something to the idea that she was the original Mycenaean Mother Goddess?

then, embraced death in battle as his choice, but he is unwilling to spend his life for that *kleos* just yet.[308]

On the other side, Odysseus separates the two qualities as diametrically opposed to each other as he debates whether to battle on alone:

> Cowards, I know, would quit the fighting now
> but the man who wants to make his mark in war
> must stand his ground and brace for all he's worth--
> suffer his wounds or wound his man to death

(Fagles, Book 11, ll. 483-6).

This formula is easy to understand even if oversimplified: cowards flee; heroes fight. In his desperation to prove he is no coward, Odysseus fights, killing five men before attacking Charops, whose brother, Socus, tries to save him, taunting Odysseus in the process, so Odysseus "plunges a spear in [Socus'] back between the shoulders" (l. 527), taking the time to taunt the dead man in return before attending to his own wound. How heroic was it for Odysseus to stab his taunter in the back?

Yet even these diametrically opposing views—hero versus coward—do not really explain heroism as the ancient Greeks saw it,

---

[308] Strangely, numerous scholars have argued about whether or not Homer was aware of or promulgating the Cult of the Hero in *The Iliad*. Lewis Richard Farnell, in his book *Cult Heroes and Ideas of Immortality*, waffled promiscuously over the idea that Homer was aware of the growing Cult of the Hero by admitting, at one point, that the poet "seems willing to admit the possibility of the mortal becoming divine" and "was probably aware of the worship of Achilles" elsewhere in the Greek world (11), but argues that the poet "while probably aware of hero-worship and the occasional deification of the mortal, did not accept **or did not wish to encourage this vain dream of the self-exaltation of man**" (emph mine, 11). In the end, Farnell chooses to see the poet as reflecting "the tone of an intellectual aristocracy, men of cool intellects, less troubled by ghostly concerns than the average Greek" (11). These arguments appear to be more wish-fulfillment on the part of Farnell, who probably saw himself this way, rather than arguable interpretation of what *The Iliad* and *The Odyssey* show us about the poet's intentions. The mere fact that Achilles embraces his chance for *kleos*, instead of living a life of fulfillment as a husband and father, demonstrates how deeply the Cult of the Hero had become by Homer's time. Ironically, acknowledging the probable "father" of each of the heroes in the poem, becomes a form of repeatable catchphrase in *The Iliad*, echoing another story written around the same era by another culture with lots of "begets" in it.

though.[309] Odysseus, like most good propagandists, could be reciting the party line.

To understand a more authentic ancient Greek point of view, we must separate out our modern concepts of a hero, which has, undoubtedly, changed over the last two and half millennia. For us, today, a hero is a person who acts selflessly, often risking death, on the behalf of others. This concept of heroism is embraced in some form or another by many cultures.

The Lakota, a warrior culture from the American plains, had a ritual for their young in order to impress upon them that they might be called upon to give the ultimate sacrifice, their own lives, for their tribe. Charles A. Eastman in his autobiography, *Indian Boyhood*, wrote about spending the first 15 years of his life c. 1860 ACE as a traditional Lakota. When he was ten, he experienced this preparation-for-the-ultimate-sacrifice ritual firsthand, writing about it, in the middle of his autobiography in third person, possibly because the memory was still so painful for him. The story is called "Hakadah's First Offering," and describes how Eastman was told to prepare for this ritual, knowing he would have to sacrifice the thing that meant the most to him in the world. He assumed he would be asked to sacrifice his beloved paint set, since he thought he most enjoyed painting images on hides, his body, his pony, and his dog with those paints, and knew they were very valuable because they were hard to come by (Eastman 104). However, as the ritual begins, he learns he must sacrifice his dog, Ohitika, his beloved companion, in order to prove he would be willing to give his own life for his people (Eastman 107). What Eastman does not include in his story, which was first printed in a boys' magazine, is the fact that, not only was his dog killed in such a way as to honor the Great Mystery, *Wakan Tanka*, but also cooked and fed to him to complete the

---

[309] Michael Lindblom and Gunnel Ekroth give this definition as what the ancient Greeks believed: "a hero is a mythical or real person who has lived and died and is worshipped after death on a more official level than the ordinary dead. This sets heroes apart both from the ordinary dead and from the gods" (235). However, these scholars never really try to apply this definition to the "heroes" of *The Iliad*, probably because all of the characters using the term are supposed to be alive at the time. Can one be a hero without being dead?

ritual. We can only imagine that he was admonished to be as "loyal" to his people, as he consumed his pet, as his dog had been to him.

Such blood sacrifices were not uncommon for many cultures, including the ancient Greeks, but an important distinction I want to make between the traditional Lakota view of heroism and the ancient Greeks' views of heroism is that facing an enemy without fear and counting coup on that enemy was far more honorable, and worthy of a warrior's reputation, than actually killing the foe was to the Lakota. While a Lakota warrior, who could be female,[310] since women were expected to know enough about weapons to at least defend their children, had to be brave enough to face the possibility of her/his/xis own death, it was culturally frowned upon to kill another human being unless it was absolutely necessary.[311]

However, we find little evidence for such a humane compunction among either the Mycenaeans or the Trojan "heroes" in Homer's epic poetry. They willingly stab their opponents in the back, use taunts and name-calling, and clearly practice deception in order to win, all of which should seem antiheroic[312] to most modern readers.

I read Cedric Whitman's book, *Homer and the Heroic Tradition*, published in 1958 by Harvard University Press, fully expecting to find at least one, solid, coherent definition of what the

---

[310] In his story, "The War Maiden," Eastman states that "in the old days it was unusual but not unheard of for a [Lakota] woman to go upon the war-path—perhaps a young girl, the last of her line, or a widow whose well-loved husband had fallen on the field—and there could be no greater incentive to feats of desperate daring on the part of the [male] warriors" (260).

[311] In order to keep young, hot-blooded men, from running straight into combat when confronted by approaching enemies, the Lakota used the Peace Pipe ritual given to them by their primary protector, Buffalo Calf Woman. As the people sat in council to hear reports from the scouts about potential approaching danger, the Peace Pipe was passed, and no one was allowed to speak until every person present partook of the sacred smoke. By then, hot blood had mostly cooled, allowing better thought and planning about how to react to potential dangers.

[312] Too many scholars to count have lauded *The Iliad* as a man's-man's book, failing to recognize the antiheroic elements which are often just blatant toxic masculinity. In fact, in recent years, the toxic masculinity exposed by the epic poem has garnered lots of discussion both by "influencer" androcentrists and those opposed to toxic masculinity. See Matthew A. Sears' and Walker Larson's online articles about the topic as it relates to *The Iliad*.

ancient Greeks believed a hero was. Sadly, there was not one clear definition of the concept because Whitman clearly expected his other scholarly readers to already believe, like he did, that what a hero is was pretty self-explanatory. Ask any scholar of ancient Greece what a hero was to the Mycenaeans, and you might get this definition: a man who is valiant in battle. Heroes to Hellenistic scholars are almost exclusively male, despite Athena's presence in both *The Iliad* and *The Odyssey*, and that concept of valor in war, which is repeated throughout *The Iliad*, often gets confused with the idea of what a hero is—something Homer struggles to convey to his readers, at least to his modern readers because the idea of being valorous sounds so much like our modern definition of hero—a person bravely facing danger to her/his/xis own life in order to help others.

The people of Homer's time, though, probably understood that a real hero, by cultural definition, was tied to divinity,[313] meaning that a hero could either hear, see, or speak directly with the gods, so had a direct connection with the divine (Soles 250), or that the hero's actions, possibly divinely inspired or enhanced, led him/her/xim to be deified in the new Cult of Heroes. For example, we can accept Achilles as a "hero," using the divine connection requirement because he is semi-divine, since his mother, Thetis, is a goddess of the sea. He can see, hear, and speak directly to her, and she willingly does his bidding. The *kleos* Achilles has already collected, prior to the siege on Troy, has earned him the reputation as the best of the Achaean warriors, something not even Agamemnon refutes. Why, then, does Achilles seek more *kleos* if he is already famous as a warrior?

However, the androcentric ideal of "hero" as being a man/male (almost exclusively) who acts bravely, honorably, or with valor **in battle** (which is "male" business, according to Hector and

---

[313] Hans van Wees (2006) claims, however, that "modern scholars…have almost unanimously denied that Homer and Hesiod used *hêrôs* in **its normal religious meaning**, mainly on the grounds that these poets always apply the word to living men, whereas in its religious sense the word normally applies to the dead" (emphasis mine, 367). van Wees points out, though, that "some men in historical Greece were, after all, recognised as *hêrôes* [sic] while still alive" (367). I explore why the Greeks tied religion to their concepts of heroism in this paragraph.

many scholars, so women have readily been excluded from this definition), **with a tendency to ignore the warrior's ties to divinity**, has steadily grown to be **the** definition of Greek heroism since at least the time when the epic began to be re-examined by scholars in the throes of androcentric cultures centuries ago.

Christopher Jones, wrangles with the very word "hero" itself, drawing mostly, though, on later sources to try to justify his ideas. He links *herós* to the word *tisiseroe*, which he claims means "thrice hero," but gets so caught up discussing the transmigration of souls later in his book that he misses a connection to "thrice hero" that Pindar provides for him, claiming that "the poet says that those who have lived **three times** free from all acts of injustice on earth and below earth 'take the path of Zeus to the tower of Cronos,' where they live in a kind of paradise, a land of cool breezes and flowering plants" (emphasis mine, Jones 10). He, too, clearly misses the connection between divinity and heroism.

Yet even Cedric Whitman struggles to make his grandiose ideas about heroism seem feasible for mere mortals:

> Heroism cannot be maintained in isolation, or it becomes nothing.[314] Though the heroic nature partakes of the absolute qualities of divinity itself, and though the higher it looks, the less it can endure the compromises of the human level, still its human connections alone justify it as a human phenomenon" (195).

Here, Whitman is trying to explain why Achilles cannot continue to stubbornly refuse to leave his tent while the battle on the plains of Troy continues, but he never clearly explains why heroism has to be public, how heroism is divine, or what those "absolute qualities of divinity itself" are,[315] although he seems to realize, halfway through

---

[314] Since women were largely forbidden in Hellenic Greece from participating in public activities, if Whitman is correct, such a prohibition would mean women could never be heroic.

[315] And readers have to wonder what those divine qualities are. They are, in short, the ability to see, hear, or communicate directly with the gods (Soles 250). Being born semi-divine, of course, helps one do those things, so many Greek figures are either characterized as semi-divine or heroic to help listeners/readers grasp the idea that those characters can recognize or commune somehow with the gods.

this grandiose claim, that this definition will not work for mere mortals, so feebly shoves the two together as though saying, well, heroes have to be humans (because the gods are not real), so heroism has to be human, too.

In short, Whitman never fairly deals with the real meaning for the ancient Greeks of the word *hērōs*, but, instead, like most of the other androcentrically biased scholars before and since him, prefers to view the concept of valor during combat (which earns a warrior *kleos*, which literally means post-death glory; similar to Vikings who go straight to Valhalla, carried by Valkyries, after dying in battle—also an androcentric concept borrowed from a previously gynocentric culture) as the heroic ideal.

For these scholars, there is no heroism without war, but few admit that there could be no *kleos* without Hero Worship.[316]

Whitman, in his book, comes really close to proving my point about *The Iliad* being a cautionary tale about how not to act, however. He admits that Achilles' desire to live, despite his mother's foretelling of his death at the battle for Troy, is part of what keeps Achilles in his tent. Whitman rushes to convince us Achilles is not afraid[317] to die; he's just not ready to die. However, after Patroclus, whom Whitman admits was Achilles' lover, not just his cousin (187), dies, Achilles' wrath "changes" for Whitman (194), when, in fact, Achilles' wrath **deepens**—it grows from being aimed merely at Agamemnon, the king of kings ruling over him who stole his "prize" Briseis from him, to a grief-based wrath so deep and so profound he goes berserk in his effort to slay the world and the gods who, together, put him in his untenable position. For Whitman, this deeper wrath is justifiable because Patroclus represented Achilles' "human side" (195), so, once Patroclus is dead, so is Achilles' humanity, at least for a time. If Achilles' humanity has died, however, surely all he is left with is his divinity, right? But why is his divine side so crazy, so single-minded, so wrath-filled? In this moment, the Cult of the Hero seems to be indicating that any angry young man wantonly killing others for revenge is acceptable, allowable, possibly even encouraged.

---

[316] Hero Worship is a cultural phenomenon I will deal with shortly.

[317] After all, it would not be masculine to be afraid, according to androcentric bias.

Whitman's idea brings us ever closer to proving my main point—*The Iliad* is a cautionary tale that demonstrates what happens when human beings allow their humanity, their ability to feel pity and to show mercy for others, to die, preferring, instead, to engage in violence with everyone who stands in their way. Achilles, himself, will come to regret his wanton violence, so should readers forgive him, too? What kind of social precedent does such forgiveness set?

Whitman struggles with these concepts, claiming:

> The heroic spirit, if it is to exist as a reality, must act in this world. The hero's head may touch Olympus, but his feet can leave the earth only when he dies" (196).

Again, Whitman admits that a hero must be in touch with the divine, but he does not explore the full meaning of what heroism truly meant in a spiritual sense to the ancient Greeks, the Mycenaeans. For Whitman, and other scholars who romanticize war with their androcentric-colored glasses, the Wrath Achilles experiences is merely Achilles' "search for himself" (197), so that Wrath is nothing serious, which seems to turn the bloody violence of the epic into nothing serious.

The normalization of these violent androcentric views means scholars have, for centuries, missed Homer's real point.

Even though Whitman claims "the path [of] heroism means the search for the dignity and meaning of the self" (193)[318] on one side, the scholar continues to have trouble justifying the brutality of battle, going so far as to claim that the battle in Book 9,

> Far from being a mere indulgence in **the more primitive aspect** of old epic, this Battle is a shapely masterpiece, a panorama in which the earthly and heavenly effects of the Wrath are deployed in piercing detail, steadily inwrought [sic] with motivations of the approaching tragedy" (emphasis mine, 194).

While Whitman admits that violence is "more primitive," his romanticism of any of the violence in the epic is a symptom of androcentrism, which firmly believes Male Might Makes Right

---

[318] How can a violent, selfish man find himself, let alone understand dignity? At what point in violent rage is a person supposed to learn "the meaning of self"?

because it believes men are "naturally" dominant, so "naturally" brutal, so we are supposed to admire this brutality when it is depicted in glorious detail, as though the beauty of the descriptions somehow makes up for the brutal, primitive acts themselves.

However, Whitman has to admit that "as the *Iliad* progresses, the compensations of **true greatness** appear less and less engaging, and the **supreme hero** is less and less the gloriously satisfied top of the heap, and more and more a lonely and haunted sojourner among **men of inconsequence** and half-hearted ideals" (emphasis mine; 185). Nowhere does Whitman describe what "true greatness" is, and he never tries to justify calling Achilles "the supreme hero." Why? Because the accepted androcentric views put forward by other scholars about this epic presume to make these hierarchical ideas, going so far as to put a ranking on the value of individuals' lives, widely accepted.

Yet, according to Lada Stevanović, hero worship[319] only became popular among the ancient Greeks as cities developed in *polis* or city-states in the Eighth Century BCE, about the time Homer or his scribe wrote down what would become *The Iliad* (8). Stevanović argues that the new *polis* used hero worship, replacing ancestor worship,[320] to accomplish several things—from regulating funeral activities, which had once allowed women to weep and wail for their dead in extravagant shows of grief, by limiting women's roles in funeral processions through laws that dictated behavior, such as forbidding women from walking in front of men (9), to using the idea of *kleos* in order to convince men to go to war for the polis (13). Through this politically encouraged hero worship, "heroic virtues ('courage,' 'sense of duty,' 'keen feeling of honor') became

---

[319] Even in 1932, Martin Nilsson admitted that "many genealogies and eponymous heroes [were] created for political purposes" (1).

[320] Ancestors in matrilineal cultures would be female, of course, because we always know who the mother is of a child. Switching to Hero Worship from ancestor worship allows men to be deified and worshipped in a manner similar to what women must have once enjoyed. While Martin Nilsson tried to trace various cities and heroes back before Mycenaean culture linguistically, he never considers that the reason "the names [which were derived] of the older series [of names] are often difficult to explain etymologically" was because prior to that period people were named for their mothers, not their fathers, even though it was well established by 1932 that the early Mycenaeans were matrilineal (16).

incorporated in the norms of civic morality" as propaganda "to mobilize people and make them eager both to die and to kill" (Stevanović 15) by praising and elaborately commemorating those who died in battle, specifically (16). This change in social mores means that the type of democracy advocated for by Athenians, for instance, valued equality as an abstract concept, but still weighed national identity as more important than individuality because "the idea that a dead child can be replaced with another one is, on one hand, indicating the absence of the idea that each individual is unique," thus important (Stevanović 17).

According to Hans van Wees,

> by the late archaic period, the Greeks saw their legendary heroes as not merely great men of the past but still-present superhuman beings who protected those who honored them with sacrifices and games. Agamemnon, Achilles and Odysseus may have started out as Mycenaean kings and nobles or they may have begun as chieftains invented by Dark Age oral tradition, but they ended up transfigured into 'demigods' (*hêmitheoi*) and worshipped as 'heroes' (*hêrôs*) in the **Greek religious sense** (emphasis mine, 2006, 363).

van Wees also examines some scholars' argument that the word "hero" is used as a form of address to the other warriors by the warriors themselves, so was just a secular term of respect, not religious in nature. Van Wees points out that "unless we are prepared to argued that 'god' also has a unique secular meaning in Homer, we will have to accept that the term *hêrôs*, even when used by the heroes themselves, can perfectly well have had its normal religious meaning" which was to view the person as divine or at least semi-divine, demigods (368). The religious idea of hêrôs arose as far back as the Bronze Age, according to two Linear B tablets (van Wees, 2006, 370), so we must conclude that the Cult of the Hero began then and still existed, perhaps reaching its peak, in Homer's time.

The change in thinking about apotheosis, which is the belief that humans can be transformed into gods, from being gynocentric to androcentric probably could not have happened if the Myceneans had not already believed that their ancestors made such a spiritual or divine transition upon death. This process of deification was probably overly simplistic in nature, but the believed spiritual transition would have taken on a religious importance that far outweighs what we modern scholars can imagine. We are simply too blinded by modern religious concepts to understand the nuances of a truly pantheistic religious view. A possible symptom of this issue is the fact that I could not locate a thorough scholarly examination of the connections, which seem obvious, between *heiros*, which indicates things that are sacred or tied to the supernatural, and *hêrôs*, even though many scholars, including Plato, bent over backwards trying to tie less obviously related words/concepts as the origins of the Greek concept of *hêrôs*.

However, prior to this drastic religious and social change by the societies and governments of the individual Greek cities, the ancestral "dead were considered to be deities" (Stevanović 18), a concept that is transposed onto the men who die in battle who begin to be seen as heroes or divine human beings or semi-divine beings who exist in "eternal glory" in a type of rebirth that guarantees they will never truly die (19) because *kleos* demands that their fame will be sung perpetually to glorify them[321] (Stevanović 7).

Imagine how attractive such an idea would be to young men who see images of "heroes" on wall frescoes and pottery; such a combination of seeing heroes on frescoes, hearing about heroes during recitations, and envisioning themselves as those heroes,

---

[321] Many scholars, such as Tobias Kienlin, argue that the "heroic" epic began with the *Epic of Gilgamesh* (21), but this claim ignores the fact that, while Gilgamesh and Enkidu do slay a giant, they are not setting a heroic standard by doing so, since the Sumerian version of the tale has Gilgamesh frightened of the giant, and the Babylonian version has Enkidu frightened under the same circumstances. The primary points of Gilgamesh's experiences in the epic are to illustrate how human beings are made from beasts (via sacred sex), how abusive kings can be (which starts the whole epic, just like in *The Iliad*), and how the most important thing is not fame or living eternally, but living in the here and now by appreciating what Gilgamesh has. This latter lesson is one Homer appears to be working through in both *The Iliad* and *The Odyssey*, although he never makes that point blatant. The only real "heroic" quality Gilgamesh exhibits is the fact that he is two-thirds divine because his mother was a goddess. The math exhibited here clearly indicates that, even as the epic was being written down, **women were still considered more divinely powerful than men**.

helped the Cult of the Hero to become one of the most successful propaganda campaigns in history. A propaganda campaign that continues today.

In this way thorough cultural saturation, "violent erotics," as Stevanović calls the descriptions of violence in *The Iliad*, turn traditional oral stories into "war propaganda" (19), which, in turn, produces epic poetry like *The Iliad*. However, we have to ask: was *The Iliad* meant to be pro-war propaganda or anti-war propaganda? After all, just because the new *polis* governments were trying to bolster their armies with new recruits does not mean the poet or poets who put together *The Iliad* in its written form agreed with the new government sanctioned propaganda.

Steve Farron in his article, "The Portrayal of Women in the *Iliad*," clearly, was not willing to credit Homer with writing the epic "to urge social reform" (31). Indeed, in many ways, especially in the way the warriors in *The Iliad* continually consider each other "heroes" in order to build their own reputations as warriors, Homer **appears** to be enforcing the new social paradigm of Hero Worship. But was he?

A mother could be forgiven for elevating her son's status among his peers, just as Thetis calls Achilles a "hero among heroes" twice in Book 18, but it is quite probable that Homer harps on the idea of the warriors each being heroes was simply because slaying anyone else in such a brutal manner would be dishonorable, much in the same way Achilles' berserker moment in the poem paints him with a zealous bloodthirst that was probably even too much for the ancient Greeks to tolerate.

In his uncontrolled rage, Achilles symbolically encompasses both the rising androcentric male indifference to the value of other people's lives, and the androcentric male arrogance that he can do anything he wants on his own terms because no one else can physically or divinely stop him. Achilles becomes both of these things—indifferent and arrogant—when he goes berserk and begins slaying the Trojan "heroes" relentlessly in Book 20.

As Achilles cleaves Iphition's[322] head in two, he "exults" over his death, "'Here you lie, Otrynteus' son—most terrible man alive!/ Here's your deathbed! Far from your birthplace, Gyge Lake" (Fagles, Book 20, l. 443-4). He not only renders Iphition's face practically unidentifiable, but he does so with needless force; Achilles then dares to arrogantly mock him for dying so far from home.

Then Achilles, who now sweeps the field like a tornado, kills Demoleon[323] by stabbing him with his "pitiless bronze" spear through "his temple," cleaving "his helmet's cheekpiece./ None of the bronze plate could hold it—boring through/ the metal and skull," splattering Demoleon's brains across the insides of his helmet (Fagles, Book 20, ll. 451-4). This second purposeful head wound is possibly a metaphor by the poet to demonstrate the removal of individual identity, defying the state of *kleos* so closely tied to the Cult of Heroes, through death.

The third "hero" Achilles slays in his frenzy is Hippodamas, who is trying to escape. Achilles has no compunction about stabbing Hippodamas in the back, making the young man bellow "like some bull/that chokes and grunts when the young boys drag him round/ the lord of Helice's shrine" (Fagles, Book 20, ll. 457-8). For Achilles, Hippodamas is a beast to sacrifice. However, for a culture of warriors who found using bows and arrows a "weak" way to fight, surely stabbing a retreating warrior in the back should be just as cowardly. Achilles' thoughts, though, shared with us by the poet, clearly indicate he sees the young man as a sacrificial animal. For what god is this sacrifice made? Perhaps it is not made for a god, but for Achilles' ego.[324] According to the Cult of the Hero, Achilles

---

[322] Since *iphios* means strong or mighty, does Iphition's death symbolize the slaughter of the mighty, possibly leading to the idiom, "how the mighty have fallen"?

[323] Demoleon's name literally means "people's lion." Is his death a warning against or for democracy?

[324] Given the fact that Odysseus pays a heavy price for denying the gods have any value in Homer's other important epic poem, *The Odyssey*, which is set post-Trojan War, Achilles' arrogant massacre, seeking only to appease his own angry grief, could symbolically be a warning from the poet that replacing the gods by worshipping "heroes" will harm the culture in the long run.

was a god or at least semi-divine, so the sacrifice might be seen as a tribute for himself.

But Achilles' battle madness does not stop there. His fourth target is King Priam's youngest son, Polydorus, but "the young fool, made to display his speed,/ went dashing along the front to meet his death" (Fagles, Book 20, ll. 467-8). As Achilles rushes past Polydorus, he

> speared him square in the back where his war-belt clasped,
> golden buckles clinching both halves of his breastplate--
> straight on through went the point and out the navel,
> down on his knees he dropped--
> screaming shrill as the world went black before him--
> clutched his bowels to his body, hunched and sank

(Fagles, Book 20, ll. 470-76).

Here, Achilles kills another "hero" by stabbing him in the back, managing to disembowel him in the process. As if this scene is not graphic enough, after Hector attempts to avenge his little brother, Achilles claims Hector only escapes his grasp because Apollo rescues him. Still inflamed with an unquenchable fury, Achilles goes on not just to disembowel Rhigmus and stab his chariot driver, Areithous, in the back, lifting him clear of the chariot on the end of Achilles' spear, Homer additionally paints an even grimmer picture for us of male anger unleashed:

> ...as the great Achilles rampaged on, his sharp-hoofed stallions
> trampled shields and corpses, axle under his chariot splashed
> with blood, blood on the handrails sweeping round the car,
> sprays of blood shooting up from the stallions' hoofs
> and churning, whirling rims—and the son of Peleus
> charioteering on to seize his glory, bloody filth
> splattering both strong arms, Achilles' invincible arms--

(Fagles, Book 20, 563-69).

Achilles is Male Might run amuck, seeking to gain more kleos[325] for himself by slaying "hero" after hero because a man endowed with divine power has an unbalanced advantage over others, so killing anyone less than a "hero" would be the equivalent of murdering helpless children. Yet there is nothing honorable in Achilles' berserk slaughter fest, so we have to be truly amazed at the sheer number of androcentric scholars who stand by their

---

[325] By this point in Achilles' career as a warrior, he should have enough kleos stored up that the Trojan War should add little to it.

interpretations of his character as a "hero," even in our own modern sense of the word. Most forgive him because his Wrath is two-fold, caused by being wronged by Agamemnon, then deepened by his loss of his lover, Patroclus, who he learns was loved more by Briseis than he himself was, adding jealousy to the dangerous mix of emotions. However, violence for violence's sake, no matter the reason, is never heroic, not even during war. For a people who bandy about the words for valor and honor so much, such actions should have been condemned by the ancient Greeks, which is probably why Homer later has Achilles repent over his frenzy.

Perhaps the one argument many Hellenic scholars make about *The Iliad* that puzzles me most, though, is the idea that "Homer [took] an extreme line" by "omitting from his picture not only hero cult but all ritual and commemorative activity at tombs" (van Wees, 2006, 372). We cannot get much closer to hero worship when Homer recites a warrior's heroic progenitors. Since ancestor worship, which would have focused on women more than on men prior to the understanding that men had a biological role in procreation, easily morphs over to hero worship, wherein males become worshipped with *kleos* for performing valiantly in battle, the easiest way to "call on" the powers of those now-divine ancestors is by reciting their names and recounting their deeds. In this form of "who's-your-daddy" commentary, *The Iliad* reinforces the rising patriarchal values and the Cult of the Hero at the same time.

Perhaps the Greeks needed the Cult of the Hero to become fully patriarchal?

Since many cult worshippers called on their ancestors, then on their heroes, to assist them, and since we see no direct examples of such pleas by the characters in *The Iliad*, many scholars believe Homer believed such worship was not valuable, despite the many details provided of Patroclus' funeral, which includes the sacrifice of 12 Trojan youths[326] and includes a very important precaution overlooked by most scholars: the ritual immolation of Patroclus'

---

[326] Cedric H. Whitman cautions readers to understand this moment in The Iliad clearly because "Sacrifice and funeral have blended, in part, and they become actually identical when the twelve Trojan youths are immolated at the funeral of Patroclus. Surely, there is a consciousness here of **an inverted order of things**" (emphasis mine, 130).

body. We know that the Greeks believed a person had to have an intact corpse to be allowed into the Underworld after death from such stories as *Medea*, wherein Medea purposely chops up her little brother and tosses the pieces into the ocean to delay her father's ship, which has to stop to retrieve the pieces in order to give him a proper burial, thus allowing her and Jason's escape. But we can also determine that the Mycenaeans already believed this idea from the fact that the Achaeans knew they could not build a true monument to Patroclus for his heroic actions (as they saw them) because, if they lost, that monument would then be destroyed by the Trojans. So, being very practical, they instead enact the Cult of the Hero by burning Patroclus' body in a funeral pyre, along with the bodies of the 12 Trojan youths, who would become Patroclus' slaves in the afterlife because of his heroic role in the Trojan War. While modern scholars might not view a cremated body as whole, the Mycenaeans and later Greeks probably did because the fire, which is an offering, remember, to Hestia, the goddess of the hearth, wafts smoke upwards toward Olympus[327] as a signal to the gods that a hero has died, and the burnt remains can then be gathered all in one pot or urn, thus making certain the mortal remains stay together.

However, van Wees cites previous scholars as he examines the idea that there was no obvious hero worship in *The Iliad* because "'there was no one to do the worshipping,'" but he does not examine the fact that there **were** people there to do the worshipping because tending to graves and decorating them "with wreaths and ribbons" (van Weeks 372) was one of the many religious tasks women saw to, and, as we can tell because of the kerfuffle over Briseis, the Mycenaean men had many women with them. However, while the women do not overtly help Achilles plan or implement the memorial service for Patroclus, they do begin wailing their grief when Patroclus' body is returned to the Mycenaean camp:

> And the women he and Patroclus carried off as captives
> caught the grief in their hearts and keened and wailed,
> out of the tents they ran to ring the great Achilles,

---

[327] Of course, Hestia was only one of the Mount Olympus gods, at least of the Dodekatheon (Council of 12) depending on different time periods and different city-state perspectives, some of which replace her with Dionysus (Baring), which could be taken as another sign of the rising androcentrism that held ancient Greece in its sway.

387

all of them beat their breasts with clenched fists,
sank to the ground, each woman's knees gave way
(Fagles, Book 18, ll. 31-35).

Even the sea nymphs join Thetis, "all in one mounting chorus beating their breasts" (Book 18, l. 58), raise their voices in lament. Apparently, all of these females adored Patroclus as much as Achilles did. Is Homer emphasizing how important women's roles are in lamenting the dead? Or is this only one way we see the Cult of the Hero enacted, in addition to the sporting events Achilles organizes, and the 12 Trojan youths[328] sacrificed at Patroclus' funeral?

I have yet to find a believable argument that Homer was not aware of the Cult of the Hero. However, just because he was aware, does not mean he endorsed the idea.

Have scholars been so eager to embrace Homer's epic poem, *The Iliad*, for the best reasons? Have they heard the poet's real pleas in regards to war and its accompanying deplorable violence?

More importantly: have they read *The Iliad* through gynocentric lenses, instead of their programmed androcentric eyes?

ΔΔΔΔ

Social Changes: The Rise of Androcentrism

Steven Farron came so close, in 1979, to arguing my theory for me that *The Iliad* is a cautionary tale written by a writer wishing to highlight the trauma of violence and oppression and how these rising political elements affected Greek society circa 800 BCE. While his article, "The Portrayal of Women in the *Iliad*," completely ignored discussing the female gods directly, Farron argued that Homer "knew that women were complete human beings and constantly emphasized how deep and intense their feelings are. By doing this, he demonstrated how brutal the conduct of the men in

---

[328] Oliver R. Gurney notes that blood "sacrifices are made to the deceased and 'the chthonic powers, to the ancestors'" for the Hittites, whose practices were similar to the ancient Greeks' (61). Like the Greeks, the Hittites then consume a ritual feast for the dead; both of these elements are present in Patroclus' burial ceremonies, which demonstrates, contrary to popular belief, that *The Iliad* does contain elements of ancestor worship, if not outright hero worship.

the *Iliad* is" (30). Farron claimed that "Homer was intent on showing the effects of the intersexual attitudes of his characters...; thus making the tragedy of the women more complete," adding, however, "I doubt whether he did this in order to urge social reform" (31).

What if Farron is wrong? What if Homer recognized that the rise in violent androcentrism through the Cult of the Hero was, in fact, a bad thing for his people? What if this oppression of women that is depicted in *The Iliad* is relatively new for the Greeks because it only came about, or got significantly worse, during the Mycenaean transition period, before we actually get to historical Greeks?

It is highly probable that the Mycenaeans were not as violent as most literary scholars and historians had once believed they were because scholars took the images painted on frescoes and pottery as historical reports rather than symbolism, so believed the violent iconography depicted real moments. We also know now that the violent iconography found from Early Egypt, especially that depicted in the Narmer Palette, which indicated that the "king" smashed heads to unify Upper and Lower Egypt, was merely symbolic[329], too. The unification of Egypt was a much more peaceful and cooperative process than what many images made archeologists believe, so it is distinctly possible that the violence depicted on murals and pottery for the Mycenaeans was also symbolic and not literal.

We also know that the transition from Mycenaeans to ancient Greeks saw a gradual change from matrilineal to patrilineal succession. In fact, we now know every major historical culture in the Eastern Mediterranean region went through the same gender-preference shift within historical times (beginning circa 2400

---

[329] Béatrix Midant-Reynes notes that the Gebel el-Arak knife-handle, which depicts a figure using a mace to strike three enemies...who are linked to him by some kind of rope...is the embryonic form of the supreme triumphal image that emerges about 200 years later on the Narmer Palette, the constant symbol of royal power for many centuries" (208). She later points out that "there is no archaeological evidence to support" the belief that violence is what united Upper and Lower Egypt in Dynasty Zero (Midant-Reynes 237).

BCE[330]) largely because humans had worked out, through the domestication of swans, the only monogamous animal[331] ever domesticated (which also happened to have a short gestation period that enabled humans to note the connections between sexual intercourse and conception) that males have a biological role in procreation.[332] This discovery began a process that would completely overturn the idea that women were sacred beings who could create new lives out of their own bodies, bleed for three to five days without dying, and could transform beasts into men[333] and men into kings through sexual intercourse.[334] As word spread, men

---

[330] William Stiebing, Jr. corroborates this date, stating that "from at least the Early Dynastic II Period [circa 2500-2330 BCE] onward, Mesopotamian culture was patriarchal" (56).

[331] This discovery possibly led to the use of eggs as explanations of cosmic origins.

[332] Prior to this discovery, semen was considered "milk" necessary to feed babies in the womb, but not seen as an important part of conception. Many different beliefs were held about pregnancy and childbirth post-confirmation of the biological connection between sexual intercourse and procreation, such as that semen coagulated women's menstrual blood to form babies (Grahn 6 & Neumann 31), an idea that is corroborated by the traditional Lakota tale, "Blood-Clot Boy" (Eastman *Wigwam* 68) or that men's spirits, often collectively, usually magically, helped women become pregnant, which explained why some children looked like some men (Grahn 48). Not until the microscope was invented in 1590 ACE did humanity realize that sperm were the active live cells inside semen that are required in order to fertilize women's ova, which were not seen for nearly another 100 years, even with microscopes. So between c. 2400 BCE and the late 17th century ACE, humanity could only guess with many different theories about how human conception actually happened—that is more than 4000 years.

[333] See the *Epic of Gilgamesh* where Enkidu is transformed from beast to man by having sacred sex, *hashadu*, with a priestess from Inanna's temple. Realize that androcentric bias leads many interpreters to call this "taboo woman" a prostitute or harlot, instead of the purveyor of sacred, healing sex that she was. Even Kristiansen and Larsson admit that, "rather than term this prostitution, we should probably understand it as being part of a different cultural perception of religion and sexuality" (351).

[334] Misnamed after the Greek concept *hieros gamos*, the *qursu* ritual for ancient Sumerians sanctified a king's rule through re-enacting sexual intercourse between the goddess Inanna and her consort Dumuzi through the High Priestess of Inanna's temple and the man who wanted to be king. Sex with Inanna was the way in which the sacred *me* from the gods was continually passed down to humanity. According to Beate Pongratz-Leisten, however, male gods had been given the ability to designate kings by 1820 BCE in Sumeria, so the texts stop referring to the *qursu* ritual (58).

came to believe that they, themselves, were the creators, planting homunculi as "seeds" in women's wombs, which became equated with fertile fields—things to be owned and controlled, planted, and harvested.

These presumptions led to the rise of androcentrism as men gradually changed cultural norms[335] and established themselves as the dominant gender—all because they came to believe they were the ones who were most god-like. In the process, the gods changed from creator-destroyer type female gods to even more iron-fisted creator-destroyer type male gods, with male gods assuming the process of procreation from within their own bodies.

As androcentrism gained more favor and was embraced by more people, male violence, used to oppress "lesser" males and women, was also becoming more widely accepted as a fact of life.

In his article, Farron calls fellow scholar Charles Beye a male chauvinist, quoting Beye's article wherein he stated that "'Andromache shows a woman's typical determination to direct her husband on matters she does not understand'" (23). Farron recognizes this view as an extremely androcentric one, so argues that Andromache's "advice is obviously well thought out and, as far as we can determine, sensible" (23), going so far as to admit that "it is impressive that Andromache, despite her emotional horror of the war, can still think about it in a calm and intelligent manner" (23).

Yet, while Farron argued that Homer "formed all his women to display the agonies of war," it never once crosses his mind that Homer might have chosen to create his human female characters this way to lament the losses both men and women endured from the rise of androcentrism, which rose steadily on the back of the

---

[335] The changes can be tracked through the written literature left to us through the changes in how females are perceived, such as removing some of Inanna's powers to give them to male gods who were invented long after Inanna was adored and worshipped by ancient Sumerians, as well as changing the Akkadian word for women, *sinnustra*, into a word indicating evil, sinister. Most of the cultures started these changes to religious views by getting rid of matrilineal descent in favor of patrilineal descent, and by favoring male gods over female gods. In fact, we know that the Sumerian god, Enki, arose after Inanna, so the myth about her having to steal the sacred *me* from him in order to give them to the people was created in order to justify why a female god had those powers to dispense.

mere threat of violence as a means to frighten and to subdue. We know, historically, that, by Homer's time, women were being bought and sold as slaves—literal "booty"—in many Greek cities around the Mediterranean, but so were "lesser" men.[336]

Farron, like so many androcentric-based scholars before him, does consciously try to view the epic poem through a female point of view, but the point of view he chooses is still androcentrically centered, largely because he still exhibits the belief that **<u>sexuality is shameful</u>**, especially uncontrolled female sexuality, which he discusses most thoroughly when examining Helen as a character who he assumes is fully conflicted by her role in the war against Troy because of her "sexual seduction" of Paris: "Helen is the fulfillment of lust which [Aphrodite] gave to Paris" (Women, 20). He tries to examine Helen and Aphrodite's relationship as though Helen has a split personality—one sexual, one rational: "a conflict between Helen's rational, moral (or at least shame-avoiding) personality and **her sexuality, which is irrational, amoral and shameless**" (emphasis mine, Farron, 19).

Why does Farron consider Helen's sexuality "irrational, amoral, and shameless"? Especially when he notes that Paris "accepts his erotic nature without a qualm" (Women, Note 13, p. 18).

As I have argued elsewhere, women have long been associated with sex because priestesses in the Sumerian goddess Inanna's temple were taught to use sexuality's liminal nature to heal and to transform, but the rise of androcentrism after humanity learned males actually do have a biological role in procreation (circa 2400 BCE), led to male assumptions that men are the primary creators, and women are just vessels to carry the men's children. By assuming then embracing male primogenitorhood and excluding women's biological contributions to procreation, males were able to shift social dynamics gradually over hundreds of years, leading us to the androcentric view that only the fathers matter, and that

---

[336] In fact, while almost impossible to prove, castration was probably the handiest means to control "lesser" men and their behavior, which could account for the drastic DNA findings from early history wherein only one man for every 27 women were passing along their genes (Diep).

women's sexuality must be controlled in order to accurately determine who the father of each child is.[337] The best way to control that sexuality is to make females' desire for sex shameful in the same way that spilling blood in combat used to be shameful (Flannery & Marcus 44).

When Helen battles with herself, she tries to stand firm and repudiate her complicity in the actions that led to the war, but then Aphrodite steps in and causes her to sleep with Paris again. For Farron, "Homer could not express divided feelings and internal conflict in a person's personality," so used a god's influence to explain inexplicable thoughts, feelings, and actions (Women, 19).

However, Helen would have been very familiar to the ancient Greek audience to whom this tale was told, but not as a mortal woman. Even Nilsson admits that "Helen is an old goddess whose origin belongs to the Mycenaean age and who always was a goddess at Sparta.... She had two temples at Sparta: one is the so-called Menelaeion" (41). So Menelaus, Helen's first husband in *The Iliad*, is a fictional male figure named after Helen's Spartan temple?

Nilsson connects Helen, who was twice kidnapped, first by old man Theseus, and later by the most beautiful human male on earth, Alexander/Paris, to Kore-Persephone because both are carried off as "brides by capture," but also are vegetation goddesses associated with the seasons (42). He notes, however, that Helen gets drawn "into the heroic mythology...[so] consequently, came to be treated as a mortal woman," sinking "still lower, to become the woman who eloped with an Asiatic prince" (42).

This demotion of females already had a long history before the Mycenaeans, since one of the earliest gods associated with being captured and sent to the Underworld, in the process causing the seasons, was Dumuzi, the shepherd god of the ancient Sumerians (circa 3000 BCE). Dumuzi was the consort of the goddess Inanna, possibly one of the most powerful gods ever conceived of by human imagination, since she was not just the goddess of sexuality, gender, knowledge, and war, she also controlled who became the divinely sanctified ruler of each Sumerian city. While Inanna is often

---

[337] Ironically, monogamy has not solved this issue, since many people have been startled to learn that the men they assumed were their fathers share no DNA with them because of recent DNA testing fads.

equated with other goddesses from the Mediterranean region, she is more like Athena[338] than Helen, although she has attributes of both.

Inanna predates the Mycenaean goddesses by approximately 1500 years, probably more. She was the wholly sexual woman to whom men, kings especially, groveled for sexual favors and in the hope she would tell them they were the best lovers.[339] Like Helen, then, Inanna was so beautiful and sexually appealing, she had men falling over themselves to gain her favor.

Like Athena, however, Inanna was also cunning. Even in a poem that reflects the growing androcentrism taking over the Sumerian cultures, Inanna gets her "father" god, Enki, drunk and steals the sacred *me*, the divine knowledge and powers that Inanna then gives to humanity.[340] When Enki sends his minister after her to stop her as she escapes on the Boat of Heaven, Inanna uses her wits to confuse and confound both the minister and Enki, so they send various monsters after her to try to take the sacred *me* from her, but she escapes every time. While Inanna repeatedly calls the male gods, who are gradually raised above her in importance, "father" out of respect,[341] Athena is literally the daughter of Zeus,

---

[338] While many casual scholars lump many goddesses from the Mediterranean region together (something they rarely do with male gods), there is another story worth pursuing about why so many of these goddesses—Athena from Greece, Inanna from Sumeria, Arinna from Hattuša, Ishtar from Babylonia, Asherah from Judea—were seen, first, as protectors of cities or regions. None of these goddesses appears to have borne children, but their citizens loved them dearly in their roles as protectors or, what many want to call, Mother Goddesses. While this association of protector goddesses with motherhood indicates that women were long seen as wielding Female Protective Magic, the fact that none of these goddesses ever bore children should not be overlooked, since it indicates that protecting children is more important than biological motherhood.

[339] All except Gilgamesh sought her sexual favors. In the *Epic of Gilgamesh*, King Gilgamesh of Uruk shuns Ishtar, arguably a later, more frightening iteration of Inanna, because all her earlier lovers died.

[340] This theft is a move copied by the Egyptians when Isis arises in their pantheon because Isis, too, steals magic from her "father." The later Greeks, however, punish Prometheus for stealing fire and giving it to mortals.

[341] Some people insist that, since the word for father was used by the ancient Sumerians so often, that they had to have known of the biological connection between sexual intercourse and procreation, but the main problem with this idea is that, even today, we have men we call "father" who are not related to us at all,

who is often called, especially when he is with Athena, the "father god."

Just as changes to Inanna's worship demonstrate how androcentrism rose in Sumeria, morphing this super powerful goddess into a more frightening, yet less powerful goddess, Ishtar, for the Babylonians, Athena's strange birth from Zeus' head is a symbolic demonstration of how religious changes from ancestor (read "female") worship to hero worship must have inverted women and men's procreative roles for the ancient Greeks. Quoting Hesiod's *Theogony*, Lucy Corcoran summarizes Athena's birth, "not from her mother, but from her father Zeus, springing fully armed from his head" (6). By swallowing whole Athena's mother, the goddess of wisdom, Metis, Zeus hoped to prevent one of the many prophecies of a son being born who would overthrow him, as he did to his own father. While Athena's "birth" through a gash cut in Zeus' head adds "wisdom" symbolism to her birth, Zeus also assumes the role of procreator—just as the men of ancient Greece, at least by Homer's time, embraced patrilineal descent for inheriting kingdoms. We know, as Homeric scholars, however, that the ancient Greeks were still largely matrilineal because Odysseus has to kill all of Penelope's suitors, since it is his wife, not their now adult son, who determines who will next take over Odysseus' kingdom through marriage to her.

Athena seems to have been Homer's favorite goddess, since she appears prominently in both *The Iliad* and *The Odyssey*, and he seems to enjoy making her the most complex personality in *The Iliad*. While she is a war goddess, "Athena is not bloodthirsty, unlike her male counterpart, Ares" (Cocoran 2). But Athena has chosen a side to support, and, unfortunately for the Trojans, she chose the Greeks, so, while the Trojan women proffer her statue and shrine within Troy with rich robes and other offerings, Athena chooses to

---

so this argument is a red herring meant to distract from the fact that Inanna calls all the male gods, except Dumuzi, "father" in many of the poems and hymns about her—and they cannot possibly all be her real "father," can they, except in the age-old concept that the men of a community "contribute" spiritually to the children women bear. More probable is the idea that such terms, which we now consider familial, were, before the connection between heterosexual intercourse and conception were made, simply social terms of respect.

ignore them, effectively ignoring her role as the city's protector. However, as Robert Luyster[342] points out in his article, "Symbolic Elements in the Cult of Athena," "Troy could not be defeated as long as [the city] possessed" the Palladium Athena statue, which is why "Diomedes carried it away with the help of Odysseus and brought it back to Argos in Greece" (156). This division within Athena's powers is fascinating because it seems to indicate that Athena, the goddess, was not in control of Athena, the guardian statue, which continued to protect Troy until it was stolen away. This dual quality **seems** to call into question, ever so subtly, whether or not the gods were believed to be as powerful as they were claimed to be, an idea that Homer explores even more fully in *The Odyssey* because he has Odysseus doubting the gods. Athena, in her role as prognosticator, sees Odysseus' doubts as the precursor to atheism among the Greeks.

However, we must remember to think like pantheists, who believed everything was embodied with divine spirit. When the Greeks built a statue to honor a god, that statue was clearly not the god, but actually had a divine energy borrowed from the god and the people who created it to become an entity all its own. While we could think of such a statue as being a telephone connecting us to the divinity we appeal to with offerings and prayers, we must recognize that the Greeks probably believed that the statue, while a part of the deity, was not, itself, powerless, which led to the fact that so many gods are referred to in the literature by their common name or by a name that linked them with a specific place. Therefore, the act of praying to Athena's statue by the Trojans all while Athena, the goddess, was acting as an agent for the Greeks on the battlefield would not have seemed to be as schizoid a belief to the ancient Greeks as it appears to us now.

Even Homer does not completely question the validity of the belief that the gods are more in control over the battle between the Greeks and the Trojans in *The Iliad*. For instance, when a Greek

---

[342] Luyster's article is a fascinating examination of the many different views the Greeks held of the goddess Athena, although his word choices are often androcentrically biased, such as when he calls Poseidon "the mate of Medusa," even after just reminding us that Poseidon raped Medusa in Athena's temple on the Acropolis (159).

warrior, Diomedes, prays to Athena for help during battle in *The Iliad*, "she draws near to him and gives him the strength of his father[343] and the power to tell gods from men" (Corcoran 3), so that, even if Diomedes was not already a hero, he becomes the kind of hero who communes with the gods, or, in this case, convinces one god to help him harm another god. Athena also deceives Hector into believing she is one of his brothers, come to help him fight Achilles, only to allow him to enter combat alone, leading to his death.

While the theory of using divine intervention to express complex human emotions can be applied quite often in *The Iliad*, Athena clearly plays favorites—keeping Achilles from killing Agamemnon in his rage, helping Diomedes to harm Aphrodite as payback for unfairly winning the contest of the golden apple for "the fairest," and outright deceiving Troy's main hope through the pitiless slaughter of Hector. Is Athena merely a convenient tool for the poet to use to justify seemingly random actions? Or is Homer's own struggle with atheism showing through?

Farron also believes, wrongly, that few women would have been involved in the siege on Troy, in order to explain why Briseis is only one of two human women Homer mentions directly in the poem on the Greek side (Women, Note 3, p 15). Farron ignores the plethora of women who had been taken as booty, many of whom he lists later as prizes exchanged among the men in the same article: "well-girdled women" and "a woman who was skilled at much handiwork" in Book 23 (Farron, Women, 27). He even goes so far as to claim that the men in the poem do not display much interest in heterosexual intercourse: "indeed, not only did these Greek warriors not consider women to be significant human beings, they were not even particularly attracted to them as sex objects"[344] (Farron, Women, 27).

---

[343] Diomedes' father was Tydeus, who was an Aetolian hero as one of the Seven Against Thebes. Are sons of heroes automatically heroes? Or do the gods have to single them out?

[344] According to Robert B. Koehl, one custom among warriors that was deeply entrenched by the third century BCE was when a *philetor* (lover) would select a *parastatheis* (companion) to initiate into manhood by abducting him, taking "into the countryside for two months of hunting and feasting" and sex, after which "the *philetor* [presented] the *parastatheis* with gifts as required by law, including

What did Farron imagine all these women were doing in the Achaean camps, then? Clearly, as aware of feminist views as he tries to be, Farron dismisses these "lesser" women's significance to the story simply because they are not visibly named.[345]

In his footnote (#40) regarding this particular claim, Farron claims "the Greek warriors had little concern for sex because they treated it so casually and generally regarded their female partners as interchangeable" (Women, 27), pointing out that even Achilles' goddess mother, Thetis, equated food and sex as equally important, completely ignoring the fact that we need food daily to survive, which would seem to mean, via Thetis' point of view, that these Greeks needed sex daily, too.[346]

What Farron misses is the possibility, undoubtedly, that the Mycenaeans probably had a much more casual regard for sex than the Greeks of Homer's time because they probably did not fully believe sex had anything to do with procreation. After all, it takes 40 weeks from the time of fertilization to birth. There are no flashing lights to tell a woman when she has conceived, and, in some women, especially malnourished women, it is difficult to detect outward physical signs of pregnancy. If the women are having sex with many men, and the men are having sex with each other, some of the men might wonder why none of them ever get pregnant if, indeed, the men are planting homunculi "seeds"[347] that become babies inside the women. It is also possible that many of the men in Homer's poem possibly view sex with women as another form of violence, both of which seem to give them pleasure because "battle thrilled them more than the journey home" (Fagles, Book 11, l. 15) and the men fantasized about "plenty of women/ crowding your lodges" (Fagles, Book 2, ll. 264-5).

---

an ox, military gear, and a drinking cup" (116). Perhaps even the poet creating *The Iliad* preferred homosexual relations to heterosexual ones?

[345] Similarly, many scholars ignore the fact that the goddess Hestia, goddess of hearth and home, is never mentioned in *The Iliad*, despite the fact that she would be "present" at every hearth and at every sacrifice.

[346] Thetis' view implies that sex **is** sustenance, indicating an age-old need for the transformative powers of sexual intercourse for both women and men.

[347] One of the most common beliefs about pregnancy after the discovery of procreation.

Yet there is still something terrifying to these men about female bodies and heterosexual intercourse, and Homer shows us these fears symbolically in two important ways.

First, Homer symbolically shows us this fear through Aphrodite, who is not truly the goddess of "love" or "compassion," since she shows neither, but is, indeed, the goddess of sexual lust; Aphrodite not only caused many of the circumstances that led to this war, demonstrating that uncontrolled lust can lead to war,[348] but also continues to meddle in the violence itself, both in saving Paris by whisking him away from dying at Menelaus' hands and by ineffectively taking part in the combat herself, which many take as a sign that proves females, not even female gods, should not participate in battle, since she sustains an injury that pulls her from the battle.[349] Aphrodite could symbolize, instead, how sexual lust blinds men, especially, to reason, the fear of which might constrain some men from following through with their lustful thoughts.

Second, sex is used as a weapon by Hera to distract and to incapacitate Zeus,[350] the "king" of the gods. More than Aphrodite's ability to suddenly make a man feel lust[351] after a woman he sees, these men fear incapacitation. They have to be in control. That need for control leads to Achilles' wrath, Patroclus' death, and Hector's arrogant ignoring of Andromache's war experience-based advice. In fact, Achilles' wrath is a direct result of his losing control over something he thought he owned, so that the Wrath that consumes him is ironic because, by indulging in it, he loses control of himself as a human being even as he believes he is taking control.

---

[348] While Paris' lust for Helen begins this particular conflict, such themes recur throughout literature, including Rama's rescue of Sita from Ravana and Lancelot's lust for Guinevere.

[349] Aphrodite's injury, however, serves as a means to explain why the gods are immortal, but still prone to pain. As a story, it clearly explains the reasons behind a spiritual belief.

[350] We know from the Linear B tablets that Zeus was not the original supreme god on Olympus. Instead, Zeus replaces a female god, Diwia, and Poseidon replaces a female god, Posidaia (Lupack 271, Serrano Laguna 288).

[351] The excuse that the woman (i.e. Aphrodite) is to blame, instead of the man raping her, is still heard today, demonstrating that some people never expect men to grow up and learn self-control.

The only male in the entire epic who does not need to be brought back into control is an old man, Priam, king of Troy, and a hero in the sense that he can hear and can recognize messages from the gods, saying, "I heard her voice with my own ears./ I looked straight at the goddess, face-to-face" (Fagles, ll. 265-6). Priam does not hesitate to take action in Book 24 when Iris, Zeus' messenger, bids him to take gifts to Achilles, willingly endangering his own life after having seen so many of his sons murdered by the Greeks by going straight into the Achaean camp to ask Achilles for the return of his oldest son's body. Priam's cart filled with gifts for Achilles is even driven by Hermes himself, leaving him at the gate to the Myrmidon camp because "it would offend us all/for a mortal man to host an immortal face-to-face" (Fagles, l. 544-5), but urging Priam to "stir" Achilles' heart.

Why does King Priam go to Achilles, a mere prince, to beg for the return of his son's body and not to Agamemnon, the Greeks' king of kings, who should be the one making such decisions?

The simplest answer is that Achilles is the one who killed Hector, so Priam could assume that he has kept Hector's body as a trophy, since Achilles has been dragging it around the city of Troy every day for twelve days as a way to defile the body, even though the gods, Apollo and Aphrodite, are keeping it preserved.

I admit that the poetic choice to have Priam appeal to the younger man, instead of to the older king, creates the most poignant scene, by many accounts, in the entire epic because Priam does everything a noble, heroic Greek (not to mention a king) should **<u>not</u>** do, given the new rising male codes of androcentrism. Priam begs, and he kisses Achilles' hands, pointing out that he has just done what no other mortal has ever done—kissed the hand that murdered his children (Fagles, Book 24, l. 591). Priam appeals to Achilles' own sense of family sorrow by asking Achilles to think of his own father and what he must feel because "no doubt the countrymen round about him plague him now,/with no one there to defend him, beat away disaster" (Fagles, Book 24, ll. 572-3). After the two of them share a cathartic cry (also highly unmasculine in today's standards), Achilles admires the old man's bravery, but he chides him:

> You must bear up now. Enough of endless tears,
> the pain that breaks the spirit.
> Grief for your son will do no good at all.

You will never bring him back to life--
sooner you must suffer something worse.
(Fagles ll. 642-6)

Is the last line a threat? Or is it a statement of inevitability? Soon, Priam, too, will be dead. It is cruel, even sadistic, for Achilles to taunt Priam in this way, but, perhaps, this statement shows us a depth of sympathy Achilles can hold for a man who will soon die.

Priam boldly tells Achilles that the gifts he's giving him are "a king's ransom" (Fagles, Book 24, l. 652), to which, ironically, Achilles angrily, possibly ironically, retorts, "don't tempt my wrath" (l. 656). Does Achilles react this way because he is so caught up in his layers and layers of anger? Or does Achilles' anger subtly point toward a breakdown in the hierarchical "palatial" system that preceded the Greek Dark Age? Clearly, Achilles feels he is a better man by almost every measure than Agamemnon, who has declared himself the king of kings. *The Iliad* demonstrates, again and again, that rank does not matter because everyone dies.

However, the poetry is providing yet another important symbol for us. Here, the old ways, symbolized by the gentle old king who cared for the women in his palace (having been one of the few to pity Helen, he also fathered 19 children on his wife, and 31 on other women), are being replaced by younger, more violent, more volatile male ways, symbolized by Achilles and his unquenchable wrath that leads him to violently murder countless men in his quest for deathless "glory" to ensure his name lives on, despite having acquired plenty of *kleos* before coming to Troy. Priam symbolically and literally appeals to what remains of Achilles' humanity in order to convince him to do the right thing. In the same way, the entire *Iliad* symbolically appeals to readers' humanity, in the hope that we, too, will recognize the symbolism embedded in this cautionary tale and choose to do the right thing(s) in our own lives.

ΩΩΩΩ

Matrilineal Descent Preceded Patrilineal Descent

It should seem obvious, by this point, that ancient humans were matrilineal and gynocentric before males ever learned they play a biological role in procreation. Even though we know the

Mycenaeans were originally matrilineal and matrilocal in their cultural practices, the pressure to make men seem superior to women must have been enormous for them because of the rising androcentrism in the eastern Mediterranean, especially by the time we get Greeks proper—by the time of Homer. First, they had to dismantle the idea that women were divine creators in order to elevate themselves into that role, something we see in the symbolic birth Zeus portrays as he gives birth to each of his children.

Cedric Wilson bravely addressed the fact that scholars knew the early Mycenaeans practiced matrilineal descent—everything passed through the mother, and, when children noted their parentage, they named their mothers, not their fathers, as their progenitors. Wilson states, "Matrilineal descent has been widely attributed to early Greece, and is known to have been the rule in Lycia also" (39). In fact, Wilson admits that "it is a striking fact that the most distinguished of Homer's Achaeans have short genealogies.[352] On the other hand, the family history of the Abantid Kings of Argos is one of the longest of all, counting six generations before Io, the antediluvian ancestress of the house" (39). The mere fact that Wilson calls the goddess Io "antediluvian" gives away his particular Judeo-Christian bias towards history, but he is honest enough to discuss how ancestry was, once, traced solely through the mothers, not the fathers.[353]

Yet Homer makes a clear point of listing as many male progenitors for the warriors on the plains of Troy as possible.

Why?

Homer hammers patrilineal descent[354] in *The Iliad* because, like hero worship, the assertion that fathers are identifiable and primary connectors to their children, more so than their mothers

---

[352] Wilson overlooks the fact that these Achaean genealogies are short because they only count back to known fathers in most cases, since doing so is a new thing at this point in Greek history. As Cedric H. Whitman admits, "Matrilinear descent has been widely attributed to early Greece, and is known to have been the rule in Lycia also" (39).

[353] The same goes for Judaism, which still traces religious ties through the mother. In fact, the Book of Ruth is an attempt to convince the women of Judea to change inheritance traditions regarding land holdings.

[354] Patrilineal descent becomes emphasized in Genesis, as well, which was probably first written near the same time as *The Iliad*.

was the new preferred social view by those in power. We can discern from this repetitive action that the poet is, at least in part, trying to please someone or some group of people with power. But even Hack, way back in 1929, believed enough in the idea that Homer could have been a dissenting voice against the Greek national propaganda that was arising around the Cult of the Hero to address the question (61).

Stephen O'Brien attempts to address previous sweeping claims about who, exactly, was in power during the Mycenaean periods, urging caution regarding some long-standing beliefs about the social structures, and arguing, for instance, that claiming the *wanax*[355] was the "king" of a particular city-state or *polis* is problematic because that claim was based solely on the idea that the largest land holding determined who ruled (29). He notes many historians who believed this biased view throughout his article but points out the information from the Linear B tablets indicates a more heterarchical society (29) wherein local landowners, probably all landowners who also employed others to work for them, were expected to raise and to equip warriors when the city or region needed them (30). Such a social and political obligation could have come about because the land had been given to them for previous service by the community or even by a "big man" ruler, but O'Brien readily agrees that the iconography depicting warriors within community buildings or on ceramics could be "read as a strategy to promote an understanding of the world in which particular forms of violence were associated with the 'palace'" or with the needs of the community as a whole (30). O'Brien also argues that new military items found in many of the excavated grave sites "should therefore be considered to represent **a change** in Mycenaean ideologies and their attendant practices, rather than a change in population" (emphasis mine 31).

The one important idea O'Brien overlooks, however, is that the iconography might be largely symbolic as mentioned earlier. He

---

[355] According to Thomas G. Palaima (2006), "the underlying meaning of the term *wanaks* is connected with 'birth' and 'generation'" and connected to "blood-lines" and "burial and ancestor cult" (57). If true, the *wanax* most likely refers to female leaders, who were both the original ancestors and from whom most Mycenaeans traced their lineage.

comes close when he examines the "Warrior Stele" from Mycenae when he emphasizes that the stele could be indicating "a collective group rather than any particular individual" warrior in the scene (33). Because the buildings in which many of the warrior frescoes appear were built over former megarons[356] (O'Brien 32-33), which were constructed for large ceremonial sacrifices with huge hearths, so are reminiscent of the gynocentric views[357] that preceded the androcentric ones, the people who constructed the buildings and painted the walls could have been consciously replacing a gynocentric world with an androcentric one, depicting war and violence as acceptable social norms for all who viewed them.

Does this increased pressure to emphasize androcentric values and patrilineal descent mean Homer embraced androcentrism, too?

Even in 1929, Leicester Holland knew "that the historic cults of Delphi and Eleusis go back without a break to prehistoric worship of the Great Mother" (60). Holland reports that "statues of the priestesses of Hera, whose consecration served as the basis of Argolic chronology, stood before the Argive Heraeum" (60), giving ancient Greek historians a measurement of time for dating events based on "the chronology of the Argive priestesses" (61). He notes

---

[356] According to Thomas G. Palaima (2008), "The canonical central megaron within Mycenaean palatial centers was a place of royal ritual. The central hearth reminds us of the importance of fire to Greek communities and of the goddess of the hearth, Hestia." (353). However, moving the central hearth indoors clearly limits who can participate in such rituals, which could be another way the rising androcentrism of the ancient Greeks sought to limit women's participation and their former powers.

[357] Hearths were originally associated with homes and homemaking, the purview of women, wherein individuals and families ritually gave sacrifices of meat (mostly fats, which both stoke the flames and create attractive smoke for the gods when burned) and libations of wine before meals. Hestia, the goddess of both hearth and home, was always present at such common rituals, with the ancient Greeks believing she received these sacrifices, so would protect their homes and families. Despite her prevalence in such rituals, Homer never mentions her in *The Iliad*. The simple act of moving ceremonial hearths from outdoors to inside "palaces," indicates that someone wanted more control over who could participate in such rituals (Kienlin 25). Were women being systematically closed out of such ceremonies of which they would have, once upon a time, been the primary leaders?

that "the chronology of Lindos, Rhodes, was similarly based on the annual priesthood of Athena" (Note 7, p. 60).

Even though many scholars and archeologists resist the idea that there was a gynocentric oriented view of humanity wherein women were divine because they wielded Female Creative and Protective Magics, historical documents, including legal changes from matrilineal descent to patrilineal descent, indicate there was a period prior to the rise of androcentrism wherein women were valued not just as human beings who were heads of their households and their families, but also considered divine or at least magical. Once males learned that they have a verifiable biological role in procreation c. 2400 BCE, they assumed they were the magical creators planting "seeds" in women's fertile "fields"; an uninformed belief held well into recent history. Among the ancient Greeks in particular, ancestor worship—which would have involved worshipping female ancestors, since men were not believed to have a biological role in procreation, so had no direct kinship ties— purposely changed to Hero Worship, so that men's brutality in battle became lauded in order not only to establish men as able to be divine or superhuman, but also to create propaganda to convince young men to die for their people. This social and political change touting Might as Right affected everyone, especially women and "lesser" men, and everything about Mycenaean culture. As a result of rising androcentrism, even favored goddesses were demoted.

The possibility that the Greek Dark Age was a time when Mycenaeans realized what was happening to their culture, possibly blaming writing on these changes because of laws that began to eliminate and to regulate women's lives in a suppressive manner, that both men and women alike who were repulsed by the denigration of females, especially their goddesses, revolted. It is interesting to compare the results of the Greek Dark Age with the rise of the patriarchal culture, Judaism, which recorded the *Torah*, aka *Pentateuch*, at about the same time Homer wrote down *The Iliad*. While controversial, the Book of Ruth, when examined gynocentrically, reveals that the Hebrews were in the process of convincing the women of their tribes to accept certain changes, particularly changes in land inheritance. But we also know from archeology that, even though the leaders of Judaism resorted to using a "lost book" to justify worshipping just one god and having just one king, that the common people of Judea continued to worship Asharah, their Mother Goddess,[358] so it is not inconceivable that the common people of this region of the world rose up against the "elites" of their cultures in order to try to stop the change from a gynocentric view to an androcentric one. Ironically, their very act of rebellion might have given stimulus to the idea that Might Makes Right.

Therefore, it is clearly conceivable that whoever finally recorded the poem in writing framed *The Iliad* as a cautionary tale, attempting to subtly warn his listeners and readers about the dangers of embracing Male Might as Right, which encourages violence against all those who do not fit the "hero" label, and even needlessly endangers those who do.

Modern heroes[359] act selflessly, not expecting praise or thanks. The need for *kleos* only comes from those who are self-centered, who need glorifying to feel important or valued. That the Cult of the Hero appealed to Greek male vanity, which had possibly been relatively unassuaged in more gynocentric times, is what sold the Cult of Heroes to the ancient Greeks, possibly indicating that the poet's efforts to elucidate and to warn his/her listeners on the negative effects of encouraging brutal male hubris went unheard or, at least, unheeded. It is almost needless to point out that a vast majority of the scholars studying *The Iliad* have also missed the importance of this cautionary tale.

---

[358] William G. Dever argues that "the biblical writers and editors knew very well what the female figurines [of Asherah] represented, and therefore they deliberately suppressed any reference to them" (emphasis his, 184). Such open suppression would have created an air of hostility between the priestly noble classes and the common folk. J. Alexander MacGillivray notes that "the double axe icon, then could be shorthand for a Minoan goddess whose name begins with 'a' and is called something like Ashera" (120).

[359] The fact that few choose to view pregnancy and childbirth as heroic events, simply because they are so common, clearly ignores the fact that death during wartime is also quite common, but it is still considered an heroic act. The only difference is the fact that one is completed by women, while most people still view war as "men's business." This paradox further demonstrates how males are believed to be more important than females.

Chapter Twenty-Six
Hero Worship, Now and Then:
Love Super Human Men?
So Did the Ancient Greeks

My son and his friends, understandably, grew up loving super heroes. As a group, these young men followed super hero franchises through various iterations, and probably talked about these more-powerful-than-human beings frequently in their conversations over the years, beginning in elementary school.

Even now, in their mid-20s, they play video games online together on weekends. The games often include super human beings, sometimes female, who can fight harder, kill more, and triumphantly claim victory over imagined hordes of enemies over and over. The odd female fighter is thrown in to make players think all this violence is not part of any form of toxic masculinity. These video games help these young males, and their occasional female friend, feel a little like super human heroes.

Joseph Campbell became famous because of his theories about heroes and the "heroic journey," largely because George Lucas credited Campbell's ideas in helping him form the basic story for Luke Skywalker in the original Star Wars movie. Campbell had noticed this "heroic journey" pattern when he learned about and read myths and legends from many cultures. Even though Campbell noted that "of all animals, [humans] remain the longest at the mother breast" (6), he saw the basic function of mythology as a psychological release "from the bonds of [our] mother-complex" (11). Campbell firmly believed that women did not have, nor did they need, child-to-adult rituals like men did because, well, women had menarche, and that was enough of a change in their lives that alerted them that they were no longer girls without adding ceremony to it. In his most famous book, *The Hero with a Thousand Faces*, the Index includes only one entry for "woman," listing them only as "symbolism in hero's adventure," "as goddess," "as temptress," as "Cosmic Woman," and "as hero's prize" (416). Of course, Campbell completely ignored the plethora of girl-to-woman's rituals in existence—from bat mitzvahs and quinceañeras to the Navajo na'ii'ees puberty ceremony. Women in Campbell's androcentric view simply were extraneous, not central, to mythology or to heroism.

Meanwhile, Judy Grahn did cover women's rituals and discussed their continuing influence over our modern lives in her not-to-be-missed book, *Blood, Bread, and Roses: How Menstruation Created the World*. In this ovular book, Grahn points out how many ordinary things have come down to us from women's menstrual rites, such as things associated with royalty—from birthing chairs that have been turned into thrones to the colors red and purple (think blood and clotting) as being colors only important people could wear during certain times in history. Like Campbell, Grahn touches on the trauma of separation, especially separation of young women at menarche from the rest of their families, acknowledging that such separation must be traumatic psychologically, even while such separation empowered women to create rites which still influence us today (15). In exploring the meaning of words, such as taboo, sacred, and sabbath, Grahn reveals how many myths were formed around women's menstrual cycles. She goes so far as to analyze one blood-soaked myth Campbell relates about sacrificing boars in Melanasia, which she sees as "an explicit depiction of early 'lunar' menstruation" (136). Where Campbell was androcentric in his views of humanity, Grahn is gynocentric, so, presumably, reality lies somewhere in the middle.

Neither Joseph Campbell nor Judy Grahn directly associated coming of age rituals with either ancestor worship or hero worship, possibly because ancestor worship has not been fully explored via archeology and anthropology, and because hero worship has long been believed to be solely a male thing. What most people do not realize is that our current enthrallment to super heroes is a direct result of a propaganda campaign, **possibly the most successful propaganda campaign on earth ever**, started by the ancient Greeks roughly 2700 years ago.

Probably during a period now called the Greek Dark Age, but possibly earlier during Mycenaean (pre-Greeks) times, a propaganda scheme was hatched to change the emerging cultures of the Aegean from its emphasis on gynocentrism, or female centered values, to androcentrism, male-centered values. So sometime between 1700 BCE and 700 BCE, someone decided that men needed to be deified the same way the ancients had deified ancestral women. Women, after all, had been long considered semi-divine, even while alive, because they could magically produce children from their own bodies, could bleed three to five days every month without dying, and could feed, cloth, and heal people. Prior to verifying procreation took two people circa 2400 BCE, women were the **<u>only</u>** verifiable ancestors people had. So ancestor worship meant that women, once they had died, became literal gods, divine entities to be propitiated in order for the living to receive their continued blessings. There is reason to believe that Greeks who trace their ancestry to Io are actually tracing an ancient ancestry to a real woman who became deified after death. DNA research of human remains found through archeology has confirmed that most ancient people buried their loved ones in communal graves, most likely centered around one female progenitor, and, later, buried individuals near their mothers before moving on to the individual tombs that aggrandized male rulers. Most anthropologists and archeologists admit that these burial practices mean that most ancient people were matrilineal, meaning they traced their descent from their mothers. Matrilineal descent makes absolute sense prior to the knowledge of procreation because we always know who the mother of a child is.

The Mycenaeans, like the Lycians and the Minoans, were matrilineal cultures as late as 1700 BCE. The Lycians remained matrilineal for as long as they were an identifiable culture, in fact, possibly as late as 500 BCE. Since we always know who the mother of a child is, matrilineal descent is **the most natural form of inheritance**—property and status passed down through the mother's line. In fact, archeological studies, especially those that use DNA to verify gender, have shown us that most of the pre-literate people around and near the Mediterranean Sea practiced matrilineal descent for millennia.

These simple facts that our ancestors lived with gynocentric world views seems shocking to those of us raised in androcentric cultures. Our shock demonstrates just how much androcentrism from the last 4300 years has buried itself deep into the world's psyche. We must, however, embrace the fact that women used to be revered, not just as mothers, but as leaders, as primary ancestors, and as gods after they died. Ancestor worship, which was pretty much practiced everywhere on earth prior to the discovery of procreation, literally means that people were revering their dead mothers as divinities, seeking supernatural aid from the deceased women in their families. So, when scholars talk about ancestor worship by the ancients, instead of assuming they were worshipping men, we need to embrace the fact that they would have been worshipping the women—those people who had, seemingly all on their own, as if by magic, created other human beings.

How can we verify this claim is fact? There are several methods.

A primary factor that lends credence to the fact that humanity was once gynocentric, one we have known about since archeology became a science, is because Birth and Rebirth were the primary themes in every culture that carefully buried its dead. The most common feature in burials around the world are not weapons or tools, but ochre—usually tinted a bloody yellow or a deep red. Ochre symbolizes birth blood—that which brings us into the world, and that which takes us into the Next World. Blood is central to the menstrual and birthing process, and was such a revered, spiritually powerful symbol that men took to circumcision in order to imitate menstrual spiritual power (Grahn 46), so we cannot overlook ochre's symbolic meaning for our ancestors.

People buried their dead in the earth because they believed they were burying them in the womb of Mother Earth, just as seeds, which rejuvenate plants, are buried in the earth. Such an act demonstrates our ancestral need to imagine that our loved ones will be reincarnated in some form somewhere. The very fact that we still refer to the earth as our mother and to Mother Nature, also demonstrates how long and how deeply this particular idea has been embedded in the human psyche. Our ancestors noted how plants seemed to die in winter but were rejuvenated in spring, all from the "womb" of Mother Earth, so assumed human beings could do the same thing, especially if we were buried within the earth or within a "mound of creation," meaning mastabas and pyramids for the ancient Egyptians, which were symbolically female pudenda "rebirthing" those buried within into the next life.

In fact, the Sumerians used the term *kur* to refer to the place where the dead go, but Judeo-Christian biased scholars translate the term, not as "to the mountain," which is what it means, but as "the nether world" or as Hades or Hell,[360] believing erroneously that the Sumerians thought, like the Greeks and later Christians will think, the next world was below ground, instead of inside a mountain peak. However, the earlier Sumerians influenced the ancient Greeks, who believed their gods resided on Mount Olympus, by believing their ancestors turned-divine-upon-death were returning to the mound of creation, in this case, a sacred place on a mountain.

Another important factor, which is not widespread knowledge yet, is that while there were, undoubtedly, men who were important in most cultures around the world prior to the discovery that males had a physical role in procreation, circa 2400 BCE, most cultures believed women were magically creating babies on their own, or, possibly, via spiritual influences by the community, since some of the offspring might look like some of the men. How did they explain how some children resembled some men, if they did not believe men had a biological role in conception? Even in parts of the world today, some cultures still hold that men influence children being born in a spiritual, not a physical manner, so that all the men in a village are responsible for the care and raising of the children because they all have a part in the spiritual shaping of the babies during pregnancy. Humans did not make the connection between heterosexual intercourse and pregnancy until after they domesticated, circa 2400 BCE, the only monogamous animal ever domesticated, swans, which allowed them to determine that coitus caused fertilized eggs (they also had to wonder why not all the eggs would be fertilized, which is another reason this scientific process took so long), thus offspring.

Prior to learning of procreation, if men were not literal fathers of children, there could have been no patriarchies, but that

---

[360] Thorkild Jacobsen makes this leap when he claims that, "for the Sumerians *kur*, "the mountains," represented Hades" (53). Clearly, he and other androcentric scholars like him are refusing to admit that the ancients viewed mountains in the same way they viewed mounds—symbols of female pudenda or rebirth mechanisms.

does not mean ancient people lived in matriarchies, either. Realize that such polar-opposite thinking is nonproductive, so no-one reading this needs to fear women will create matriarchies in order to emasculate men, despite the fact that men have imposed egynovi[361] on women for millennia. A world run by women would probably not be nearly as oppressive a place to live as the one still run, largely, by men is, since believing women should dominate men runs counter to egalitarianism. However, there is ample evidence from archeology[362] that most societies, while possibly hierarchical in nature, were still more egalitarian in most places around the world prior to the rise of androcentric biases.

But we do need to recognize that the ancient Greeks seized the opportunity either during or immediately after the Greek Dark Age, presumably at the time of or directly after the palaces had been sacked, to rebuild over those palaces using pictorial propaganda to sway people's thinking toward androcentrism. While some of the new structures kept megarons—the huge hearths around which important religious ceremonies took place—many of the new structures replaced previous gynocentric symbols embedded in the hearths and painted public rooms with murals depicting warriors in full battle regalia, instead. Furthermore, they began incorporating the wonders of "heroes" everywhere they could—from pottery and jewelry to the names of cities.

Just in case you were unaware, the hearth of a home, in most cultures, not just for the ancient Greeks, was tended by the women, who owned the homes. In fact, we get the term "homemaker" from this ancient fact—the women usually built the homes most families lived in, from tipis and huts to yurts, hogans, and wattle-and-daub houses. In fact, Hopi women, many Asian, African and Middle Eastern women are still responsible for building and maintaining the structures of the homes in which their families live. So having

---

[361] There has been no equivalent word for "emasculate" until now. "Egynovi" means to disempower women based solely on their gender; to unempower the feminine.

[362] We know, for instance, that the ancient Sumerians, like the Iroquois/Haudenosaunee Confederacy, functioned with elected officials in at least two councils (e.g. the Matrons Council and the Warrior Council) that made decisions for whole communities, so that democracy is far older than most people believe it is.

ancient Greek males assume control of the giant ritual hearths and the megarons in which they were housed would have been a tremendous undertaking, so we should not assume they were able to make this transition overnight.

Scholars still argue about what brought about the Greek Dark Age, but none that I have read have considered the possibility that because new laws, written down for the first time, were geared toward reducing women's status, including putting favored goddesses below male gods and regulating ancestor worship, putting women in the back of funeral processions and limiting their roles in such ceremonies, so that both men and women rebelled against this refocus of values from gynocentrism to androcentrism. After all, if Campbell, who references Carl Jung, is correct, separating human beings from their mothers is traumatic, so that most humans seek the circle of comfort that is "full circle, from the tomb of the womb to the womb of the tomb" (12).

Signs that the ancient Greeks were becoming more dominated by hierarchical rather than communal views come from the fact that the ancient hearths were no longer outside but had been transported inside the megaron rooms often featured in early Greek palaces. The ceremonies held at these more isolated hearths were for "elite" eyes only, although undoubtedly a chosen individual or perhaps half a dozen youths from the more common elements of the society were allowed to witness such ceremonies in order to keep the legends of what happened at those ceremonies alive. In many of these rituals, it was usually young women from "good" homes who only recently reached menarche or who were about to reach menarche who were chosen to participate, demonstrating that the presence of young, nubile women was an important element to many of these ceremonies, possibly because of the belief that the ability to bleed monthly and not die, all while experiencing sometimes crippling menstrual cramps, imbued them with special spiritual powers.

Besides wanting males to be worshipped like the ancestral females had been, as divine beings with powers to alter people's destinies who could be called up for assistance, the ancient Greeks needed young males willing to fight and to die for their culture in order to build a fighting force so that ambitious leaders could attack various Greek and non-Greek cities scattered around the Aegean in order to build their empires.

Now, we know from the Spartans and the Minoans that the females were trained in physical combat, too, so why did the ancient Greeks prefer male combatants?

Possibly earlier, but also possibly around the same time that humans determined that males have a biological role in procreation, humans also discovered that castrated males, especially when castrated at the right moment in life, can not only grow bigger and stronger than intact males (at least among cattle), but also are more manageable than intact males, which is why so many males who had been captured in warfare were castrated—to make them manageable. Their captors wanted their technical skills, but not their belligerence, so castration reduced most threats these men might pose.

So, if a culture can afford to castrate a great number of males in order to keep them docile without losing the ability to produce enough children, they can also afford to lose men in battle. My husband's generation called such young men "cannon fodder."

But how do you convince young, possibly self-centered men to willingly throw themselves into battle, possibly to die most gruesome and painful deaths just to expand one man's or one city's sphere of power?

We create **propaganda that makes being a warrior seem attractive, super human**. Ergo the ancient Greeks created myths about such super human feats by seemingly super human men, like the warriors described in famous ancient Greek epic poem, *The Iliad*, Achilles and Hector.

Images of warriors in full combat gear began appearing not only as murals on important public walls, but also on pottery, on personal seals (which stood as a person's signature in those days), on jewelry, and, more easily portable, in stories.

As the Cult of the Hero grew in stages (Whitman 77), burial practices changed from group or family entombment or even from cremation (which happens to Patroclus' remains in *The Iliad*), to individual burials wherein the warriors are gifted with more than just their armor and weapons, but also with items made from precious metals and stones as befitted their new status in the society.

Since funerals of such "heroes" would have been public spectacles that lasted for days, such celebrations for the dead would have convinced many young men, just as Achilles is persuaded in *The Iliad*, that dying in battle is glorious, so much so that their names would live on long after they were dead through *kleos*. If their deaths were glorious enough, where they died might become local shrines. If their fighting skills were super human enough, they could even get cities named after them.

This kind of adulation had been reserved, previously, for revered, semi-divine ancestors. With the rise of the Cult of the Hero, that reverence, that worship that was once reserved for revered female ancestors (such as the goddess Io, to whom some Greeks still trace their ancestry), was transferred to male warriors—the more brutal their conquests, the better.

And, if these particular "heroes" were lucky, not only were their graves worshipped and visited by people from far and wide who admired their violent life choices, but also important points of geography, such as hills or valleys and even cities, were named after them.

What greater motivation could there be to risk one's life in battle than to be immortalized? The Greeks called this immortalization, *kleos*, which translates into "what others say about you."

Scholars still debate about when the Cult of the Hero arose in various places around the Aegean Sea, but the writer of *The Iliad* (circa 750 BCE) was definitely familiar with it because we get elements of Hero Worship in the scenes describing Patroclus' and Hector's funerals, including 12 young Trojan men sacrificed to Patroclus through immolation on his funeral pyre, and having Achilles speak to his "honored ghost" in the hope of appeasing him. Patroclus' spirit appears to Achilles in a dream and admonishes him for neglecting his corpse, claiming he needs a proper funeral because "the spirits, the shades of the dead, are keeping me out" of the Underworld (Green, *The Iliad*, Book 23, line 72), where he needs to reside before he can be reincarnated. This dream-plea prompts Achilles to plan an elaborate funeral for his dead friend.

What is clear is that violent, murderous men, who stab other warriors in the back, taunting them as they fall dead, became the "norm" in *The Iliad*, which intentionally or not helped usher in an age wherein many males felt entitled just by possessing their male genitalia (or not, as the case would be if they were castrates) to use any form of weapon to suppress others for their own monetary or honorific gains. **That kind of attitude and action is the exact definition of toxic masculinity: using force, including threats of force, to overpower or to suppress others for personal gain.**

What we must do now is to reflect on whether or not we want to live in the type of world that admires and even worships super human violence because, frankly, we have no real super human enemies to defeat, just our personal arrogance and greed.

The kind of heroes we really need are selfless, not selfish, helpful, not domineering, and are full spectrum humans, not dispassionately super human. We need people to be humane, compassionate, and cooperative.

As Brian Hare and Vanessa Woods put it in their book, *Survival of the Friendliest*, "cooperation is the key to our survival as a species because it increases our evolutionary fitness. But somewhere along the way, 'fitness' became synonymous with physical fitness" (xvi), warning us that believing "being the biggest, strongest, and meanest animal [is necessary for success] can set you up for a lifetime of stress" (xvii), which ends up making us weaker emotionally and socially, not stronger.

We now know how and why humanity began to embrace super human violence as being a desirable trait, and we know approximately when this still operating propaganda machine first started--at the outset of the Cult of the Hero some 3000 years ago.

It is time we all throw off the ancient Greeks' propaganda scheme, so we can be free from toxic masculinity, and, instead, nurture healthier masculine and feminine values to choose from.

Chapter Twenty-Seven
The Translation Conundrum

As I began my research into *The Iliad* as a lamentation over the loss of a gynocentric world view, I realized I had a dilemma. For someone like me who doesn't read much Greek, let alone ancient Greek, finding a reliable, trustworthy English translation of *The Iliad* has been problematic, largely because there is not one that does not exhibit some androcentric bias.

Every single translator has carried into the poem her/his own biases while trying to find, as one of the most recent translators puts it, "the *mot juste*"—which I take to mean the best word that does justice to the original meaning (Homer, Green, Preface, xii). Translator Peter Green admits that some ancient Greek ideas would seem "difficult and often alien" (Preface, xii) to modern readers, but he never addresses the precautions he must take to prevent his modern bias from creeping into the translation. Nor does he bother to explain why he cannot take the time in his translation to explain those "difficult and often alien" ideas to his readers.

What I have discovered from examining several different translations of the epic poem is that none of the translators are as well informed about ancient Greek culture as they should be.

None of the translators of the different scholarly editions of *The Iliad* that I consulted[363] address the very important issue of androcentric bias directly in the Prefaces or Introductions to their books. This bias-towards-males colors their interpretations considerably, including undervaluing the many goddesses, especially the minor ones, mentioned in the poem, leading most of them, importantly, to misinterpret that important first line. While their androcentric biases undoubtedly mean they do not fully consider the importance of the goddesses, the focus on rituals, and the irony embedded in the poem that only appears to applaud violence and male bravado, it also means their translations lose

---

[363] Because I will be quoting from several different translations of *The Iliad*, I will maintain each translator's choice of spelling within the quotes.

many of the nuances that only the ancient Greeks would have understood.

All attempt to do their best, but most, with the exception of Robert Fagles (Homer, Preface, xii), seem fully unaware that they carry any biases with them, let alone androcentric ones.

Peter Green gushes in his Introduction to the epic that *The Iliad* focuses "not only [on] gods and monsters, but [also] on the great deeds of warriors long dead, its enskyment of heroic fame, *kléos* in Greek—the only claim that evanescent individuals could make to any kind of immortality" (Homer, Green, Introduction, p. 8). Ironically, he immediately then refers to the *Epic of Gilgamesh*, not to admit that Gilgamesh is not enskied, even though evanescent, because of his warrior prowess, but because he sought immortality and learned to live in the moment instead—a fact that completely undermines Green's claim that only warriors can do great deeds or be heroes. Green conveniently overlooks Gilgamesh's human faults—such as being frightened and literally running home to momma often (something he has in common with Achilles)—in favor of lauding "great deeds of warriors" (Homer, Green, Introduction, p. 8) as his definition of what a hero is.[364] Green, alternatively, uses *Gilgamesh* as a segue into discussing poetic formulae but misses entirely an important point about poetics— they are born from ritual, and ritual undoubtedly arose from socially, publicly produced ceremony to invoke, to understand, or to

---

[364] While Green's definition of what a hero is alludes that only brutal acts by warriors can be heroic, he is like many androcentric biased scholars when he ignores the actual ancient Greek definition of *kléos*, which is having the person's name remembered long after her/his death, although remembering the person does not have to be from acts of violence or warfare. A hero was, for the ancient Greeks, someone who is so favored by the gods that it seems s/he could speak directly with a god or gods (Soles 2016, 250), and should be the definition used, since *hieros*, knowing or recognizing the sacred or hidden spiritual things, is clearly a precursor to the Latinate term *hero* used to only describe males since the 1700s, according to the *Oxford English Dictionary*. This concept of the sacred, "hieros," is extremely common, still, so that we study *hiero*glyphics, or the sacred writing of Egypt, or know about *hiero*dules, or the slaves who served in temples, or even argue about hierogamy, what many call "sacred sex." The fact that this term (hiero to hero) has been transformed from sacred knowledge to a man who wields war effectively is another demonstration of how the rise of androcentrism discolored word definitions.

connect with the spirit world.[365] For most cultures, failing to recite ritual poetics in the appropriate manner could result in catastrophe, which is one reason most poetic forms shared orally have meter and rhyme—to make them easier to remember and to recite from memory. For the ancient Greeks, rituals arose from gynocentric powers that linked women to earth and moon magic and growing androcentric views linking males to sky, especially sun, powers.

Green admits that "dawn is always rosy-fingered, and Achilles swift-footed. Before a feast the heroes regularly stretch out their hands to the good things set before them, and only move on to other activities when they have satisfied their desire for food and drink" (Introduction, 8). But, while he admits "this formulaic phraseology is not as all-pervasive as has sometimes been alleged" (Introduction, 9), he conveniently overlooks poetry's undeniable connections to ritual, to spirituality, to connections with the divine, which, for the ancient Greeks, meant a connection with the earth itself and all of its myriad spiritual forms that such a pantheistic culture would embrace.

I had sincerely hoped for more awareness of androcentric bias from Caroline Alexander's translation, also published in 2015, simply because she is female, but was sorely disappointed.

Alexander's androcentrism goes deeper than just lauding male bravado and machismo; it's clearly monotheistically biased, probably toward Christianity, making frequent singular references to a god or "the god," such as "the ramparts build by god" (Alexander, Book 21, l. 526) and at least once capitalized the word in a very Christian sounding sentence, "God accomplishes all things to fulfillment" (Alexander, Book 19, l. 90). Alexander recalls Christian versions of hell when she mentions "demonic fire" (Alexander, Book 20, l. 490), a concept that had not been adopted by Homer's time, and confuses readers when she fails to name the god of the Scamander River, but merely says, "the great god did not leave off" (Alexander, Book 21, l. 248).

Androcentric bias clearly clouds Alexander's interpretations when she refers to individual adult women or goddesses as "girl."

---

[365] Ritual, of course, becomes important in *The Iliad*, with one invisible goddess present at every single offering made by either the Greeks or the Trojans. See below.

Briseis, who is obviously no girl since she was captured as another man's wife and has been with Achilles for some time, is diminished in importance, first, by Agamemnon, who wants to take her from Achilles, "so I and Achilles have fought over a girl" (Alexander, Book 2, l. 377), but Alexander continues to diminish her importance by calling her a girl, repeatedly—even when she's probably not being diminished in importance by the actual speakers in the poem. Helen, too, is both diminished and faulted when Achilles, who has finally found some pity in his heart, at least for an old man, Priam, tells Atreus, "we two…raged in life-devouring strife for the sake of a girl" (Alexander, Book 19, ll. 57-8).

While there are undoubtedly some instances. like the two above, when the speakers in the poem might be using a Greek word indicating a young girl, not a woman, in order to diminish the specific woman's importance to the characters a speaker is addressing in the story, Alexander does use the word "woman" a few times when most other interpreters still use the word "girl," such as when Thersites dares to upbraid Agamemnon in Book 2, line 226—"your huts are full of bronze, many choice women"—and lines 232-3—"or a new woman so you can join in fornications, a woman you can possess apart?" (Alexander). Clearly, she's uncomfortable, as a translator, talking about grown men having sex with mere girls, as she should be, but so are Fagles and Green, who also use the words "woman" and "women" in these same moments. However, Mitchell and Fitzgerald awkwardly use "one more beautiful girl to screw in your hut" (Mitchell, Book 2, l. 219) and even more modern slang when he writes, "and the hottest girls—we hand them over" (Fitzgerald, Book 2, l. 194). Clearly, this particular passage is being influenced by the translators' choices, tainted by their points of view about women. Either Thersites is as liberal as he sounds as he berates Agamemnon for his hierarchical greed (Thersites is pro-democracy so should consider adult women to be women, not girls) or he's a mixed bag, depending on the interpreter's views.

Alexander makes another interesting choice in Book 8, when Hector derides Nestor, calling him a woman (line 164), which to male warriors was a decided demotion, but, where some interpreters have Hector actually call Nestor a girl when he tells him to "be off!" (Butler, Classics, l. 164) or a "sissy" (Mitchell, Book 8, L.

154), Fagles uses all three concepts, "now they will disgrace you, a woman after all. Away with you, girl, glittering little puppet" (Book 8, ll. 185-86). Alexander uses the worse diminution of the three used by calling Nestor a "poor puppet" (Book 8, L. 165), which must be close to the original Greek word used because Lattimore jeers "down with you, you poor doll," (Book 8, L. 164), Green sneers "on your way, craven dolly!" (Book 8, L. 164), and Fitzgerald pities him with "you empty doll" (Book 8, L. 138). As a critical reader, I come away wondering exactly what the ancient Greek word was that gives us either "doll(y)" or "puppet," but wonder if the ancient Greek idea might actually be closer to "poppet," a simulacrum used in magic—not just manipulated by strings like a puppet, but actually used to bring harm to others. Such a meaning would deepen the accusation Hector projects at Nestor, since it would imply Nestor is not in control of his decisions, but is acting based on manipulative magic, instead of just as a woman was supposed to, only upon orders from someone intending fatal harm on Hector and his family.

Consider how much more powerful the jibe is if Hector is subtly accusing Nestor of being a bearer of curses or other magic meant to harm. Even with the gods intervening in the battle, such additional magic would be unseemly to warriors who frown on arrows being used because they are dishonorable, since they remove the necessity of face-to-face combat. Realize that I am not claiming that the ancient Mycenaean warriors did not believe in divine forms of magic, but there is no evidence in *The Iliad* that the combatants resorted to magical means, other than the occasional direct appeal to a god, to win the battle.

This confusing lack of preciseness in interpretation is frustrating for scholars who do not read ancient Greek because we have to guess which version, if any, is closer to being accurate. While we can acknowledge that these male characters are, undoubtedly, putting each other down by calling each other some form of feminized figure, whether they are using the word "girl" or "woman" makes a decided difference to modern readers, since girls are females younger than 12 years old (or pre-menarche) and women are females old enough to bear children. Even though the ancient Greeks probably did not think in these distinct age limits, they did have different terms for pre-menarche girls, *parthenoi*, and post-menarche women, *kanephoroi* (Connelly 18) or *gynaikes* (Pence-Brown), because *kanephoroi* were treated differently and assigned different religious roles in processions and rituals, dressing and behaving differently both in the home and in public. In public, *kanephoroi*, young women who were considered marriageable because they had passed menarche, which occurred around the age of 14 (Pence-Brown), dressed in "more modest *chiton* or *peplos*" (Pence-Brown), led processions carrying sacred items particular to each festival to display their marriageability (Connelly 33). Girls, for the ancient Greeks, dressed like Artemis, or rather Artemis, stuck in perpetual maidenhood, dresses like young girls in a short dress called a *chitoniskos* (Pence-Brown), and girls propitiated Artemis with votive statues "hoping to aid in the coming menstrual flow and subsequent marriage" (Pence-Brown). While most androcentric scholars call Artemis a "virgin," she's actually frozen in "eternal youth by her father Zeus, meaning that she never reached menarche" (Pence-Brown), which explains her dress. The fact that none of the interpreters of *The Iliad* make such distinctions clear when they mention the "virgin" goddesses demonstrates that they do not fully understand ancient Greek thinking, which focused more on where women are in relation to menstruation rather than on whether the woman had had sexual intercourse, which is a patriarchal preoccupation, even though Agamemnon hastens to assure Achilles that he has not slept with Briseis.

While some scholars gush about how Homer, or whoever finally wrote down *The Iliad* and *The Odyssey*, was "seeing...untapped potential for an examination, in depth and at length, of **the whole human condition**" (emphasis mine; Green, Introduction, 11), how can two epic poems that seem to largely focus solely on male perspectives, given the androcentric interpretations we have thus far, give us a window into the "whole human condition" at the time of the Mycenaeans?

Avoiding addressing this important question demonstrates just how androcentric all these interpreters, even the woman, Alexander, are. This pro-male cultural bias inserts itself even deeper into the poem than most readers realize, even though Fagles quotes Maynard Mack in his "Translator's Preface," who put this conundrum benignly, stating that interpreters are "wearing the spectacles of our time" (xii). If this claim is accurate, the society Fagles reflects in his translation is still extremely sexist. Admittedly, while the rise of androcentrism demonstrates growing sexism among the ancient Greeks, the issue becomes separating ancient sexist views from modern ones. When is Homer actually using a word denoting a pre-menstrual female? When is the use of the word "girl" simple sexism on a translator's part?

Think carefully about how the various editions—even from the earliest known Greek manuscripts which Bernard Knox claims existed in physical form in sixth century Greece "for we hear of official recitations at Athens and find echoes of Homer in sixth-century poets" (Fagles, Introduction, p. 6)—and many translations of the "original" Greek text published by the Oxford University Press in 1902 (Knox in Fagles, p. 5) have ever so slightly, or even whole heartedly, altered the poem's original wording.

Fagles, for instance, discusses in some depth how scholars can recognize the addition of Phoenix as part of Agamemnon's ambassadors sent to appeal to Achilles to an original edition of the poem because the writer "forgot not only to explain [Phoenix's] presence in Agamemnon's council but also to amend the dual forms" of Greek that indicate only two people approached Achilles (22). How do we, then, ferret out what Homer or his scribe copying his words dutifully down as the (or "a") great poet(s) spoke them aloud truly meant, so that we truly know what was really being conveyed to us over the millennia?

In other words, how many millennia of growing androcentric bias has tainted the poem?

Fagles argues that "there seems to be only one possible explanation of the survival of these dual forms in the text: that the text was regarded as authentic, the exact words of Homer himself. And that can only mean that [the source] was a written copy" (22).

While many scholars have scoured the poem for authenticity, translators continue to take liberties with the wording, so that those of us who do not read Greek must dissect the poem's translations, first to note and to possibly remove the androcentric biases translators have carried into the text on purpose or by accident.

Some scholars will undoubtedly argue that such a process is not necessary. After all, the poem is clearly about war and, for many still locked in androcentrism's embrace, war is only something men are concerned with. They would be wrong and obviously biased to make such a claim because *The Iliad* takes pains to illustrate what women's lives are like because of the rising rage of men, especially those women who do not choose to take up arms, including brief mentions of the Amazons, "women who were the peers of men" (Butler, Book 6) because they did not shy away from combat.

What we must acknowledge, first, is that modern androcentric prejudices, perhaps more heavily steeped than in the ancient Grecian times, have colored *The Iliad*—how it is taught, how it is read, and even who lauds it most. The darker side of *The Iliad*'s praises come from those men who laud it as evidence of the superiority of the European people, especially as an indication of the superiority of men.

Once we recognize how modern androcentrism has affected the perceptions of the epic poem, we then have to reckon with that bigotry's effects on translations, which perpetuate those beliefs.

For instance, only Green's interpretation admits, from that first important line, that Wrath is a goddess: "Wrath, goddess, sing of Achilles" (Green, Book 1, l. 1). Also known as Lyssa, Furor, and even Rabies, Wrath was the goddess of rage and crazed fury (GreekBoston 2017). While Caroline Alexander might be implying the same identity, "Wrath—sing, goddess, of the ruinous wrath of Peleus' son Achilles" (Alexander, Book 1, l. 1), her phrasing is not as clear as Green's; she could simply be doubly referring to the wrath she wants an abstract goddess to sing about.

The other five translators--Samuel Butler, Stephen Mitchell, Robert Fitzgerald, Robert Fagles, and Richmond Lattimore--are similarly unclear about the fact that the poem begins by addressing Wrath, the goddess, not necessarily wrath as a theme.

"Sing, O goddess, the anger of Achilles son of Peleus" (Butler, Classics Book 1, L. 1) begins with a clear request for an unidentified goddess to sing but makes no attempt to clarify that the goddess being asked to sing is the goddess of Anger/Rage/Wrath/Lyssa.

"The rage of Achilles—sing it now, goddess, sing *through* me" (Mitchell, emphasis his, Book 1, l. 1) does not name the goddess the first line refers to nor does it tie her to the concept of rage but, instead, begins the metatextual self-referencing that the poet, Homer, sneaks into the poem repeatedly.

"Anger be now your song, immortal one" (Fitzgerald, Book 1, l. 1)does not bother to name who the "immortal one" is, either, but it is clear from his wording that Fitzgerald does not know the goddess to whom the poem is referring is Anger/Wrath/Lyssa.

"Rage—Goddess, sing the rage of Peleus' son Achilles" (Fagles, Book 1, l. 1) confuses readers because both Rage and Goddess are capitalized, signaling that they are both sentence starters, discoloring the fact that the goddess being referred to is also known as Rage.

"Sing, goddess, the anger of Peleus' son Achilles" (Lattimore, Book 1, l. 1) likewise doesn't bother to identify to which goddess the speaker is referring, also making the line sound as though some unnamed goddess (perhaps a muse?) is being asked to sing about Achilles' rage, not bring that rage to fruition.

Except for Green, all the translators demonstrate an inherent bias against females in their interpretations right from the start—that tone setting first line.. Either that, or they simply did not know a fundamental aspect of Greek religion—that most of the emotions, if not all of them, are female gods or *daimona*, spirits who "existed in ancient times before the Olympians took control of the world" (GreekBoston 2017).

As an agent of the madness caused by rage, Lyssa/Wrath/Rage was said to show more astute maturity than many of the other *daimona* or gods because she once refused to inflict Heracles with madness, even though ordered to do so by Hera, who hated him and most of Zeus' other bastard children irrationally[366] (GreekBoston 2017). In some accounts,

---

[366] Hera's hatred of most of Zeus' bastard children is a clear foil for the patriarchal views being embraced by the ancient Greeks by Homer's time. Prior to realizing males have a biological role in procreation, any child produced by women that they chose to keep and to raise were honored because most cultures viewed them as evidence of women's powers to create. Once men can prove they are a part of procreation, they begin to control women's bodies, largely through laws and violent methods, but also by stimulating other women's jealousy over men, to ensure they could identify who fathered what child. The fact that Hera appears quite irrational about Zeus' bastards potentially demonstrates how far women had been brainwashed to believe men are contributing more than just half of their

Lyssa/Wrath/Rage only cursed Heracles with madness because the goddess Iris forced her to do so (GreekBoston 2017).

In fact, many of the translators, while calling most of the *daimona* goddesses, seem to avoid admitting that these particular gods are older and wiser than most of the Olympus gods.

The first *daimona* or minor goddess--after Lyssa/Wrath and other than Achilles' mother, Thetis, who appears in Book 1--to make an overt appearance in *The Iliad* is Eos, goddess of the dawn. Alexander writes, "Dawn the goddess set her foot on high Olympus,/ heralding light of day to Zeus and the immortals" (Book 2, ll. 48-9), clearly acknowledging, even if not naming, Eos as a goddess. Green's wording is similar, "the goddess Dawn drew near to high Olympos, bringing/ her announcement of light to Zeus and the other immortals" (Book 2, 48-9), although the lack of modern punctuation to set off a proper name as an appositive could create some confusion. Fagles doesn't depart far from similar wording, "Now the goddess Dawn climbed up to Olympus heights,/ declaring the light of day to Zeus and the deathless gods" (Book 2, 57-8), clearly naming the goddess and describing her role in this moment. Fitzgerald uses extra breaks in his poetic lines to write, "Pure Dawn/ had reached Olympos' mighty side,/ heralding day for Zeus and all the gods" (Book 2, 54-5), clearly expecting readers to know he is using Dawn as a name of some sort of entity who is reaching the side of Mount Olympus, but she seems more natural phenomenon here rather than liminal, which is how the ancient Greeks would have seen Eos—the divinity that defines night from day: the personification of a magical or spiritual moment in time between worlds. Lattimore writes, "Now the goddess Dawn drew close to tall Olympos/ with her message of light to Zeus and the other immortals" (Book 2, 48-9), clearly, again, naming the goddess and describing her as a more humanoid divine being. Mitchell is more prosaic, "As dawn spread out and arrived at the peak of Olympus,/ announcing the light to Zeus and the other gods" (Book 2, 46-7), fully eliminating any capitalization of a proper name for the

---

biological or spiritual force to these other women, so that wives should be jealous of their husband's sexual contact outside legal marriage. Alternatively, the writers of the times simply wanted listeners and readers to believe women were irrational.

goddess, so that, here, her divinity has become nothing more than a time period in the day's cycle. Butler, also more prosaic because he uses actual prose, says, "The goddess Dawn now wended her way to vast Olympus that she might herald the day to Jove and to the other immortals" (Classics, Book 2), once again making her a human-like divinity who is actively "heralding" in the new day. Of these, only Mitchell ignores the fact that he's speaking of a goddess, even if she is a minor god who does not reside on Olympus but de-emphasizing a human-like divinity whose "rosy fingers" are replete throughout the poem.

Iris, goddess of the rainbow and primary messenger god in *The Iliad*, also appears frequently in the poem, usually sent by Zeus. While Hector is spoken of in fearful or reverent tones by many of the characters in the epic, it is not until Book 2 that, like Achilles and Agamemnon, we realize he fits the most basic definition of a hero—he can hear and speak to the gods, thus is favored by them (Soles 2016, 250)--in this case, Iris, even though she is disguised as his brother, Polites. In this guise, she advises Hector to organize his troops by nationality, since his allies speak different languages. In this way, Hector can ensure that every warrior understands what is being asked of him as they all march into battle. We are told, by Alexander, that "Hector did not fail to recognize the word of a goddess" (Book 2, 807). Mitchell is, again, more prosaic, when he writes, "When Hector heard her, he knew that she was a goddess" (Book 2, 747). More formally, Fitzgerald writes, "Hektor punctiliously obeyed the goddess" (Book 2, 807). Lattimore states, "She spoke, nor did Hektor fail to mark the word of the goddess" (Book 2, 807). Fagles notes, "Hector missed nothing—that was a goddess' call" (Book 2, 917). Green concurs, "So she spoke, and Hektor did not mistake the goddess' voice" (Book 2, 807). Hector, like most heroes, recognizes a divine voice, even when it is in disguise—a true hero.

Hector's punctiliousness in obeying the war advice of Athena draws an important distinction in *The Iliad* because, later, when his wife, Andromache, offers him sound advice about how to best protect the city of Troy, he dismisses her advice. She wisely suggests, "Take your stand on the rampart here" (Fagles, Book 6, l. 511), then "draw your armies up where the wild fig tree stands,/

there, where the city lies most open to assault,/ the walls lower, easily overrun. Three times/ they have tried that point, hoping to storm Troy" (ll. 513-16). Green is a little more generous to Andromache who advises:

> Station your troops by the fig tree, where the city's most open
> To scaling, where the ramparts are vulnerable to assault—
> Three times has a picked force come there to make the attempt,
> Led by both Aiases, and by famous Idomeneus,
> With the sons of Atreus and Tydeus's valiant offspring;
> Someone, perhaps, well skilled in soothsaying told them,
> Or else it's their own spirit that drives them, urges them on
> (Book 6, ll.433-9).

Aggravatingly, Mitchell's interpretation does not contain Andromache's sage advice at all, so that, when Hector calls her "my foolish darling" (Book 6, l. 485), he's not so much being a smarmy, arrogant male mansplaining war to his wife, even though he still orders her to command her women "to go about *their* work" (emphasis Mitchell), so "the men [can] take care of the fighting" (Book 6, l. 492).

Hector's response to Andromache's sage war advice is to frighten her with a narrative he envisions will happen once she's captured and taken away after his death, as though she hasn't already contemplated that fate. As Simone Weil points out, "such a destiny [for her child] is more awful for [the mother] than death itself" (49). Weil also notes that "the husband hopes to die before seeing his wife reduced to" her ultimate post-war fate (49), but the critic fails to consider that this selfish desire on Hector's part demonstrates cowardice or even cruelty. As Fewell and Gunn point out about biblical war celebrations, the "victors" force "the Canaanite women to approve unconsciously [of] their own imminent rapes," should the war end less favorably (408). By biblical times, the women are so used to such threats that Sisera's mother lists his expected rapes as war spoils, "a womb, two wombs, per head" (Fewell and Gunn, 407). *The Iliad* is just one step toward that total female submission that enables male domination.

Hector dismisses Andromache's sage advice with, "as for the fighting,/ men will see to that" (Fagles, Book 6, ll. 587-8). Green has Hector as dismissive with "but warfare shall be the business of men" (Book 6, l. 492). Clearly, Hector is willing to listen to the goddess of war, but not to a mortal woman's observations about what their enemies have tried to accomplish.

Lattimore's treatment of this vital moment is more even handed, and he gives Andromache more reason to give war advice to Hector because she witnessed Achilles kill her father and her seven brothers (ll. 414-423), then saw her mother captured and ransomed only to see her shot down by Artemis "in the halls of her [mother's] father" (Book 6, ll. 428). As in Green's interpretation, Lattimore lists the warriors Andromache has witnessed trying to scale the Trojan wall near the fig tree, and has Hector speak to her somewhat more kindly, "'Poor Andromache! Why does your heart sorrow so much for me?" (Book 6, l. 486), admonishing her to "take up your own work,/ the loom and the distaff, and see to it that your handmaidens/ ply their work also; but the men must see to the fighting" (Book 6, ll. 490-2).

Fitzgerald's interpretation of this moment is much like Lattimore's, although more poetic so somewhat more challenging to read. He gives us Andromache's full lament over the loss of her father and her seven brothers at Achilles' hands when they were "amid their shambling cattle and silvery sheep" (Book 6, l. 420). Fitzgerald has Hector caressing Andromache when he reminds her, "unquiet soul, do not be too distressed/ by thoughts of me. You know no man dispatches me/ into the undergloom against my fate" (Book 6, ll. 517-19), then commanding Andromache to attend to her womanly work because "as for the war,/ that is for men" (Book 6, ll. 524-25).

Surprisingly, Alexander has Hector belittling Andromache as badly as the worst of the male interpreters. After allowing Andromache to demonstrate that she knows the worst of war from the murders of her own family members, so knows enough to give her husband some sage battle advice, Hector, of course, brags about how after Andromache is captured, someone will note her tears and think of him, Hector, because she "is the wife of Hector, who used to be best of the horse-breaking Trojans/ in waging battle" (Book 6, ll. 460-1), blatantly lecturing Andromache that "war is the concern of men," not women (Book 6, l. 492). To Alexander's credit, Hector comes across as a fully self-centered egotist who fails to see other people's fear and anguish —even that of the people he supposedly loves most. All of the interpretations have Hector laughing when his son, a mere babe in arms, is frightened of his father in his full battle gear.

Interestingly, Fitzgerald, Lattimore, and Fagles admit that, after Hector's son is frightened by daddy in war gear, Hector laughs first, then Andromache follows, but Butler, Green, Alexander, and Mitchell all have her laugh **at the same time** Hector does. This minor detail speaks volumes. Is Andromache genuinely amused by her baby's terror of a man in full battle armor or is she merely following her husband's lead, relieved to have something to laugh about? Hector has been much less sorrowful than Andromache about his approaching death, even attempting to be philosophical about it, but he's also tried to terrify his wife with details about what her life will be like once he's dead and Troy falls as though he has no capacity to truly worry about his family's fate and then completely dismisses her sage war advice to both his and Troy's demise. War is terror, Homer is clearly saying, and the aftereffects of war are just as terrible as war itself, since women and children, whom the interpreters of the poem seem to see as equally minor concerns, become **its continued victims**. Certainly the poet seems to be making a statement about the callousness of men in their cold disregard, not only for women's sage advice, but also for women's welfare, which points to a drastic change in human attitudes toward women, who have by this time in history just become chattel to exchange between winning brutes—a theme repeated in the dialog the various men in the poem have about the spoils of war—unless they are goddesses.

Another lesser known goddess, Dione, who is not a goddess of Olympus status, but is possibly a Titan, possibly one of the ancient female versions of Zeus (Serrano Laguna 2016, 289), appears quite notably in Book Five when she tends to her daughter Aphrodite's wound inflicted with pitiless bronze by Diomedes, who attempted to stop her from spiriting her own son, Aeneas, away from battle. Green writes:

> He sprang at [Aphrodite], did high-spirited Tydeus's son, and lunged
> And sliced into the flesh of her hand with his keen-edged bronze--
> That delicate hand! The spear drove straight into her flesh—
> Clean through the fragrant robe toiled on by the Graces themselves—
> At the base of her palm: out flowed the goddess's blood, immortal
> Ichor, such as flows in the veins of the blessed gods,

> For they neither eat bread nor drink fire-bright wine, and so
> Are bloodless, and come thus to be called immortals.

(Book 5, 334-41)

For most modern readers, the news that the gods can be harmed, even if not killed, will come as a surprise, but it is interesting that Homer felt obliged to add information about why the gods are considered immortal—it is not because they cannot be wounded, but because they have ichor (imagine liquid gold) flowing through their veins instead of blood. Because the gods have ichor flowing through their veins, they are not only more physically powerful—able to survive conditions that would kill mortals and imbued with super strength—but also able to use magic to change their appearances and to move from one physical spot to another in an instant. So it is interesting that they can still feel pain and can be wounded by that pitiless bronze.

Earlier though, Green's interpretation belittles Aphrodite's war efforts—largely overlooking her tremendous part in the whole conflict—by telling us that Diomedes "had taken his pitiless bronze/ in pursuit of Kypris, aware what *a weakling goddess* she was" (emphasis mine, Book 5, ll. 330-1), having Diomedes, after he wounds the goddess of love, try to provoke her with a weak taunt, "Does it not suffice you to mislead weakling women?" (l. 349), overlooking the fact that it was Paris she misled; Paris she tricked not only into giving her the golden apple, but also of stealing Helen from her king-husband.

This portion of the poem is fascinating when we compare Aphrodite to her progenitor, Inanna, the Sumerian goddess of both sex and warfare because, in order to make war a sacred duty, the rising androcentrists had to equate the bloodletting of war with the natural moon cycles of women, which had long tied women to everything sacred. Having Aphrodite wounded in battle is ironic in this light, especially when her sexual manipulations of Paris and Helen, which brought the conflict and its subsequent voluminous blood loss to life, contrast with her ichor blood and temporary pain. Like women during their moon cycles, she bleeds, but does not die. This blood magic is what the men in *The Iliad* attempt to recreate, but, because they are not as magical, they fail, bringing only destruction and misery.

We have to wonder how many Grecian women listening to the poem recited aloud tittered about Diomedes' refusal to accept that Aphrodite affects more than just women or even realized that she, above all, was the divine influence that most guided the conflict. While many blame Helen for the Trojan War, calling her "the face that launched a thousand ships," shouldn't we really be blaming Aphrodite's cunning manipulations?

Of Aphrodite, Fitzgerald translates that Diomedes "knew her to be weak, not one of those/ divine mistresses of the wars of men" (Book 5, ll. 332-3); by this time in history, goddesses are no longer as powerful as Inanna once was, since all of her previous powers have been divided. Lattimore is not as direct, but is possibly more accurate, since he more closely quotes the ancient Greek, describing thus that Diomedes "swung the pitiless bronze at the lady of Kypros, knowing her for *a god without warcraft*, not of those who,/ goddesses, range in order the ranks of men in the fighting" (emphasis mine, Book 5, ll. 330-2). Lattimore seems to be attempting to point out that not only were their goddesses who practiced warcraft, but who were, possibly, more powerful than men, but seems reluctant to go that far.

Fagles is not so generous in his description of Aphrodite because he states that Diomedes, "knowing her for the *coward* goddess she is" (Book 5, l. 371) was "hunting Aphrodite" with "his ruthless bronze" (emphasis mine, Book 5, l. 370). Alexander is more prosaic with "the son of Tydeus...ranging after Cyprian Aphrodite/ with his pitiless bronze spear, knowing that she was not a fighting god,/ nor of those goddesses who hold sway *through the war of men*"[367] (emphasis mine, Book 5, ll. 330-3). Mitchell's translation, though, is the easiest and most direct—the least like ancient Greek: "Diomedes had gone after Aphrodite,/ his spear in his hand, knowing that she was fragile/ and not a goddess triumphant in human warfare" (Book 5, ll. 309-11).

Homer must have had Diomedes taunt Aphrodite in much the same way that Green's interpretation has him doing, since all of the interpreters mention this ironic taunt in some way, but

---

[367] Clearly, Alexander sees war as only for men in this phrasing, but she also seems to indicate that some goddesses are only effective when they act during human conflicts.

Lattimore's wording is most intriguing because he has Diomedes ask Aphrodite the question in a different way: "It is/ not then enough that you lead astray women without warcraft?" (Book 5, l. 349). Lattimore felt no need to call women pejoratives here, but does, perhaps unknowingly or unintentionally, emphasize the kind of battle Aphrodite does regularly engage in—lust. Of all the interpreters of this passage, Lattimore comes closest to touching on the fear that the men truly hold of the women—they are afraid they can be "led astray" by methods other than overt conflict; in other words, the men clearly fear and desire to control women's bodies in order to control their worst fears of losing control.

Controlling our emotions comes to us in many forms in *The Iliad*. Dione's message to her wounded daughter should not completely be unfamiliar to modern readers since many mothers still admonish their children to "suck it up":

> Bear up, my child, and endure, despite what you're suffering;
> For many of us suffer likewise—we, the Olympos-dwellers—
> At the hands of men, and bring harsh grief on one another.

(Green, Book 5, 381-3)

Dione then provides a list of ways many of the gods have suffered either at human hands or by torture from other gods, including a clue as to why Hera might really hate Hercules—

> Thus Hērē suffered, when Hēraklēs, Amphitryōn's strong son,
> Shot her in the right breast with a three-barbed arrow;
> Then she too was seized by unquenchable agony.

(Green, Book 5, 391-3).

The ancient Greek audience listening to this portion of the poem would have understood something that is not explained by most of the translators about this passage (although Zeus explains it later)— Zeus, in disguise as Amphitryon, seduced Alcmene, and it was from that sexual union she bore Hercules/Herakles. This moment in the poem is important because it not only introduces us to Dione, a goddess more ancient than most of the prominent goddesses mentioned in *The Iliad*, but also demonstrates **the primary struggle of the patriarchy—establishing paternity**.[368] Who really fathered Hercules? Zeus or Amphitryon?

---

[368] Prior to DNA testing, the patriarchy could not exist if men did not control women's bodies in order to determine who fathered what child, which is another reason so many interpreters of ancient Greek poetry often misunderstand ideas

If our basic definition of hero is someone who can communicate with the gods directly, or at least through dreams, wouldn't all the men and women, like Alcmene, and who are seduced by various goddesses and gods in the many Grecian myths be heroes, too?

In all the translations, Diomedes misjudges Aphrodite by weakly comparing her to Enyo, yet another *daimona*, who is notorious for causing the pillaging of cities—the primary purpose of war for the ancient Greeks. While Enyo, whom Homer briefly mentions here, is a minor goddess usually seen accompanying the primary war god, Ares, she will, in a later poem, Virgil's first century poem *The Aeneid*, take part in the sacking of Troy along with three other minor goddesses: Eris (Strife), Phobos (Fear) and Deimos (Dread). In *The Iliad*, however, she appears again in Book 5 with Ares when "Enyo, queenly goddess:/ she brought with her the reckless tumult of battle" (Green, ll. 592-3). For Fagles, Enyo is "Strife flaring on, headlong on" (Book 5, 599). Lattimore expounds that "Ares led them with the goddess Enyo,/ she carrying with her the turmoil of shameless hatred" (Book 5, 592-3). For Fitzgerald, the Trojan push forward is "impelled by Arês and by cold Enýô/ who brings the shameless butchery of war" (Book 5, 594-5). Mitchell inserts the passage several lines earlier than the rest, stating that the Trojans are "led by Ares and fierce Enýo,/ who always comes with the ruthless carnage of war" (Book 5, 536-7), implying that, like Inanna, Enyo represents the blood spilled both in war and by women monthly. For Butler, "Mars and dread Enyo led them on, she fraught with ruthless turmoil of battle, while Mars wielded a monstrous spear" (Classics, Book 5). Alexander writes, "Ares led them and lady Enyo, spirit of war,/ bringing Panic with her; the ruthless turmoil of war" (Book 5, 592-3), so that we get not just the *daimona* Enyo, but also the *daimona* Deimos, goddess of panic and fear.

Completely counter to these goddesses of pillaging and fear is the goddess, Themis, goddess of divine law, who resides on

---

about virginity for the ancients. According to Joan Connelly, "Greeks defined virginity (*parthenos*) ... [as] the condition of a maiden who had passed through puberty but was not yet married. Emphasis was not focused on a state of intactness, which the modern definition requires" (18).

Olympus, but is not one of the ruling twelve. Like Dione, Themis is a Titan, and "one of [Zeus'] earliest brides...together they gave birth to Hours and Fates who together represented the establishment of natural law and order" (GreekGods.org 2013-2018). Homer brings Themis into the epic in Book 15 where she plies Hera with drink after Hera's infuriating argument with Zeus. The Titan returns in Book 20 when Zeus orders her to assemble all the gods in order to let them all know that he is willing, finally, to allow them to take sides in the conflict. Perhaps what we do not see in *The Iliad* that is most important about Themis' role in the Trojan War--her plotting with Zeus to arrange for the "for the fairest" golden apple to be the catalyst for the conflict, which Green claims in his explanation for the Judgment of Paris is described in "Proclus's synopsis of the *Cypria* supplemented by Apollodorus in Epitome 3.2" (Glossary p. 529). In *The Iliad*, Themis does no overt plotting, instead exhibits concern for Hera but follows Zeus' orders.

Eileithyia, goddess of childbirth and daughter of the two chief gods, Zeus and Hera, is mentioned twice in *The Iliad*. The first time is in Book 16 when Homer describes the captains of Achilles' ships, so that we learn that Eudoros was fathered by the god Hermes, but "when finally Eileithyia, goddess of childbirth, brought him/ out to the light, and he saw the sun's rays for the first time" (Green, Book 16, ll. 187-88) he is willingly raised by a human as his own son. Interestingly, Alexander emphasizes Eileithyia's role differently, saying she is "the goddess who causes pain in birth" (Book 16, l. 187). Eileithyia remains unnamed in Mitchell's interpretation for whom she is "the pain-dealing goddess of childbirth" (Book 16, l. 167). Lattimore at least names her "Eileithyia of the hard pains" (Book 16, l. 187), but Fitzgerald is more poetic, "Eileithyia, sending pangs of labor,/ brought him forth to see the sun-rays" (Book 16, ll. 183-4).

The most dramatic description of Eudoros' conception and birth, though, comes from Fagles, even if he refrains from naming Eileithyia directly in the process:

> Hermes, the giant-killer
> Lusted for her once—she ravished the god's bright eyes,
> Swaying among the dancers singing...
> And straightway up to her chamber Hermes climbed,
> The Healer, in secret, lay in her arms in love

And the woman bore the god a radiant son, Eudorus—
lightning on his feet and a crack man of war.
But soon as the Lady of Labor's birthing pangs
Brought him to light and he saw the blaze of day...
(Book 16, ll. 214-23).

While all the other translators have Hermes lust after Polymela, the dancer, Fagles has Polymela actively ravishing "the god's bright eyes," as though the god is a passive victim to her beauty; like Agamemnon, Hermes just couldn't help himself.

The goddess of childbirth is mentioned again in Book 19 when Agamemnon, hoping to deflect blame for his own poor choices, compares his deception by the gods with Hera's deception of Zeus after he boasts that Hercules' birth, with the help of "'Eileithyia, spirit of birth pangs, today will bring to the light a man who will rule over all those dwelling about him, one of that line of men whose ancestry is my own'" (Green, Book 19, ll. 102-5). As Hera's daughter, Eileithyia is kept, at least twice, by Hera from attending the births of Zeus' children to other women.

Fagles, once again, fails to name Eileithyia, simply stating that Zeus proclaims that "today the goddess of birth pangs and labor/ will bring to light a human child, a man-child/ born of the stock of men who spring from *my* blood" (emphasis his, Book 19, 119-21). Paternity is clearly important to Fagle's Zeus. Green names the goddess, "Eileithyia, spirit of birth pangs" (Book 19, l. 103), but Mitchell, unsurprisingly, does not; she is merely "the goddess of childbirth" (Book 19, l. 101).

The entire passage in Book 19 is particularly illuminating about how some men will go to great lengths not to shoulder blame for their own actions, and, if possible, to blame women for the men's own weaknesses. Alexander confronts this misogyny head on, specifically taking the time to name what appear to be at least three daimona—"Fate and the Fury who walks in darkness" (l. 87), "the elder daughter of Zeus is Delusion, who infatuates all men" (l. 91), and Eileithyia--in Agamemnon's attempt to remove any question of his blame for his actions. However, Lattimore seems to give us two more daimona names in the same passage when Agamemnon claims, "yet I am not responsible/ but Zeus is, and Destiny, and Erinys the mist-walking/ who in assembly caught my heart in the savage delusion./ On that day I myself stripped from him the prize of Achelleus" (ll. 86-9). However, the Fates were also known as the Furies or the Erynys who determined every human being's Destiny. This passage must give most interpreters trouble, since Alexander attempts to be as clear as possible when she names "Zeus and Fate and the Fury who walks in the darkness,/ they who in the assembly cast savage Delusion in my mind" (l. 87-8), but she still fails to recognize that Fate and the Fury are one and the same set of goddesses. Delusion is also a goddess' name: Atê, like the Fates, was a daughter of Zeus. As a *daimona*, she instigated reckless impulsive behavior (Delusion) that often led to ruin (Theoi Project, Ate). None of the translators use Atê's Greek name, though, probably in an attempt to be as clear as possible to their English readers.

In Book 5, Leto, another minor goddess, helps Artemis heal Aeneas near Apollo's shrine. Leto, "a daughter of the Titans Coeüs and Phoebe" was one of Zeus' lovers before he married Hera, but Hera, in her raging jealousy, had pregnant Leto pursued to prevent her from giving birth, but Leto eventually gives birth to the twin gods Apollo and Artemis (Tripp 1974, 344).

Using Roman names, Butler translates that "Apollo took Aeneas out of the crowd and set him in sacred Pergamus, where his temple stood. There, within the mighty sanctuary, Latona and Diana healed him and made him glorious to behold" (Book 5). The Titan, Latona is called Leto by Green, who reports that

> Apollo the archer, who then
> took Aineias out of the turmoil, set him down far away
> in sacred Pergamon, where his own shrine had been founded:
> there Leto together with bow-hunter Artemis
> healed him in the great sanctuary, shed splendor on him"

(Book 5, ll. 444-8).

Mitchell is more prosaic: "the god took Aeneas out of the crowd and set him apart from the fighting/ on Pergamus, where his glorious temple stood./ There Leto and Artemis, inside the great inner sanctum,/ healed his wound and made him more splendid" (Book 5, ll. 410-15).

In contrast, Fagles is more dramatic:

> And Apollo swept Aeneas up from the onslaught
> And set him down on the sacred heights of Pergamus,
> The crest where the god's own temple had been built.
> There in the depts of the dark forbidden chamber
> Leto and Artemis who showers flights of arrows

Healed the man and brought him back to glory.
(Book 5, ll. 512-7)

Lattimore's translation, though, makes Leto sound like a casual secondary insertion, not Artemis and Apollo's mother or a Titan:

...Apollo,
Who caught Aineias now away from the onslaught, and set him
in the sacred keep of Pergamos where was built his own temple.
There Artemis of the showering arrows and Leto within
the great and secret chamber healed his wound and care for him.
(Book 5, l. 444-8).

But then neither does Fitzgerald acknowledge either their relationships or Leto's importance:

Diomedes backed away a step or two
Before Apollo's terrible anger, and
The god caught up Aineias and set him down
On Troy's high citadel of Pergamos
Where his own shrine was built.
There in that noble room Leto and Artemis
Tended the man and honored him
(Book 5, ll. 442-8).

Similarly, Alexander paints a brief picture of the event without drawing any significance to the fact that a Titan, Leto, is helping her children, Apollo and Artemis, to heal this one warrior—a warrior who, the Romans believed,[369] went on to found Rome:

And Apollo set Aeneas apart from the throng of battle
In holy Pergamos, where his temple stood.
Here Leto and Artemis who showers arrows
Healed and glorified him within the great inner shrine.
(Book 5, ll. 445-8).

Later, in Book 24, Achilles obliquely refers to Leto when he attempts to get Priam to eat, saying that "even fair-haired Niobē was minded to eat,/ though all her twelve children had perished there in her halls" (Green, Book 24, l. 602) because "Niobē had praised herself over fair-cheeked Lētō" (Green, Book 24, l. 607), which caused Leto's children, Apollo and Artemis, to avenge the slight on their mother—"Her brood Lētō's twins then slaughtered, every last one" Green, Book 24, l. 609).

---

[369] Have the later Romans then tampered with the poem to make this portion more important than it was originally?

Alexander is more direct in her interpretation of this passage, which, once again, emphasizes the importance of males over the importance of females, similar to Butler's "lusty sons" (Butler, Book 24):

> For even Niobe of the lovely hair did not forget her food,
> She whose twelve children were destroyed in her halls,
> Six daughters, and six sons in the prime of manhood.
> The sons Apollo slew with his silver bow
> In his anger with Niobe, and Artemis who showers arrows slew
>   the daughters,
> Because Niobe equaled herself to Leto of the lovely cheeks—
> For she would boast that Leto bore two children, while she
>   herself had borne many.
> So two only though they were, they destroyed the many.
> (Alexander, Book 24, ll. 602-9)

More poetic, Fitzgerald makes the gruesome scene seem beautiful, even though Achilles is rather ham-handedly tempting the deeply grieving Priam to eat:

> Now let us think of supper. We are told
> That even Niobê in her extremity
> Took thought for bread—though all her brood had perished,
> Her six young girls and six tall sons. Apollo,
> Making his silver longbow whip and sing,
> Shot the lads down, and Artemis with raining
> Arrows killed the daughters—all after
> Niobê had compared herself with Lêto,
> The smooth-cheeked goddess.
> (Book 24, ll. 599-606).

Lattimore provides a similar, if oddly worded translation (the dangling modifier makes it sound as though Niobe ate her own children):

> Now you and I must remember our supper.
> For even Niobe, she of the lovely tresses, remembered
> To eat, whose twelve children were destroyed in her palace,
> Six daughters, and six sons in the pride of their youth, whom Apollo
> Killed with arrows from his silver bow, being angered
> With Niobe, and shaft-showering Artemis killed the daughters;
> Because Niobe likened herself to Leto of the Fair colouring
> And said Leto had borne only two, she herself had borne many;
> But the two, though they were only two, destroyed all those others.
> (Book 24, ll. 601-9)

More business-like and direct, Mitchell's translation sounds less ominous, but doesn't make Achilles seem contrite for being a murderer at all:

> But for now, let us think of supper.
> For even Niobē came to the point of eating,
> And that was after all twelve of her children were killed,
> Six daughters and six sons, in the prime of their youth.
> Apollo shot down the sons with his silver bow,
> And Artemis shot down the daughters. They were both angry
> At Niobe for boasting that she was the equal
> Of Leto, because the goddess had borne just two,
> While she herself was the mother of many children;
> So the pair, though only two of them, slaughtered her many.

(Book 24, ll. 595-605)

Perhaps to passive-aggressively strike back at Priam for listing the deaths of most of his fifty sons, Fagles' Achilles indicates that Niobe's loss is comparable to Priam's, taking the time to actually give Niobe a voice:

> Now, at last, let us turn our thoughts to supper.
> Even Niobe with her lustrous hair remembered food,
> Though she saw a dozen children killed in her own halls,
> Six daughters and six sons in the pride and prime of youth.
> True, lord Apollo killed the sons with his silver bow
> And Artemis showering arrows killed the daughters,
> Both gods were enraged at Niobe. Time and again
> She placed herself on a par with their own mother,
> 'All you have borne is two, but I have borne so many!'
> So, two as they were, they slaughtered all her children.

(Book 24, ll. 707-17)

While Leto, a Titan, takes an active role with her children in Book 5, she is used as a cautionary example in Book 24, since Priam has taken a clear risk in entering the Mycenaean camp to retrieve Hector's body from Achilles. First time readers of Book 24 might shudder with dread for Priam's safety after Achilles feeds him then puts him to bed, since, in our sleep, we are most vulnerable. Is it irony that brings Leto into Book 24? After all, she is one of the goddesses of motherhood, and here we have a father appealing to another man's son for the return of his own son's lifeless body. However, Leto is also a goddess of modesty (Project Theoi, Leto), and Priam is so modest that he becomes obsequious to Achilles' power in order to wheedle back his son's remains.

Several other lesser known goddesses are named throughout *The Iliad*. Panic (Deimos), Strife (Eris), Prowess (Alke), Pursuit (Ioke), and the Gorgon (Medusa) are listed as appearing on "the tasseled aegis" that Athena wraps over her shoulders (Green, Book

5, 739-42). In Book 18, we get a roll call of the daughters of the Old Man of the Sea, Nereus, who all respond to Thetis' siren call:

> Thither came Glauke and Thaleia, Kymodoke,
> Nesaia, Speio, and Thoe, and ox-eyed Halie,
> Kymothoe and Aktaia, along with Limnoreia,
> Melite and Iaira, Agaue, Amphithoe,
> Doto and Proto, Dyamene and Pherousa,
> Dexamine and Amphinome and Killianeira,
> Dors and Panope and far-famed Galateia,
> Nemertes, Aspeudes, and Kallianassa.
> With these also came Klymene, Ianeira, and Ianassa,
> Maira and Oreithyia and fair-tressed Amatheia,
> And other Nereids from elsewhere in the sea's depths.

(Green, Book 18, 39-49)

Thirty-four sea goddesses, counting Thetis, respond to Achilles' grief over the loss of Patroclus.

Fagles gives the Nereids English names:

> Glitter, blossoming Spray and the swells' Embrace,
> Fair-Isle and shadowy Cavern, Mist and Spindrift,
>
> ...
> Race-with-the-Waves and Headland's Hope and Safe Haven,
> Glimmer of Honey, Suave-and-Soothing, Whirlpool, Brilliance,
> Bounty and First Light and Speeder of Ships and buoyant Power,
> Welcome Home and Bather of Meadows and Master's Lovey Consort,
> Gift of the Sea, Eyes of the World, and the famous milk-white Calm
> And Truth and Never-Wrong and the queen who rules the tides in beauty
> And in rushed Glory and Healer of Men and the one who rescues kings
> And Sparker, Down-from-the-cliffs, sleek-haired Strands of Sand

(Book 18, ll. 45-55).

Only Mitchell, of all the translators, completely skips over this long list of goddesses, seemingly unimpressed by the number of daughters Nereus fathered.

Yet one mighty Olympian goddess, Hestia, one of the original 12 Olympian gods, remains unnamed throughout the poem by all the translators, despite the fact that she would have been "given her fair portion" of each and every meal and each and every ritual sacrifice. We have to assume Homer overlooked Hestia on purpose, but why?

Hestia is not mentioned when Agamemnon sacrifices a hecatomb of oxen[370] to Apollo to appease the god in order to rid the Mycenaean camp of the plague kills men and animals in any of the translations:

> to Apollo they made perfect sacrificial hecatombs
> of bulls and goats alone the shore of the murmuring sea
> and the savor rose to heaven amid a swirl of smoke
> (Alexander, Book 1, 315-17);

> "they offered up to Apollo
> hundreds of oxen and goats, and they skinned them and burned
>   them
> on the shore of the restless sea, and the broiled fat's savor
> rose with the black smoke, spiraling up to heaven"
> (Mitchell, Book 1, 316-19);

> then offered up to Apollo their perfect sacrifices
> of bulls and goats by the shore of the unharvested sea,
> and the savor went up to heaven, circling around the smoke
> (Green, Book 1, 315-17);

> Then they accomplished perfect hecatombs to Apollo,
> of bulls and goats along the beach of the barren salt sea.
> The savour of the burning swept in circles up to the bright sky
> (Lattimore, Book 1, 315-17);

> then to Apollo by the barren surf
> they carried out full-tally hekatombs,
> and the savor curled in crooked smoke toward heaven
> (Fitzgerald, Book 1, 315-17).

Hestia, Zeus' older sister and the oldest of the Olympian gods, was the goddess of fire, of the hearth, and, because she chose to remain unmarried, Zeus had pledged that she would be afforded

---

[370] Homer might have included oxen as the sacrificial animal here for many reasons, but, if the poet wrote the poem based on what was visibly left from the earlier palatial age of Mycenaean Greece, he would have undoubtedly seen the many images of bulls being led to their sacrificial slaughter, which , according to Thomas G. Palaima, were more represented through images on walls and on official seals than the real animals sacrificed more frequently, according to texts left behind, such as sheep, goats, pigs, and even deer (354).

her due portion of every sacrifice. So we know she must be there and that that "savor" that comes from burning fat in a fire, one method for keeping a fire blazing enough to use it for cooking and for warmth, that rises with visible smoke into the air—that attractive scent was meant to entice her protection, to ensure her presence at the fire in order to maintain the fire (adding the fat to the fire helps a great deal, so was a common ritual for many cultures who believed they were giving "the best" part of the meat to the spirits), and it rises upward, not toward the heavens—because the ancient Greeks did not place all of their gods in the heavens, only those immortalized as constellations; instead, their gods resided high on Mount Olympus—but to Hestia high on that mountain.

Two other goddesses that appear to go unmentioned by name, at least judging by the translations we have at hand, are Achlys, the goddess who brings the mists of death, and Lethe, the goddess of forgetting and forgetfulness, linked with the River Lethe in the Underworld where "the [ordinary] dead who descended to Hades drank" its waters in order to forget "their former lives" (Tripp 1974, 344). While Achlys could be (should be?) being alluded to each time a warrior dies in the poem, Lethe might be being obliquely mentioned in Book 2 when Zeus sends Morpheus, the god of dreams to Agamemnon in order to influence him to stay to fight the Trojans further because Morpheus tells Agamemnon, "Keep this in your heart, don't permit/ **forgetfulness** to seize you when honey-sweet sleep lets you go" (emphasis mine, Green, ll. 33-4). Four other translators—Alexander, Fitzgerald, Lattimer, and Mitchell--also use the word "forgetfulness," but Butler simply says, "Remember this, and when you wake see that it does not escape you" (Classics, Book 2), while Fagles writes, "But keep this message firmly in your mind./ Remember—let no loss of memory overcome you/ when the sweet grip of slumber sets you free" (Book 2, 38-40). Why would five out of seven translators obliquely refer to Lethe, if not because the ancient Greek writer uses a term that alludes to the goddess and the forgetfulness that comes with waking?

The fact that every interpretation of *The Iliad* to date is tainted by androcentrism is a fact none can dispute. How much modern androcentrism affects each translation can be measured by how fair each translator is toward the women in the poem, especially given the fact that others, including the Romans, could have layered on more androcentrism than what Homer and his colleagues carried into the poem themselves. While many would excuse such male-biased interpretations as a natural flaw in scholarly views regarding this nearly 3000-year-old poem, we owe it to the women of ancient Mycenae and ancient Greece to uncover the truths about their society's attitudes toward a growing male fondness for war, rape, and booty—what some euphemistically call "brides by capture." That Homer names more goddesses than gods in *The Iliad*, despite the androcentric biases of its translators and those of its creators, is our first clue toward the gynocentric views that preceded the rise of androcentrism.

Chapter Twenty-Eight
Gynophobia and Fear of Goddess Worshippers:
Why Many Feminist Scholars Are Ignored

Gynophobia is the fear of anything "feminine."

Ironically, patriarchies tend to stoke both the fear of the feminine in everyone, as well as the fear of the masculine, androphobia, because oppressors use fear for optimum control. Patriarchists use fear of women "out of control" and of men who rape in order to control. They also stoke fears of gender difference, trying to make people believe trans women are really just males waiting to rape women in restrooms or locker rooms, when, in fact, few rapists ever disguise who they are, and many rapists willingly infiltrate private spaces where people live.

In fact, homophobia is a complex intertwining of both gynophobia and androphobia because men, especially, are taught to be afraid of other men, just as women are told to be, in sexually compromising situations, such as the old dropped-soap-in-the-communal-shower scenario often inveigled by men to incite fear in other men because getting raped by a man is supposed to only be something that happens to women or to the effeminate.

If you happen to be one of the few people on the planet who does not know or who refuses to believe that women are oppressed in many wannabe patriarchal cultures around the world, what follows will be startling news to you.

While many conservatives in America attack institutions of higher education for being "woke" (the current slang term for being aware that other people have rights, especially the right not to have their rights infringed upon; the term attempts to make fun of the idea of being "enlightened" or well educated), they fail to recognize the number of times the people who run such educational institutions are actually quite bigoted against particular people and ideas, such as against the idea that women were once believed to be capable of magic and once believed to be such powerful magicians that whole swathes of humanity believed we could create whole human beings simply out of our physical bodies, magical abilities which are directly connected to the divine or spiritual powers of the cosmos through our ties between the moon and our menstrual cycles. Most college and university administrators have changed from being people who support liberal arts education to being conservatives who cut programs, usually the liberal arts, allegedly to save money. Only outspoken critics of such systems have kept anything close to equality at the forefront of some institutions.

Beliefs in female power, especially magical or spiritual ones, aside, many scholars—most paid by institutions of higher education—deny, deny, deny the basic logical idea that, **before humanity learned that procreation required heterosexual intercourse of some kind to happen, women were valued as creators, leaders, and community organizers** (aka kings, to use the scholarly preferred term) for many cultures and their communities around the world.

Why is it so frightening to some people to envision women as natural leaders?

That is a rhetorical question because we know what causes bigotry like gynophobia—social programming.

Social programming is usually, but not always, quite subtle, but it can be so pervasive that people who belong to the groups being oppressed can actually adopt those prejudices toward others like themselves.

I know because I have witnessed it firsthand. As a college professor, I had a male student of color who argued with other male students of color in class after he said he would never hire a male person of color because "you can't trust them." At first, I thought he was joking around, but I learned he was serious. He had been so programmed by the culture he grew up hearing, seeing, and experiencing every day that, despite his being of the demographic population he was prejudiced against, he vehemently believed the mythology about that specific group of Americans. As happens so frequently, the person exuding the bias believes her/him/ximself separate from or somehow different than those s/he/xi is bigoted against.

I have witnessed this same kind of self-loathing-not-applied-to-the-self in many Midwestern women who firmly believe women "caught" having had sex by getting pregnant should be "punished" by being forced to carry unwanted fetuses to term, even if those fetuses are dead or dying.

Because many wannabe patriarchal cultures in the world today are based on the Abrahamic religions, there has been a concerted effort by many scholars raised in cultures steeped in those religions to deny, negate, and actively suppress information about what had to be gynocentric cultures early in human history. Such suppression denies active discussion about women's place in ancient societies, as well as our place in modern ones.

I have examined several of these scholars and their prevailing bigotry in this book.

However, I did not discover Elinor W. Gadon's *The Once & Future Goddess* scholarly book until very late in my research for this book, but it sounded like a valuable predecessor to Judy Grahn's *Blood, Bread, and Roses* scholarly book that I located quite by accident early on in my research, as well as to Gerda Lerner's *The Creation of Patriarchy*. I only discovered Charlotte Perkins Gilman's treatise, *The Man-Made World* because I looked up a reference for a friend online about Gilman, even though I regularly taught her story, "The Yellow Wallpaper," in my college classes.

The fact that I discovered all four of these books, not through examining the references lists from the other sources I read, but sheerly **by accident**, either stumbling through a library or the internet, demonstrates exactly how much feminist scholars with books that support either the idea that there was a discernable start to the patriarchy or that the original primary divinity for most cultures was a Creator Goddess have been purposely overlooked by other scholars.

If you are not a scholar, you might not care that some scholarship is being overlooked on purpose because of androcentric biases.

If you are a scholar, you know how egregious this oversight is.

Scholarship is **an ongoing conversation** between people who are exploring the whys and hows about things, whether that thing we are discussing is particle physics or ancient Sumeria.

Therefore, purposely excluding scholarly works because of a bias against a particular perspective presented in those works is **NOT** scholarship. It is pure bigotry by omission.

A dedicated scholar reads every other scholar's writing about the topic (as much as they can locate within their budget because some scholarly services, especially when the scholar is not associated with a college or university, are expensive), even if the work is never used directly within her/his/xis own article or book, which is why many scholarly books have Works Consulted pages.

Realize that I am not arguing that either Gadon's or Grahn's or Lerner's arguments are wholly perfect, since no scholar's argument is flawless, but all three women rely on the same sorts of scholarly sources and create their best arguments based on that available scholarship. All three books are worthy of reading and discussing.

However, Gadon's book has two important mistakes that I need to address: 1) she assumes that figures carved on a plaque found at Çatalhoyük demonstrate that humanity knew about pregnancy being caused by coitus circa 7500 BCE (36), and 2) that women had to wrest power from men through controlling food supplies through agriculture, presuming that men had previously held more power than women (37).

First, the four figures in bed from Çatalhoyük (see image below), according to Gadon, demonstrate that "the causal relationship between intercourse and procreation was now fully understood" by 7000 BCE (36) because the two figures on the left appear to be a couple having sex, while the two figures on the right appear to be a mother and child. For Gadon, this pairing is enough to pronounce that the plaque demonstrates the knowledge of procreation by these ancient peoples. For her, the couple has sex on the one side, and the product of that coitus appears on the other.

Drawing of Four Prone Figures carved in stone, Çatalhoyük, c. 7000 BCE
Fuller

However, the carving could simply represent four people sharing the same bed. Instead of demonstrating a causal determination for the results of coitus, the plaque could simply be demonstrating how public sex was, so that coitus happens even when other people are in the same bed. Since this type of communal sexual activity was conducted as late as American history when many families lived in one-room cabins, my interpretation is much more likely than Gadon's, especially when we consider that the typical room with "platforms" for sleeping[371] in a Çatalhoyük house was only about 20 square feet in size. Since most of the people living there were also engaging in coitus on top of the dead remains of their ancestors, I doubt they had many qualms about having sex near the children. Anxiety about public sexuality did not arise until androcentric cultures began controlling women's sexuality, thus began describing sex as a "dirty" activity.

Second, Gadon, like so many scholars, assumes that men were "natural" leaders because of their presumed superior physical strength.[372] However, most tribal cultures exercised shame on

---

[371] These platforms were dirt filled graves over the people's ancestors or dead children, probably topped with layers of reed mats and furs.

[372] Charlotte Perkins Gilman points out that "superior physical strength does not make the poorer wealthy, nor gives even the soldier a general" (113). Remember, the current record holders for the most weight lifted show that women can lift

449

individuals who aggrandized their own achievements or greedily tried to acquire more than their fair share of goods. In fact, most ancient cultures, prior to adopting androcentrism, demonstrate that they elected their leaders because they were good speakers, not because of their gender or because they possessed particular battle prowess or intimidated their peers more. The Aztec word for leader literally means "speaker," for instance. The might-makes-right attitude developed after androcentrism's rise because anthropologists have also identified cultures that believe killing taints the soul, so they enact blood purification rituals after such acts. Humans are, by and large, more cooperative than we are combative.

Gadon does admit that the people of Çatalhoyük appear to have worshipped a goddess, but she does not question another scholar's theory that women had to **regain** any form of power from dominating males as she continues to struggle, through her discussion of scholarly ideas in the book, with how the patriarchy rose to suppress women. Ironically, she stated earlier in the book that "women are now generally credited by prehistorians with the discovery of agriculture," waxing religious when she claims that "a strong link was forged connecting women as the cultivators of grain, to grain as the bounty of the Goddess, and to bread as the staff of life" (22). Her mistake was to assume that c. 7000 BCE was a reasonable time for humanity to have learned about procreation. She too easily gives up on believing women had social or spiritual power during the Bronze Age.

I have yet to read any other scholar's specific theory for how human beings would have actually determined that heterosexual intercourse was the sole cause of pregnancy, as I have done in this book.

Gadon correctly questions and identifies many things in her book. She questions scholarly use of the word "hierodule," meaning sacred prostitute (138), to discuss Inanna and Dumuzi's *qursu* ritual encounter, but she never openly questions calling the ritual a "sacred marriage," merely pointing out that Inanna "must be

---

four times their weight, while men only lift 2.8 times their own weight. Strength is relative.

properly and amply loved" (136). The Greek term hierodule is translator bigotry writ large in such tales as the *Epic of Gilgamesh.*

Gadon duly observes that "from the vantage of the Goddess from the underground is the primal womb, the matrix of all being, but to patriarchy the underworld is an alien place of horror and dread" (139), correctly identifying Inanna as the Queen of Heaven and her sister, Ereshkigal, as the Queen of the Nether World, but, like many androcentric scholars does not question the idea that the land of the dead must actually be located below ground, usually called the netherworld, when the Sumerians considered it "returning to the mountain." While, yes, caves on mountains are "below ground" and while many cultures buried their dead below ground or at least beneath mounds or pyramids, their imitations of the "mound of creation," the concept was more complex than just Above and Below and included Female Magic—the ability to give birth or rebirth, in particular.

In fact, Gadon borrows popular psychological archetypes when she claims the two goddesses are "a bipolar wholeness of the archetypal feminine" (139), largely because she believes the patriarchy was in full ascendency in ancient Sumeria, when it actually did not begin its rise until Sargon, the Akkadian who styles himself as "the Great," begins his ruthless, brutal military campaign to conquer 67 Sumerian cities, after having obtained Inanna's (thus the gods') blessing through sexual intercourse with his own daughter, High Priestess Enheduanna. However, after describing how "the sacred marriage ritual has become politicized," she admits that "the paradigm shift from a feminine to a masculine consciousness is under way" (141), signaling that she realized that the patriarchy had not yet full risen to power in ancient Sumeria. In discussing how Mesopotamia is often called the "'cradle of civilization,'" she claims, mimicking male scholars before her, that some first "achievements" that come with that label are "war, class-structured society, and slavery" (141).

Gadon in 1989, like Gerda Lerner two years earlier (1987) and Judy Grahn in 1993, understands that the patriarchy saw a meteoric, seemingly inexorable increase in power during the Bronze Age, but they all assumed that the patriarchy was already in place, even though they all demonstrate, repeatedly, throughout their books that the main events leading to the full on concept and implementation of the patriarchy occurred during, not before, the Bronze Age.

More modern scholars, like Eric Cline, Marguerite Rigoglioso, David Wengrow, Kara Cooney, and Jan Driessen, know that the Bronze Age was the age of radical social and technological change in and around the Mediterranean Sea, even though **none** of these scholars seem to grasp the fact that the main event that catalyzed all the other changes was encapsulated in one important discovery: procreation—that it takes both the male and the female to create offspring. This one piece of knowledge gives men the spiritual power they desire in order to elevate themselves above women in importance, spurring them on to take over nearly everything—from spiritual leadership to the manufacture of ceramics and bricks (construction).

YYY

The discovery of procreation led to the patriarchy, which could not exist without it.

Humanity has been more gynocentric than androcentric for most of our existence because women exhibited traits that tied them to the magic and spirituality of the cosmos more closely.

YYY

Because of this one fundamental truth, **that there was a discernable time before the patriarchy existed**, scholars still deeply immersed in the patriarchal Abrahamic religions or other androcentric world views continually overlook and underestimate the scholarship exhibited in these feminist scholars' works and the wisdom expressed by our ancestors themselves.

In fact, there are whole books on magic, often written or edited by male scholars, that completely ignore the magic of sex and Female Creative Magic, let alone the spiritual power of menstruation. While scholarly books like *Magic in the Ancient Greek World* by Derek Collins and *Magic and Ritual in the Ancient*

World, edited by Paul Mirecki and Marvin Meyer, discuss many forms of magic, including sympathetic magic (but make no mention of circumcision) and ancestor worship or propitiation (making no mention of women as the primary ancestor in question), they present little or no discussion about the magic of menstruation or birth, except for a discussion by Collins about how quotes from *The Iliad* were used to bring a woman's menstrual cycle back (110).

These clear biases against gynocentric views of the world, which undoubtedly predated androcentric ones, means most scholars are still missing half the story. A story I have only begun exploring here.

By shedding our assumptions that men are better than women at everything, by realizing that this **mythology** was created to make men seem transcendent spiritually thus more powerful socially than women, even though only enforceable via violence and oppression, and that this mythology arose out of the discovery of procreation circa 2400 BCE, we can begin to examine the symbols our gynocentric ancestors left us, better realizing how important, how vital, how creative our ancestral mothers were, and how likely it was that they led communities, nurtured their people, created religions, created cities, manufactured most goods, created healing, created communication, and instigated learning far more often than most scholars have yet to acknowledge.

The world was once gynocentric, our only known blood relatives our mothers. It makes perfect sense that once our ancestors turned those beloved women into divine beings in the Next Realm who could assist human beings in this one, they became goddesses, like Io, primary ancestresses who brought the divine to the people through various means, including sex and sacred rituals.

What is most clear is that a tendency toward androcentrism transformed into full on patriarchal form in the centuries following the discovery of procreation—that males have a biological role in creating babies. Armed with this new, verifiable knowledge thanks to domesticated swans, ancient males sought ascendency over women by using written language—laws—to counter feminine spiritual, sacred, and social powers by gradually changing positive Female Magic demonstrating women's cosmic connections into sullied, even dirty, images by making menstruation unclean, sex dirty, and allowing men to legally buy and sell women like chattel or stone women to death for having sex with more than one man. Women had to be oppressed for men's status in society to rise, so women were denied educations, despite the fact that they probably were responsible for creating writing systems as the priestesses controlling goods gathered at temples. Sex became something only men enjoy, with their favorite pejoratives alluding to women's sexual infidelity to just one man. Women were denied freedom of movement, both by the kinds of clothing they were forced to wear and by requiring they refrain from public excursions, shutting them off from exercise as well as knowledge. Women were paid less food as wages than men, becoming smaller, thinner, and weaker than their ancestors had been. In short, men did everything they could to make certain women could not compete with them on any level with that oppression solidifying during the Bronze Age.

Homer, possibly remembering the conflicts that led to the Greek Dark Age when writing nearly ceased to exist for those of Greek culture because it was through written laws that women and poor men were being oppressed and goddesses being abandoned, does his best to warn us about male-power gone mad, wherein young men were encouraged to abuse women, to oppress weaker men or men they could take advantage of, to kill wantonly men, women, and children of "other" cultures in order to secure for themselves *kleos*—immortal fame.

By honoring our female ancient ancestors' sense of spiritual power and magic, solely based on their interactions and connections with the cosmos, we can begin to rectify over 2000 years of gender oppression.

It is certainly way past time for women to ascend, once again, and take our rightful places as leaders. While we may never have a female pope (Latin for "father") leading the Catholic Church, it is high time there was a nurturing *Mother* leading humanity, showing us we can be better than we are.

Achrati, Ahmed. "Womanhood Without the Bull: Venus of Laussel, Inanna, and the Lady of Tin Tilizaghen." *SAHARA* (2010): 4-6.

Adovasio, J. M. Olga Soffer & Jake Page. *The Invisible Sex: Uncovering the True Roles of Women in Prehistory*. Routledge, 2009.

Ahdad, Karima. 2019. "The Fading Pride of Amazigh Tattoos." *TRT World*. April 4. Accessed August 22, 2020.

Alba-Lois, Luisa & Claudia Segal-Kischinevsky. "Yeast Fermentation and the Making of Beer and Wine." Scitable: *Nature Education*. 2010. https://www.nature.com/scitable/topicpage/yeast-fermentation-and-the-making-of-beer-14372813/

"A Look at the History of the Weaving Loom." Fashion Archives. April 11, 2015. https://startupfashion.com/fashion-archives-history-of-the-weaving-loom/

Ancient1580. "Prehistoric Copper Smelting in a Pit!" YouTube. May 25, 2012. https://www.youtube.com/watch?v=8uHc4Hirexc&t=15s.

Ancient Architects. Matt Sibson. "New Hypothosis: The 11,000-Year-Old Göbekli Tepe Bone Plaque." YouTube. January 26, 2022. https://www.youtube.com/watch?v=sYcc-dJC9hY&t=2s

"Ancient Pottery." *Encyclopedia of Art*. 2021. http://www.visual-arts-cork.com/pottery.htm

Animalia. "Resplendent Quetzal." 2021. https://animalia.bio/resplendent-quetzal Accessed August 14, 2021.

Antonaccio, Carla. "Religion, Basileis and Heroes." in *Ancient Greece: From the Mycenaean Palaces to the Age of Homer*. Eds. Sigrid Deger-Jalkotzy and Irene S. Lemos. Edinburgh UP, 2006. pp. 381-95.

The Archaeologist. "The 'Moche' Erotic Pottery: Ancient Pornography or Something Else?" August 24, 2021. https://www.thearchaeologist.org/blog/the-moche-erotic-pottery-ancient-pornography-or-something-else

Aridi, Rasha. "California Condors Surprise Scientists with Two 'Virgin Births.'" Smithsonian Magazine. November 2, 2021. https://www.smithsonianmag.com/smart-news/california-condors-surprise-scientists-with-two-virgin-births-180978983/

Assante, Julia. "Sex, Magic & the Liminal Body in the Erotic Art and Texts of the Old Babylonian Period." *Sex & Gender in the Ancient Near East*. Proceedings of the 47th Rencountre Assyriologique Internationale, Helsinki, July 2-6, 2001. Eds. S. Parpola & RM Whiting. 2002. 27-57.

Aveni, Anthony. *Stairways to the Stars: Skywatching in the Three Great Ancient Cultures*. John Wiley & Sons, 1997.

Babe, Ann. "Object of Intrigue: Moche Sex Pots." AtlasObscura. March 8, 2016. https://www.atlasobscura.com/articles/object-of-intrigue-moche-sex-pots

Bachofen, J.J. *Myth, Religion, and Mother Right: Selected Writings of J.J. Bachofen.* Princeton, NJ: Princeton University Press, 1967.

Bacon, Bennett. Azadeh Khatiri, James Palmer, Tony Freeth, Paul Pettitt, and Robert Kentridge. "An Upper Palaeolithic Proto-writing System and Phenological Calendar." *Cambridge Archaeological Journal.* January 5, 2023. https://www.cambridge.org/core/journals/cambridge-archaeological-journal/article/an-upper-palaeolithic-protowriting-system-and-phenological-calendar/6F2AD8A705888F2226FE857840B4FE19

Ball, Jennifer. "About This Website, the Research, and the Team." https://originofalphabet.com/about/

-----. "Breasts Depicted in Sumerian Cuneiform: a Database." *Cross Analysis of Ancient Written Languages.* https://originofalphabet.com/breasts-depicted-in-sumerian-cuneiform-a-database/ May 6, 2021.

Balter, Michael. "How Sheep Became Livestock." *Science.* April 29, 2014. https://www.sciencemag.org/news/2014/04/how-sheep-became-livestock

Baring the Aegis. "Hestia Versus Dionysos." March 4, 2013. http://baringtheaegis.blogspot.com/2013/03/hestia-versus-dionysos.html

Barrett-Morison, James. "Sumerian Language: Known Profanity Words in Sumerian." July 6, 2018. https://sumerianlanguage.tumblr.com/post/175618372191/are-there-any-known-profanity-words-in-sumerian

Barroqueiro, Silvério A. "The Aztecs: A Pre-Columbian History." Section 3. Yale-New Haven Teachers Institute. 2022. https://teachersinstitute.yale.edu/curriculum/units/1999/2/99.02.01/3

Bartosiewicz, Laszlo. "Animals in Bronze Age Europe." In *The Oxford Handbook of the European Bronze Age.* Eds. Harry Fokkens & Anthony Harding. Oxford UP, 2013. 328-47.

Beckman, Gary. "The Hittite Assembly." *Journal of the American Oriental Society*, vol. 102, no. 3, American Oriental Society, 1982, pp. 435–42, https://doi.org/10.2307/602295.

Betz, Eric. "Nabta Playa: The World's First Astronomical Site Was Built in Africa and Is Older Than Stonehenge." *Discover.* June 20, 2020. https://www.discovermagazine.com/the-sciences/nabta-playa-the-worlds-first-astronomical-site-was-built-in-africa-and-is

Biørnstad, Lasse. "Researchers Have Examined the Burial Mound Where the Gokstad Viking Ship Was Found." Sciencenorway.no. September 15, 2020. https://sciencenorway.no/archaeology-viking-age-vikings/researchers-have-examined-the-burial-mound-where-the-gokstad-viking-ship-was-found-what-they-found-surprised-them/1741928.

Black, Jeremy A., Graham Cunningham, Eleanor Robson, and Gabor Zolyomi. *The Literature of Ancient Sumer.* Oxford UP, 2004.

Bower, Bruce. "Ancient 'Smellscapes' Are Wafting Out of Artifacts and Old Texts." ScienceNews. May 4, 2022. https://www.sciencenews.org/article/ancient-smell-odor-artifacts-texts-egypt-archaeology?fbclid=IwAR3egneWXDI6BVhyGT5z8phIRWrkSrWPfJ7uWz8zdE9kjUnSK1nqgFqHxlg

Bray, Peter. "Before Cu_{29} Became Copper: Tracing the Recognition and Invention of Metalleity in Britain and Ireland During the 3rd Millennium BC." In *Is There a British Chalcolithic?: People, Place and Polity in the Later Third Millennium*. Eds. Michael J. Allen, Julie Gardiner, and Alison Sheridan. Prehistoric Society Research Paper #4. Oxford: Oxbow Books, 2012. 56-70.

Breton Connelly, Joan. *Portrait of a Priestess: Women and Ritual in Ancient Greece*. Princeton, NJ: Princeton UP, 2007.

British Museum. "Curator's Tour of Tantra." YouTube. December 11, 2020. https://www.youtube.com/watch?v=p5jOGh6j7T0

-----. "Tantric Practice and Divine Feminine Power." YouTube. November 30, 2020. https://www.youtube.com/watch?v=AU3eXB33OHs .

Brochmann, Nina Dølvik & Ellen Støkken Dahl. "The Virginity Fraud." TEDxOslo. YouTube. May 2017. https://www.ted.com/talks/nina_dolvik_brochmann_and_ellen_stokken_dahl_the_virginity_fraud?language=en

Brown, Wally. "What Happened at This Popular Anasazi Ruin?" Navajo Traditional Teaching. YouTube. October 4, 2020. https://www.youtube.com/watch?v=LY8h7pkpa-w

Bryce, Trevor R. "Ḫattušili i and the Problems of the Royal Succession in the Hittite Kingdom." *Anatolian Studies*, vol. 31, British Institute at Ankara, 1981, pp. 9–17, https://doi.org/10.2307/3642754.

Buckley, Thomas & Alma Gottlieb. *Blood Magic: The Anthropology of Menstruation*. Berkeley, CA: University of California, 1988.

Budge, Ernest Alfred Wallis. 1920. *An Egyptian Hieroglyphic Dictionary, Vol. 1*. London: John Murray.

Budin, Stephanie L. "Maternity, Children, and 'Mother Goddesses' in Minoan Iconography." *Journal of Prehistoric Religion*, Volume XXII. Aströms Förlag Sävadelen, 2010. Pp. 6-38.

-----. "Mother or Sister? Finding Adolescent Girls in Minoan Figural Art." In *Girls in Antiquity*. Susanne Moraw and Anna Kieburg, eds. Münster: Waxmann, 2014. 105-16.

Cameron, Dorothy. "The Symbolism of the Ancestors." *ReVision* 20.3 (Winter 1998). 1-6. Gale Academic OneFile. Accessed May 20, 2022.

Campbell, Joseph. *The Hero with a Thousand Faces*. NY: Barnes & Noble, 1949.

-----. *Historical Atlas of World Mythology. Volume 1: The Way of the Animal Powers, Part 2: Mythologies of the Great Hunt*. NY: Harper & Row, 1988.

Carmody, Denise Lardner. *The Oldest God: Archaic Religion Yesterday and Today*. Nashville, TN: Abingdon, 1981.

Carter, Robert. "Boat Remains and Maritime Trade in the Persian Gulf During the Sixth and Fifth Millenai BC." *Antiquity.* Vol. 80, No. 307. March 2006. 52-63. https://www.academia.edu/173149

Cartwright, Mark." Pulque." *World History Encyclopedia.* 2009-2021. https://www.worldhistory.org/Pulque/

CenterFireMade. "Casting an Egyptian Bronze Age Dagger." YouTube. August 10, 2017. https://www.youtube.com/watch?v=ngsrzXTiFjA.

Chambers, Ian. "The Stand as Ancient World Tree." February 20, 2018. https://www.patheos.com/blogs/bythepalemoonlight/2018/02/the-stang-as-ancient-world-tree/.

Clark, Janet Mulroney. "Burial Beliefs of the Aztecs of Mexico." Classroom. November 9, 2016. https://classroom.synonym.com/good-world-history-research-paper-topics-4900.html

Clark, Megan. "Paddle Dolls in Ancient Egypt: Gaudy or Godly?" Recorded Zoom Lecture for the Egypt Centre. https://www.youtube.com/watch?v=E1qodHVG__8 Posted May 12, 2020. Accessed February 9, 2021.

Clendinnen, Inga. *Aztecs.* Cambridge: Cambridge UP, 1991.

Cloth Roads. "Maguey, Ancient Wonder Plant." 2018. https://www.clothroads.com/maguey-ancient-wonder-plant/

Collins, Billie Jean. *The Hittites and Their World.* Atlanta: Society of Biblical Literature, 2007.

Collins, Derek. *Magic in the Ancient Greek World.* Malden, MA: Blackwell Publishing, 2008.

The Compact Oxford English Dictionary. 2nd edition. Oxford: Clarendon Press, 1987.

Connelly, Joan Breton. *Portrait of a Priestess: Woman and Ritual in Ancient Greece.* Princeton, NJ: Princeton UP, 2007.

Cook, Jill. "Mistaken About Eve? Finding and Representing Women in the Old Stone Age." In *Women in Industry and Technology from Prehistory to the Present* Day. Amanda Devonshire and Barbara Wood, eds. London: Museum of London, 1996.

Cooney, Kara. *The Good Kings: Absolute Power in Ancient Egypt and the Modern World.* Washington, DC: National Geographic, 2021.

-----. *When Women Ruled the World.* Washington, DC: National Geographic, 2018.

Corcoran, Lucy. "Athena: Goddess of War." University College Dublin. July 12, 2001. https://www.ucd.ie/pages/99/articles/corcoran.pdf

Cucuzza, Nicola. "Minoan Nativity Scenes? The Ayia Triada Swing Model and the Three-Dimensional Representation of Minoan Divine Epiphany." In *Estratto Da Annuario.* Volume XCI, Series III, 13, 2013. SAIA, 2015. Pp. 175-207.

Culbertson, "Women and Household Dependents in Ur III Court Records." In *Structures of Power: Law and Gender Across the Ancient Near East and Beyond.* Ed. Ilan Peled. Chicago: Oriental Institute of the University of Chicago, 2017. 115-29.

Curry, Andrew. "Mystery of the Varna Gold: What Caused These Ancient Societies to Disappear?" *Smithsonian Magazine*. April 18, 2016. https://www.smithsonianmag.com/travel/varna-bulgaria-gold-graves-social-hierarchy-prehistoric-archaelogy-smithsonian-journeys-travel-quarterly-180958733/

"Cylinder Seal with Kneeling Nude Heroes, c. 2220-2159 BCE." Metropolitan Museum of Art. http://smarthistory.org/cylinder-seals/. May 10, 2021.

Diaz, Bernal. *The Conquest of New Spain*. Trans. J.M. Cohen. New York: Penguin Classics, 1963.

Derr, Mark. "New Views of Neanderthal Are Reshaping Prehistory." *Psychology Today*. December 11, 2022. https://www.psychologytoday.com/us/blog/dogs-best-friend/202212/new-views-neanderthal-are-reshaping-prehistory?fbclid=IwAR0JIi2UVcZiLh6lU4sqW906lVgS1ZH3YSsvtY7LGp-RA2HAD4kJTEDWLEI

Dever, William G. *Did God Have a Wife? Archaeology and Folk Religion in Ancient Israel*. Grand Rapids, MI: Wm. B. Eerdmans Publishing Co., 2005.

Diep, Francie. "8000 Years Ago, 17 Women Reproduced for Every One Man." *Pacific Standard*. June 14, 2017. https://psmag.com/environment/17-to-1-reproductive-success

Dizon, Euseblo Z. & Armand Salvador B. Mijares. "Archaeological Evidence of a Baranganic Cultures in Batanes." *Philippine Quarterly of Culture and Society*. Vol. 27, #1/2. March/June 1999. Pp. 1-10. https://www.jstor.org/stable/29792430

Dodds Pennock, C. *Bonds of Blood: Gender, Lifestyle, and Sacrifice in Aztec Culture*. London: Palgrave-McMillen, 2011.

-----. "Gender and Aztec Life Cycles." In *The Oxford Handbook of the Aztecs*. Eds. D.L. Nichols and E. Rodriquez-Alegria. Oxford: Oxford UP, 2017. 387-398. http://eprints.whiterose.ac.uk/109181/

Driessen, Jan. "The Birth of a God? Cults and Crises on Minoan Crete." January 1, 2015. Brepols Online. https://www.brepolsonline.net/doi/10.1484/M.HR-EB.5.108417

-----. "Chercher la Femme: Identifying Minoan Gender Relations in the Built Environment." In *Minoan Realities: Approaches to Image, Architecture, and Society in the Aegean Bronze Age*. Eds. D. Panagiotopoulos and U. Günkel-Maschek, eds. Aegis 5, Louvain-laoNeuve, 2012. Pp. 141-63.

-----. "History and Hierarchy: Preliminary Observations on the Settlement Pattern of Minoan Crete." In *Urbanism in the Aegean Bronze Age*. Keith Branigan, ed. Bloomsbury Publishing, 2001. Pp. 51-70

-----. "A Matrilocal House Society in Pre- and Postpalatial Crete?" In *Back to the Beginning: Reassessing Social and Political Complexity in Crete During the Early and Middle Bronze Age*. I. Schoep, P. Tomkins & J. Driessen, eds. Oxford & Oakville, 2012. 358-83.

Driessen, Jan & Charlotte Langohr. "Recent Developments in the Archaeology of Minoan Crete." *Pharos*. 2014. Vol. 20, No. 1, pp. 75-115.

https://www.academia.edu/2065348/Recent_Developments_in_the_
Archaeology_of_Minoan_Crete

Dunphy-Lelii, Sarah. "The Mystery of Why Female Chimps Leave Their
Groups." Psychology Today. December 15, 2022.
https://www.psychologytoday.com/intl/blog/wild-
cousins/202212/the-mystery-of-why-female-chimps-leave-their-
groups?fbclid=IwAR3oFriMk-a-3dP0aC-
jYWSdZiONYdrt9aOnsuHvuJ6GVfcl8cLYESI58_8

DW Documentary. "The Mosuo in China—Where Women Rule." YouTube.
January 23, 2022. https://www.youtube.com/watch?v=7_2sus4Ku1g

Earle, Timothy. *How Chiefs Come to Power: The Political Economy in
Prehistory.* Stanford: Stanford UP, 1997.

Eastman, Charles A. *Indian Boyhood.* New York: McClure, Phillips & Co., 1902.

-----. *Old Indian Days.* Reprint. Lincoln: U of Nebraska P, 1991.

----- and Elaine Goodale Eastman. *Wigwam Evenings.* Lincoln, NE: University
of Nebraska P, 1990.

Easy Crafts DIY. "Amazing Two Women Building the Most Beautiful House in
Jungle." YouTube. February 25, 2020.
https://www.youtube.com/watch?v=HGr8yizWhg8

Edmonds, Susan T. "Picturing Homeric Weaving." The Center for Hellenic
Studies. November 2, 2020. https://chs.harvard.edu/susan-t-edmunds-
picturing-homeric-weaving/

Elsevier. "High-tech Imaging Reveals Precolonial Mexican Manuscript Hidden
from View for 500 Years." EurekAlert! August 18, 2016.
https://www.eurekalert.org/news-releases/505573

Eller, Cynthia. *The Myth of Matriarchal Prehistory.* Boston: Beacon Press, 2000.

Emery, Katy Meyers. "Investigating Red-Colored Bones in Mesoamerica."
Spartan Ideas. Michigan State University. June 7, 2016.
https://spartanideas.msu.edu/2014/05/22/investigating-red-colored-
bones-in-mesoamerica/

English Heritage. "How to Make Prehistoric Pottery." YouTube. April 3, 2017.
https://www.youtube.com/watch?v=nrI1LJbKIvk.

Epstin. H. "Awassi Sheep." *World Animal Review.*
http://www.fao.org/3/p8550e/P8550E01.htm. Accessed February 10,
2021.

Faculty of Oriental Studies, University of Oxford. "Inana and Ebih." The
Electronic Text Corpus of Sumerian Literature. September 10, 2020.
http://163.1.185.89/section1/tr132.htm .

-----. "Inana and Enki." The Electronic Text Corpus of Sumerian Literature.
December 19, 2006. http://etcsl.orinst.ox.ac.uk/cgi-
bin/etcsl.cgi?text=t.1.3.1#.

-----. "Inana's Descent to the Nether World." The Electronic Text Corpus of
Sumerian Literature. September 12, 2020.
http://etcsl.orinst.ox.ac.uk/section1/tr141.htm.

Farnell, Lewis Richard. *Greek Hero Cults and Ideas of Immortality.* Oxford:
Clarendon P, 1921.

Farron, Steve. "The Portrayal of Women in the Iliad." *Acta Classics: Proceedings of the Classical Association of South Africa*. 1979. Vol. 22, pp. 15-31.

Fewell, Danna Nolan and David M. Gunn. "Controlling Perspectives: Women, Men and the Authority of Violence in Judges 4 & 5." *Journal of the American Academy of Religion* (1990 (58.3)): 389-411. Print.

Fisberg, Mauro & Rachel Machado. "History of Yogurt and Current Patterns of Consumption." *Nutrition Reviews*. Vol. 73, #1, August 2015, pp. 4-7 https://academic.oup.com/nutritionreviews/article/73/suppl_1/4/181 9293

Flannery, Kent V. & Joyce Marcus. *The Creation of Inequality: How Our Prehistoric Ancestors Set the Stage for Monarchy, Slavery, and Empire*. Cambridge, Mass.: Harvard UP, 2012.

Flood, Julia. "Goddesses of the Month: Tzitzimeme." Aztecs. August 17, 2021. https://www.mexicolore.co.uk/aztecs/gods/goddesses-of-the-month-tzitzimime

-----." Goddess of the Month: Xochiquetzal ('Quetzal Flower')." Mexicolore. N.D. https://www.mexicolore.co.uk/aztecs/gods/goddess-of-the-month-xochiquetzal

Ford, Dixon and Lee Kreutzer. "Oxen: Engines of the Overland Emigration." *Overland Journal*. Vol. 33 #1: Spring 2015. Pp. 4-29. www.nps.gov/cali/learn/history/culture/upload/OJ-spring2015-oxen.pdf Accessed July 21, 2017.

Friedman, David. M. *A Mind of Its Own: A Cultural History of the Penis*. NY: The Free Press, 2001.

Fuller, Michael and Neathery. http://users.stlcc.edu/mfuller/catalhuyuk.html

Gadon, Elinor W. *The Once & Future Goddess*. San Francisco: Harper & Row, 1989.

Gelb, I.J. "Prisoners of War in Early Mesopotamia." *Journal of Near Eastern Studies*. Jan-Apr. 1973, Vol. 32. No. 1/2. pp. 70-98.

Georgiadis, Mercourios. "Metaphysical Beliefs and Leska." In *Metaphysis: Ritual, Myth and Symbolism in the Aegean Bronze Age*. Eds. Eva Alram-Stern, Fritz Blakolmer, Sigrid Deger-Jalkotzy, Robert Laffineur, and Jörg Weilhartner. Peeters Leuven-Liege, 2016. Pp. 295-302.

Gershon, Livia. "The Aztec Constructed This Tower out of Hundreds of Human Skulls." *Smithsonian Magazine*. December 14, 2020. https://www.smithsonianmag.com/smart-news/new-find-brings-skulls-discovered-aztec-tower-over-600-180976543/

Gill, Dan. "When Should Fig Trees Start to Produce Ripened Fruit?" Nola.com. July 7, 2021. https://www.nola.com/entertainment_life/home_garden/article_90bd9d35-3d7a-52c7-8961-635eac7d3400.html

Gillespie, Susan D. "Rethinking Ancient Maya Social Organization: Replacing 'Lineage' with 'House.'" *American Anthropologist*. September 2000, Vol. 102, No. 3. Pp. 467-84.

Glendinnen, Inga. *Aztecs*. Cambridge UP, 1991.

Goldade, Jenny. "Cultural Spotlight: Ancient Aztec Funeral Traditions." Frazer Consultants. February 9, 2018. https://web.frazerconsultants.com/2018/02/cultural-spotlight-ancient-aztec-funeral-traditions/

Goode, Starr. *Sheela Na Gig: The Dark Goddess of Sacred Power*. Rochester, VT: Inner Traditions, 2016.

Graeber, David and David Wengrow. *The Dawn of Everything: A New History of Humanity*. Farrar, Straus & Giroux, 2021.

Graeber, David. *Debt: The First 5000 Years*. New York: Melville House, 2011.

Grahn, Judy. *Blood, Bread, and Roses: How Menstruation Created the World*. Boston: Beacon P, 1993.

Gray, Peter B. & Kermyt G. Anderson. *Fatherhood: Evolution and Human Paternal Behavior*. Cambridge, Mass: Harvard UP, 2010.

GreekBoston. *About the Greek Goddess Lyssa*. 14 April 2017. https://www.greekboston.com/culture/mythology/lyssa/. 15 May 2020.

GreekGods.org. *Themis*. 2013-2018. https://www.greek-gods.org/titans/themis.php. 30 May 2020.

Green, Peter, Trans. *The Iliad*. Oakland, U of California P, 2015.

Guilhou, Nadine. "Myth of the Heavenly Cow." *UCLA Encyclopedia of Egyptology*. August 12, 2010. https://escholarship.org/uc/item/2vh551hn

-----. "Myth of the Heavenly Cow." *UCLA Encyclopedia of Egyptology*. September 2010. Los Angeles. http://digital2.library.ucla.edu/viewItem.do?ark=21198/zz002311pm

Günkel-Maschek, Ute. "Reflections on the Symbolic Meaning of the Olive Branch as Head-Ornament in the Wall Paintings of Building Xeste 3, Akrotiri." In Kosmos: Jewellry, Adornment and Textiles in the Aegean Bronze Age." Eds. Marie-Louise Nosch & Robert Laffineur. Leuven-Liege, Peeters, 2012. 361-7.

-----. "Spirals, Bulls, and Sacred Landscapes." In *Minoan Realities: Approaches to Images, Architecture, and Society in the Aegean Bronze Age*. Eds. Diamantis Panagiotopoulos and Ute Günkel-Maschek. Presses Universitaires de Louvain, 2012. Pp. 115-140.

-----. "Time to Grow up, Girl! Childhood and Adolescence in Bronze Age Akrotiri, Thera." In *Girls in Antiquity*. Susanne Moraw and Anna Kieburg, eds. Münster: Waxmann, 2014. 117-133.

Gurney, Oliver R. *Some Aspects of Hittite Religion*. Oxford UP, 1977.

Güterbock, Hans G. "The Composition of Hittite Prayers to the Sun." *Journal of the American Oriental Society*, vol. 78, no. 4, American Oriental Society, 1958, pp. 237–45, https://doi.org/10.2307/595787.

Hack, Roy. "Homer and the Cult of Heroes." *Transactions and Proceedings of the American Philological Association*. Vol. 60 (1929): 57-74.

Hall, Henrich. "Another Thing: Here's Looking at You—A Ship's Eyes in Bodrum." Peter Sommer Travels. October 22, 2015. https://www.petersommer.com/blog/archaeology-history/ships-eyes-bodrum

Hardy, Sabrina. "All Shall Fade: Homer's Foreshadowing of the Ending of the Heroic Age in *The Iliad*." *The Kabod*. Vol. 1, January 1, 2014.

Hare, Brian and Vanessa Woods. *Survival of the Friendliest: Understanding Our Origins and Rediscovering Our Common Humanity*. New York: Random House, 2020.

Harris, Karen. "The History of Ship Figureheads." *History Daily*. June 18, 2019. https://historydaily.org/the-history-of-ship-figureheads

Hartland, Edwin Sidney. *Primitive Paternity: The Myth of Supernatural Birth in Relation to the History of the Family*, Vol. 2. London: David Nutt, 1910.

"Hatti—the Hittite Empire." Luwian Studies. 2019. https://luwianstudies.org/hatti-the-hittite-empire/

Helmer, Daniel, Emilie Blaise, Lionel Gourichon, & Maria Saňa-Seguí. "Using Cattle for Traction and Transport During the Neolithic Period: Contribution of the Study of the First and Second Phalanxes." *Bulletin de la Société Préhistorique Française*. 116.1 (Janvier-Mars 2018). 71-98.

Hernandez, Vladimir. "The Country Where Exorcisms Are on the Rise." *BBC News*. November 26, 2013. https://www.bbc.com/news/magazine-25032305 .

High Touch High Tech. "Equinox—Oreo Phases of the Moon." ScienceMadeFun.net. March 24, 2022. Pdf.

Hoffner, Jr. Harry A. *Hittite Myths*, 2nd ed. Atlanta, Georgia: Scholars Press, 1990.

Hollager, Erik. "Chapter 11: Crete." In *The Oxford Handbook of the Bronze Age Aegean*. Ed. Eric Cline. Oxford: Oxford UP, 2010. 149-59.

Holland, Leicester B. "The Danaoi." *Harvard Studies in Classical Philology*. Vol. 39 (1928):59-92.

Homer. "Classics." 1994-2009. *The Iliad by Homer*. Ed. Trans. Samuel Butler. http://classics.mit.edu/Homer/iliad.html. 17 May 2020. <http://classics.mit.edu/Homer/iliad.html>.

-----. *The Iliad*. Trans. Caroline Alexander. New York: HarperCollins, 2015.

-----. *The Iliad*. Trans. Samuel Butler. 1994-2009. http://classics.mit.edu/Homer/iliad.html.

-----. *The Iliad*. Trans. Robert Fagles. New York: Penguin Books, 1990.

-----. *The Iliad*. Trans. Robert Fitzgerald. New York: Alfred A. Knopf, 1910.

-----. *The Iliad*. Trans. Peter Green. Oakland, California: University of California Press, 2015.

-----. *The Iliad*. Trans. Richmond Lattimore. Chicago: University of Chicago Press, 1951.

-----. *The Iliad*. Trans. Stephen Mitchell. New York: Free Press, 2011.

Horne, Jayme. "Worlds Collide: Aztec Gender Parallelism and Spanish Patriarchy." Massachusetts College of Art and Design. N.D.

Hornung, Eric. *Conceptions of God in Ancient Egypt*. Trans. John Baines. Ithaca, NY: Cornell University Press, 1982.

Houser Wegner, Jennifer. "Ancient Egyptian Creation Myths: From Watery Chaos to Cosmic Egg." Glencairn Museum News, #5. July 13, 2021.

https://glencairnmuseum.org/newsletter/2021/7/13/ancient-egyptian-creation-myths-from-watery-chaos-to-cosmic-egg

"How Is Flax Turned into Linen?" New England Flax & Linen. 2019. https://www.newenglandflaxandlinen.org/resources/faq/

Jacobsen, Thorkild. *The Treasures of Darkness: History of Mesopotamian Religion*. New Haven: Yale, 1976.

Jansen, Maarten. "Mixtec Pictography: Conventions and Contents." In *Epigraphy: Supplement to the Handbook of Middle American Indians*, Vol. 5. Austin, TX: University of Texas Press, 1992. Pp. 20-33.

Jastrow, Morris, Jr. "Sumerian and Akkadian Views of Beginnings." *Journal of American Oriental Society*. 1916. Vol. 36. Pp. 274-299. https://www.jstor.org/stable/592686.

Jarvis, Dennis. Coyolxāuhqui Stone Photo. From https://en.wikipedia.org/wiki/Coyolxauhqui_Stone#/media/File:Mexico-3980_-_Coyolxauhqui_Stone_(2508259597).jpg

Jones, Christopher P. *New Heroes in Antiquity: From Achilles to Antimoos*. Cambridge, MA: Harvard UP, 2010.

Jones, Peter. *Homer's Odyssey: a Companion to the English Translation of Richmond Lattimore*. Bristol: Bristol Classical Press, 1988.

Kellogg, Susan. "Aztec Inheritance in Sixteenth-Century Mexico City: Colonial Patterns, Prehispanic Influences." *Ethnohistory*. 33:3 (summer 1986). 313-30.

Kenoyer, Mark. "Meluhha: The Indus Civilization and Its Contacts with Mesopotamia." The Oriental Institute. October 7, 2010. https://www.youtube.com/watch?v=8zcGL1LEbmI.

Khan Academy. "Templo Mayor at Tenochtitlan, the Coyolxāuhqui Stone, and an Olmec Mask." https://www.khanacademy.org/humanities/ap-art-history/indigenous-americas-apah/north-america-apah/a/templo-mayor-at-tenochtitlan-the-coyolxauhqui-stone-and-an-olmec-mask

Kienlin, Tobias. "A Hero Is a Hero Is a...? On Homer and Bronze Age Social Modeling." in *Bronze Age Connectivity in the Carpathian Basin*. Eds. Botond Rezi and Rita E. Németh. Mureş County Museum, 2018. pp. 19-32.

Killgrove, Kristina. "20,000-year-old Cave Painting 'Dots' Are the Earliest Written Language, Study Claims. But Not Everyone Agrees." LiveScience. January 4, 2023. https://www.livescience.com/ice-age-cave-art-proto-writing-claim

King, Bruce M. "Average-Size Erect Penis: Fiction, Fact, and the Need for Counseling." *Journal of Sex & Marital Therapy*. (2021) 47:1, https://www.tandfonline.com/action/showCitFormats?doi=10.1080%2F0092623X.2020.1787279

Knapik, Joseph J., James E. Wright, Dennis M. Kowal, and James A Vogel. "The Influence of US Army Basic Initial Training on the Muscular Strength of Men and Women." Natick, MA: Army Research Institute of Environmental Medicine, 1980.

Koehl, Robert. B. "Beyond the 'Chieftain Cup': More Images Relating to Minoan Male 'Rites of Passage.'" In *Studies in Aegean Art and Culture: A New York Aegean Bronze Age Colloquium in Memory of Ellen N. Davis*. Ed. Robert B. Koehl. Philadelphia: INSTAP Academic Press, 2016. Pp. 113-32.

Konsolaki-Yannopoulou, Eleni. "The Symbolic Significance of the Terracottas from the Mycenaean Sanctuary at Ayios Konstantinos, Methana." In *Metaphysis: Ritual, Myth, and Symbolism in the Aegean Bronze Age*. Eds. Eva Alram-Stern, Fritz Blakolmer, Sigrid Deger-Jolkotzy, Robert Laffineur, and Jorg Weilhartner. Peeters Leuven-Liege, 2016. 49-59.

Kramer, Samuel Noah. "The Biblical 'Song of Songs' and the Sumerian Love Songs." Expedition Magazine (Penn Museum) 5.1 https://www.pen.museum/sites/expedition/the-biblical-song-of-songs-and-the-sumerian-love-songs/ .

-----. *History Begins at Sumer: 39 First in Recorded History*. Philadelphia: U of Pennsylvania P, 1956. Revised, 1981.

-----. *The Sumerians: Their History, Culture, and Character*. Chicago: U of Chicago P, 1963.

Kristiansen, Kristian & Thomas B. Larsson. *The Rise of Bronze Age Society: Travels, Transmissions and Transformations*. Cambridge UP, 2005.

Kriwaczek, Paul. *Babylon: Mesopotamia and the Birth of Civilization*. NY: St. Martin's Press, 2010.

Kvilhaug, Maria. Lady of the Labyrinth: Mother Goddess Series. YouTube.

Lalueza-Fox, Carles. "The Archeology of Inequality." From *Inequality: A Genetic History*. The MIT Press Reader, 2022. https://thereader.mitpress.mit.edu/the-archaeology-of-inequality/?utm_source=pocket-newtab

Larson, Walker. "Hector, Achilles, and Toxic Masculinity." *The Epoch Times*: Bright. January 2, 2023. https://www.theepochtimes.com/hector-achilles-and-toxic-masculinity_4942520.html?fbclid=IwAR0MRZUzPu6EZcu22G7GptHnUZyyqtzfSqMNchhD_PezhwHxbQkP6w1NM24&

Lawler, Andrew. *Why Did the Chicken Cross the World?* NY: Atria, 2014.

Lawson-Tancred, Jo. "An Amateur Archaeologist Has Deciphered a Cave Art Code Used by Ice Age Hunter-Gatherers." ArtNet. January 6, 2023. https://news.artnet.com/art-world/an-amateur-archaeologist-has-deciphered-a-cave-art-code-used-by-ice-age-hunter-gatherers-2239199

Lay, Alejandra Andrea Roman, Ana Pereira, & Maria Luisa Garmendia Miguel. "Association Between Obesity with Pattern and Length of Menstrual Cycle: The Role of Metabolic and Hormonal Markers." *European Journal of Obstetrics, Gynecology and Reproductive Biology*. Volume 260, May 2021. Pp. 225-231.

Leick, Gwendolyn. *Sex & Eroticism in Mesopotamian Literature*. London: Routledge, 1994.

Lerner, Adrienne Wilmoth. "The Palace at Knossos: The Archaeological Discovery of Minoan Civilization." Encyclopedia.com. N.D.

https://www.encyclopedia.com/science/encyclopedias-almanacs-transcripts-and-maps/palace-knossos-archaeological-discovery-minoan-civilization

Lerner, Gerda. *The Creation of Patriarchy*. Oxford: Oxford UP, 1986.

LesleytheBirdNerd. "7 Fascinating & Unusual Woodpeckers of North America." YouTube. June 18, 2022. https://www.youtube.com/watch?v=EibBVozFyC8

-----. "Black-capped Chickadee Nest Box." YouTube. August 1, 2021. https://www.youtube.com/watch?v=CgO_zDGgTjA

Lindblom, Michael & Gunnel Ekroth. "Heroes, Ancestors or Just Any Old Bones? Contextualizing the Consecration of Human Remains from the Mycenaean Shaft Graves at Lerna in the Argolid." in *Metaphysis: Ritual, Myth, and Symbolism in the Aegean Bronze Age*. Eds. Eva Alram-Stern, Fritz Blakolmer, Sigrid Deger-Jokotzy, Robert Laffineur, & Jorg Weilhartner. Peeters Leuven-Liege, 2016. pp. 235-43.

Linglin, Marie, Romain Amiot, Pascale Richardin, Stéphanie Porcier, Ingrid Antheaume, Didier Berthet, Vincent Grossi, François Fourel, Jean-Pierre Flandrois, Antoine Louchart, Jeremy E. Martin, & Christophe Lécuyer. "Isotopic Systematics Point to Wild Origin of Mummified Birds in Ancient Egypt." *Scientific Reports*. 10: 15463 (2020). n.p. https://doi.org/10.1038/s41598-020-72326-7 .

Lull, Vincente, Cristina Rihuete-Herrada, Roberto Risch, Bárbara Bonora, Eva Celdrán-Beltrán, Maria Inés Fegeiro, Claudia Molero, Adrià Moreno, Camila Olieart, and Carlos Velasco-Felipe. "Emblems and Spaces of Power During the Argaric Bronze Age at La Almoloya, Murcia." Antiquity. 95.380 (April 2021): 329-348. https://www.cambridge.org/core/journals/antiquity/article/emblems-and-spaces-of-power-during-the-argaric-bronze-age-at-la-almoloya-murcia/B27A3C7AD23625DD39C6D4F2C3981C2F

Luna, Jennie & Martha Galeana. "Remembering Coyolxāuhqui as a Birthing Text." In *Regeneracion Lacuilolli*. https://escholarship.org/uc/item/2dt752tn . 7-32

Lundstrom, Peter. 2011-2020. *Abydos King List*. Accessed August 20, 2020.

Lupack, Susan. "Minoan Religion." In *The Oxford Handbook of the Bronze Age Aegean*. Eric Cline, ed. Oxford: Oxford UP, 2010. 251-62.

-----. "Mycenaean Religion." in *The Oxford Handbook of the Bronze Age Aegean*. Ed. Eric Cline. Oxford: Oxford UP, 2010. pp. 263-76.

Luwian Studies. "Who Are the Luwians?" 2019. https://luwianstudies.org/who-are-the-luwians/

Luyster, Robert. "Symbolic Elements of the Cult of Athena." *History of Religions*. Summer 1965. Vol. 5.1. pp. 133-63.

MacGillivray, J. Alexander. "The Minoan Double Axe Goddess and Her Astral Realm." In *Athanasia: The Earthly, the Celestial and the Underworld in the Mediterranean from the Late Bronze and the Early Iron Age.*. Eds. Nicholas Chr. Stampolidis & Angeliki Giannikouri. Rhodes: International Archaeological Conference of 2009, 2012. 117-28.

Mark, Joshua J. "Enheduanna." *Ancient History Encyclopedia*. March 24, 2014. Https://www.ancient.eu/Enheduanna/.

Madore, Jonathon David. "When Does a Fig Tree Produce Fruit? (4 Things to Know)." Green Upside. 2022. https://greenupside.com/when-does-a-fig-tree-produce-fruit/

Maeir, Aren. "New Light on the Biblical Philistines: Recent Study on the Frenemies of Ancient Israel." Oriental Institute. May 9, 2014. https://www.youtube.com/watch?v=IAZPJRtdjmk&t=3391s

Maestri, Nicoletta. "Tlaltecuhtli—The Monstrous Aztec Goddess of the Earth." July 3, 2019. https://www.thoughtco.com/tlaltecuhtli-the-monstrous-aztec-goddess-169344

MacGillivray, J. Alexander. "The Minoan Double Axe Goddess and Her Astral Realm." In *Athanasia: The Earthly, the Celestial and the Underworld in the Mediterranean from the Late Bronze and the Early Iron Age.* . Eds. Nicholas Chr. Stampolidis & Angeliki Giannikouri. Rhodes: International Archaeological Conference of 2009, 2012. 117-28.

MacQueen, J. G. "Hattian Mythology and Hittite Monarchy." *Anatolian Studies*, vol. 9, [British Institute at Ankara, Cambridge University Press], 1959, pp. 171–88, https://doi.org/10.2307/3642338.

Makkay, J. "The Origins of the 'Temple Economy' as Seen in the Light of Prehistoric Evidence." *Iraq.* Vol. 45, No. 1. (Spring 1983). Pp. 1-6.

Mark, Joshua J. "The Five Gifts of Hathor: Gratitude in Ancient Egypt." *World History Encyclopedia.* May 6, 2020. https://www.worldhistory.org/article/58/the-five-gifts-of-hathor-gratitude-in-ancient-egyp/

-----. "God's Wife of Amun." *World History Encyclopedia*. September 15, 2016. https://www.worldhistory.org/Ma%27at/. December 28, 2021.

-----. "Ma'at." *Ancient History Encyclopedia*. March 24. Accessed September 10, 2020. https://www.ancient.eu/Enheduanna/.

Martin, Simon. "The Old Man of the Maya Universe: A Unitary Dimension to Ancient Maya Religion." *Maya Archaeology 3*. 2015. Pp. 186-227. San Francisco: Precolumbia Mesoweb Press.

Marvell, Andrew. "To His Coy Mistress." Poetry Foundation. https://www.poetryfoundation.org/poems/44688/to-his-coy-mistress

Mayo Clinic. "Vaginal Agenesis." https://www.mayoclinic.org/diseases-conditions/vaginal-agenesis/symptoms-causes/syc-20355737 .

McCartney, Eugene S. "Spontaneous Generation and Kindred Notions in Antiquity." https://penelope.uchicago.edu/Thayer/E/Journals/TAPA/51/Spontaneous_Generation*.html#note10

McGivney, Beatrice A., Haige Han, Leanne R. Corduff, Lisa M. Katz, Teruaki Tozaki, David E. MacHugh & Emmeline W. Hill. "Genomic Inbreeding Trends, Influential Sire Lines and Selection in the Global Thoroughbred Horse Population." *Scientific Reports* 10, 466 (2020). https://www.nature.com/articles/s41598-019-57389-5#citeas

Meghji, Shafik. "The Innovative Technology that Powered the Inca." BBC Travel. December 13, 2021. https://www.bbc.com/travel/article/20211212-the-innovative-technology-that-powered-the-inca

Merlan, Francesca. "Australian Aboriginal Conception Beliefs Revisited." *Man, New Series*. Vol. 21, No. 3 (Sept 1986). 474-93.

Midant-Reynes, Béatrix. *The Prehistory of Egypt: From the First Egyptians to the First Pharaohs*. Trans. Ian Shaw. Oxford: Blackwell Publishing, 2000.

Milbrath, Susan. "The Moon in Meso-America." November 19, 2020. https://oxfordre.com/planetaryscience/view/10.1093/acrefore/9780190647926.001.0001/acrefore-9780190647926-e-200

Miller, Madelaine. "The Boat—A Sacred Border Crosser in Between Land and the Sea." In *Metaphysis: Ritual, Myth & Symbolism in the Aegean Bronze Age*. Eds. Eva Alram-Stern, Fritz Blakolmer, Sigrid Deger-Jolkotzy, Robert Laffineur, and Jorg Weilhartner. Peeters Leuven-Liege, 2016. 543-6.

Miller, Mary & Karl Taube. *The Gods and Symbols of Ancient Mexico and the Maya: An Illustrated Dictionary of Mesoamerican Religion*. London: Thames and Hudson, 1993.

"Men More Obsessed with Penis Size Than Women." *Hindustan Times*. August 22, 2013. https://www.hindustantimes.com/sex-and-relationships/men-more-obsessed-with-penis-size-than-women/story-S4NyOdWOvkFtSiZublg7JP.html

Midant-Reynes, Béatrix. *The Prehistory of Egypt: From the First Egyptians to the First Pharaohs*. Trans. Ian Shaw. Oxford: Blackwell Publishing, 2000.

Mirecki, Paul & Marvin Meyer, eds. *Magic and Ritual in the Ancient World*. Leiden: Brill, 2002.

Mõttus, Siim. "The Edict of Telepinu and Hittite Royal Succession." Institute of History and Archaeology. Tartu: University of Tartu Master's Thesis, 2018.

Mursell, Ian. "Mummy Bundle Masks." Mexicolore. https://www.mexicolore.co.uk/aztecs/aztec-life/mummy-bundle-masks

New Fire Ceremony. Aztecs at Mexicolore. N.D. https://www.mexicolore.co.uk/aztecs/stories/new-fire-ceremony

Nash, Stephen E. "Skeleton Sex Pots." *Sapiens*. March 31, 2016. https://www.sapiens.org/column/curiosities/moche-skeleton-sex-pots/

Needham, Stuart, Andrew J. Lawson, and Ann Woodward. "'A Noble Group of Barrows': Bush Barrow and the Normanton Down Early Bronze Age Cemetery Two Centuries On." *The Antiquarian Journal*. 90. 2010. Pp. 1-39.

Neumann, Erich. *The Great Mother*. Trans. Ralph Manheim. NY: Bollingen Foundation, 1955. Reprint 1963.

New York University. "After 5,000 Year Voyage, World's Oldest Built Boats Deliver—Archaeologist's First Look Confirms Existence of

Earliest Royal boats at Abydos." *ScienceDaily*. https://www.sciencedaily.com/releases/2000/11/001101065713.htm

Nilsson, Martin Persson. *The Mycenaean Origin of Greek Mythology*. 1932. Reprint by Forgotten Books, 2007.

Nowak, Nick. "Josiah's Lost 'Book of the Law': Deuteronomy?" March 5, 2012. The Strange Truth of the Lamb. https://strangetriumph.wordpress.com/2012/03/05/josiahs-lost-book-of-the-law-deuteronomy/

Nowak, Troy Joseph. *Archaeological Evidence for Ship Eyes: An Analysis of Their Form and Function*. Texas A&M University Masters Thesis. May 2006.

O'Brien, Stephen. "The Development of Warfare and Society in 'Mycenaean' Greece." In *Warfare and Society in the Ancient Eastern Mediterranean*. Eds. Stephen O'Brien and Daniel Boatright. BAR International Series 2583, 2013. Archaeopress.

Occipital Horn Syndrome. Genetic and Rare Diseases Information Center. https://rarediseases.info.nih.gov/diseases/4017/occipital-horn-syndrome

Ogden, Lesley Evans. "Female Dolphins Have a Clitoris Much Like Humans'." *Science News*. January 10, 2022. https://www.sciencenews.org/article/dolphin-female-clitoris-sexual-anatomy

Ogilvie, Edward. "8 Mind Blowing Facts to Know About the Morpho Butterfly." Voyagers. March 21, 2019. https://amazoncruise.net/8-mind-blowing-facts-to-know-about-the-morpho-butterfly/

Olsen, Debbie. "Iniskim Umaapi: Is This Canada's 'Stonehenge'?" BBC Travel. January 10, 2022. https://www.bbc.com/travel/article/20220109-iniskim-umaapi-is-this-canadas-stonehenge

Oriental Institute. "Robert Ritner: A Game of Thrones and Coffins: The Death and Resurrection of Osiris." February 11, 2015. YouTube. https://www.youtube.com/watch?v=AG3Or8SMYho

Ostergard, Robert. "Are Female Leaders Better During a Pandemic?" *Nevada Today*. January 19, 2021. https://www.unr.edu/nevada-today/news/2021/atp-female-leaders-better

Palaima, Thomas G. "The Metaphysical Mind in Mycenaean Times and in Homer." In *Metaphysis: Ritual, Myth, and Symbolism in the Aegean Bronze Age*. Eds. Eva Alram-Stern, Fritz Blakolmer, Sigrid Deger-Jolkotzy, Robert Laffineur, and Jorg Weilhartner. Peeters Leuven-Liege, 2016. 479-84.

-----. "Mycenaean Religion." In *The Cambridge Companion to the Aegean Bronze Age*. Ed. Cynthia W. Shelmerdine. Cambridge: Cambridge UP, 2008. 342-61.

-----. "Wanaks and Related Power terms in Mycenaean and Later Greek." In *Ancient Greece: From the Mycenaean Palaces to the Age of Homer*. Eds. Sigrid Deger-Jalkotzy and Irene S. Lemos. pp. 53-72. Edinburgh UP, 2006.

Pandey, Sahir. "Oldest Narrative Scene: A Man, Holding His Penis, and Fighting Leopards!" Ancient Origins. December 8, 2022. https://www.ancient-origins.net/news-history-archaeology/oldest-narrative-scene-0017632?fbclid=IwAR0Khzimu5omAZozSJpV_rb0b23tpsGshTn4c7ThOYx2jY9f8N5dNGDUa8U

Pattanaik, Devdutt. "Do Hindus Worship the Phallus?" *DailyO*. August 15, 2016. https://www.dailyo.in/lifestyle/devdutt-pattanaik-phallus-shiva-hinduism-mythology-gods-fertility/story/1/12380.html

Pavid, Katie. "Beauty of the Dual-Gender Butterfly." Natural History Museum. 2021. https://www.nhm.ac.uk/discover/beauty-dual-gender-butterfly.html Penn Museum. "Fieldwork at South Abydos, Egypt." YouTube. January 17, 2014. https://www.youtube.com/watch?v=cSswHM354HY .

"The Peaks and Troughs of Hittite." May 2, 2006. University of Leiden. https://web.archive.org/web/20170203061604/http://www.leidenuniv.nl/en/researcharchive/index.php3-c=178.html

Pence-Brown, Amy. "Dress, Gender and the Menstrual Culture of Ancient Greece." July 23, 2020. http://www.mum.org/greekmen.thm.

Pinch, Geraldine. *Egyptian Mythology*. Oxford: Oxford UP, 2002.

Pongratz-Leisten, Beate. "Sacred Marriage and the Transfer of Divine Knowledge: Alliances Between the Gods and the King in Ancient Mesopotamia." In Sacred Marriages. Eds. Martin Nissinen and Risto Uro. Winona Lake, IN: Eisenbrauns, 2008. 43-73.

Prindiville, Mary & David C. Grove. "The Settlement and Its Architecture." In Ancient Chalcatzingo. Ed. David C. Grove. Austin, TX: University of Texas Press, 1987. Pp. 63-81.

Perkins-Gilman, Charlotte. *The Man-Made World*. Published in 1911. Reprint: Mint Editions, 2020.

Pinch, Geraldine. *Egyptian Mythology*. Oxford: Oxford UP, 2002.

Pitt, Daniel, Natalia Sevane, Ezequiel L. Nicolazzi, David E. MacHugh, Stephen D. E. Park, Licia Colli, Rodrigo Martinez, Michael W. Bruford, Pablo Orozco-Wengel. "Domestication of Cattle: Two or Three Events?" *Evolutionary Applications*. Special Issue: Genomics of Domestication. 12.1 (January 2019): 123-136. https://onlinelibrary.wiley.com/doi/full/10.1111/eva.12674

Popova, Maria. "How Kepler Invented Science Fiction and Defended His Mother in a Witchcraft Trial While Revolutionizing Our Understanding of the Universe. *The Marginalian*. 2022. https://www.themarginalian.org/2019/12/26/katharina-kepler-witchcraft-dream/?fbclid=IwAR1-bidRmgoy85MojUTDaxME_RSa277SbGmEIVxKvSfPWTeiQGUmqoHez4Q

Powers, Marla N. *Oglala Women: Myth, Ritual, and Reality*. Chicago: U of Chicago P, 1986.

Project, Theoi. "Ate." July 26, 2020. Https://www.theoi.com/Daimon/Ate.html.

-----. "Leto." July 26, 2020. https://www.theoi.com/Titan/TitanisLeto.html.

Pruszewicz, Marek. "The WWI Poet Kids Are Taught to Dislike." BBC News. May 1, 2015. bbc.com/news/magazine-32298697.

Puglisi, Dario. "Rites of Passage in Minoan Palatial Crete and Their Role in Structuring a House Society." In *Oikos: Archaeological Approaches to House Societies in Aegean Prehistory*. Eds. Maria Relaki & Jan Driessen. Paris: Presses Universitaires de Louvain, 2020. 63-80.

Pull Brouwers, Josho. "Chariots and Horses in the Homeric World." 19 August 2013. *Ancient Warfare Magazine*. https://www.karwansaraypublishers.com/awblog/chariots-and-horses-in-the-homeric-world/. 15 April 2018.

Quah, Jacey. "The Allure of Curses and Magic in Ancient Greece." Hellenic Museum. October 6, 2021. https://www.hellenic.org.au/post/the-allure-of-curses-and-magic-in-ancient-greece

Raevsky, D.S. "Ancient Astronomy as the Mirror of the History of Culture." Astronomical and Astrophysical Transactions. Vol. 15. https://articles.adsabs.harvard.edu//full/1998A%26AT...15..299R/0000303.000.html 299-304.

Reeves, Ed. "Is the Volvo Logo Sexist?" MotorBiscuit. April 8, 2021. https://www.motorbiscuit.com/is-volvo-logo-sexist/

Reusch, Kathryn. *"That Which Was Missing": The Archaeology of Castration*. Dissertation. St. Hugh's College, University of Oxford, 2013.

Rigoglioso, Marguerite. *The Cult of Divine Birth in Ancient Greece*. NY: Palgrave-Macmillan, 2009.

-----. *Virgin Mother Goddesses of Antiquity*. NY: Palgrave-Macmillan, 2010.

"Rock Stars: How a Group of Scientists in South Africa Rescued a Rare 500kg Chunk of Human History." The Conversation. October 17, 2022. https://theconversation.com/rock-stars-how-a-group-of-scientists-in-south-africa-rescued-a-rare-500kg-chunk-of-human-history-192508?fbclid=IwAR36q5uy7BGm-k5UGwVlpV7ZprwmswUfem8O2YDyXLXOZ-ty7UW6_y6upq4

Roos, Dave. "Human Sacrifice: Why the Aztecs Practiced This Gory Ritual." History. October 11, 2018. https://www.history.com/news/aztec-human-sacrifice-religion

Rossel, Stine, Fiona Marshall, Joris Peters, Tom Pilgram, Matthew D. Adams, and David O'Connor. "Domestication of the Donkey: Timing, Processes, and Indicators." Proceedings of the National Academy of Sciences of the United States of America (PNAS). March 11, 2008. 105 (10): 3715-20. https://www.pnas.org/content/105/10/3715

Ruether, Rosemary Radford. *Goddesses and the Divine Feminine*. Berkeley, CA: University of California P, 2005.

Russell, Sean. "What Kind of Tree Has Red Sap?" Garden Guides. July 21, 2017. https://www.gardenguides.com/13429173-what-kind-of-tree-has-red-sap.html.

Sanders, N.K., Translator. *The Epic of Gilgamesh*. Assyrian International News Agency. www.aina.org.eog.pdf. September 23, 2020.

Santos-Longhurst, Adrienne. "If You Have Questions About Your Penis Size, Read This." Healthline. January 27, 2022. https://www.healthline.com/health/mens-health/average-penis-size .

Scammell, J. M. "The Capture of Troy by Heracles." *The Classical Journal*, vol. 29, no. 6,

1934, pp. 418–428. *JSTOR*, www.jstor.org/stable/3289866. Accessed 4 Dec. 2020.

Schotsmans, Eline MJ, G. Busacca, Sarn C Lin, Milena Vasić, Ashley M Lingle, Rena Veropoulidou, Camilla Mazzucato, Belinda Tibbetts, Scott D Haddow, Mehmet Somel, F. Toksoy-Köksal, Christopher J Knüsel, & Marco Milella. "New Insights on Commemoration of the Dead Through Mortuary and Architecutral Use of Pigments at Neolithic Chatalhoyuk, Turkey." *Scientific Reports*. March 2022. 12(1): 4055.

Schroeder, Susan. *Tlacaelel Remembered: Mastermind of the Aztec Empire*. Norman, OK: U of Oklahoma P, 2016.

SciShow. "Science Proves There Are More Than Two Sexes." November 13, 2019. https://www.youtube.com/watch?v=kToHJkr1jj4

Scribe, Sesh Kemet Egyptian. 2013. *Kemetic Dictionary*. Accessed August 21, 2020.

Sears, Matthew A. "Toxic Masculinity Fostered by Misreading of the Classics." *The Conversation*. November 28, 2017. https://theconversation.com/toxic-masculinity-fostered-by-misreadings-of-the-classics-88118

Seawright, Caroline. "Min, God of Fertility, Power and the Eastern Desert." Tour Egypt. August 2, 2011. http://www.touregypt.net/featurestories/min.htm. Accessed February 10, 2021.

"A Seeress from Fyrkat?" National Museum of Denmark. https://en.natmus.dk/historical-knowledge/denmark/prehistoric-period-until-1050-ad/the-viking-age/religion-magic-death-and-rituals/a-seeress-from-fyrkat/

Serrano Laguna, Irene. "Di-U-Ja." *Metaphysis: Ritual, Myth, and Symbolism in the Aegean Bronze Age*. Eds. Eva Fritz Blakolmer, Sigrid Deger-Jolkotzy, Robert Laffineur, & Jorg Weilhartner. Alram-Stern/Peeters Lueven-Liege, 2016. 285-91.

Shady, Ruth M. "Living Conditions, Social System and Cultural Expressions of the Caral and Chinchorro Populations During the Archaic Period. In *The Chinchorro Culture: A Comparative Perspective, the Archaeology of the Earliest Human Mummification*. Sanz, Nuria, Bernardo T, Arriaza, , Vivien G. Standen, eds. UNESCO, 2014. 71-106. https://unesdoc.unesco.org/ark:/48223/pf0000227775

Shokeir, A.A. and M.I. Hussein. "Sexual Life in Pharaonic Egypt: Towards a Urological View." *International Journal of Impotence Research*. (2004) 16, 385-388.

Silver, Carly. "Romans Used to Ward off Sickness with Flying Penis Amulets." Atlas Obscura. December 28, 2016.

https://www.atlasobscura.com/articles/romans-used-to-ward-off-sickness-with-flying-penis-amulets

Simek, Rudolf. *Dictionary of Northern Mythology*. Boydell & Brewer, 1996.

Smarthistory. "Standard of Ur from the Royal Tombs at Ur." June 2, 2012. https://www.youtube.com/watch?v=Nok4cBtoV6w.

Smith, Jean. "Tapu Removal in Māori Religion." *The Journal of the Polynesian Society*. Memoir No. 40, 1974. http://www.jps.auckland.ac.nz/document//Volume_84_1975/Memoir_No._40%3A_Tapu_removal_in_Maori_religion%2C_by_Jean_Smith%2C_p_43-96/p1

Smith, K. Annabelle. "When Lettuce Was a Sacred Sex Symbol." *SmithsonianMagazine*. July 16, 2013. https://www.smithsonianmag.com/arts-culture/when-lettuce-was-a-sacred-sex-symbol-12271795/ Accessed February 10, 2021.

Smith, Michael. "The Role of Social Stratification in the Aztec Empire: A View from the Provinces." *American Anthropologist*. Vol. 88, No. 1 (March 1986): 70-91.

-----. "Life in the Provinces of the Aztec Empire." *Scientific American*. Vol. 277, No. 3 (September 1997), pp. 76-83.

Smith, K. Annabelle. "When Lettuce Was a Sacred Sex Symbol." *SmithsonianMagazine*. July 16, 2013. https://www.smithsonianmag.com/arts-culture/when-lettuce-was-a-sacred-sex-symbol-12271795/ Accessed February 10, 2021.

Snell, Daniel C. *Life in the Ancient Near East, 3100-332 BCE*. New Haven: Yale UP, 1997.

Sokol, Joshua. "The Stargazers." *Science*. June 2, 2022. https://www.science.org/content/article/what-did-ancient-maya-see-in-stars-their-descendants-team-with-scientists-find-out?fbclid=IwAR1pQSLmFmLlJKGLeH_6G1hwb7cKBAud8GfvRk6uHEj2kksuxWOI1-cwEs4

Soles, Jeffrey S. "Hero, Goddess, Priestess: New Evidence for Minoan Religion and Social Organization." Metaphysis: Ritual, Myth, and Symbolism in the Aegean Bronze Age. Eds. Fritz Blakolmer, Sigrid Deger-Jolkotzy, Robert Laffineur, Jorg Weilhartner, & Eva Alram-Stern. Peeters Leuven-Liege, 2016. 247-53.

Soustelle, Jacques. *Daily Life of the Aztecs*. Trans. Patrick O'Brian. Weidenfeld & Nicholson, 1961.

Standing Bear, Luther. *My People the Sioux*. 1928.

Stephany, Timothy J. *The Gilgamesh Cycle*. CreateSpace, 2013. Print.

Stiebing, William H. Jr. *Ancient Near Eastern History and Culture*, 2nd ed. New York: Pearson : Longman, 2009.

Stevanović, Lada. "Human or Superhuman: the Concept of Hero in Ancient Greek Religion and/in Politics." *Bulletin of the Institute of Ethnography*. SASA LVI (2). pp. 7-20.

Stewart, James. "Timeline: Galas, Sumerian Musicians." Vermont Public Radio. 2020. https://www.vpr.org/ost/timeline-galas-sumerian-musicians#stream/0 .

Stiebing, Jr. William H. *Ancient Near Eastern History and Culture*, 2nd Ed. NY: Pearson Longman, 2009.

Stone, Merlin. *When God Was a Woman*. San Diego, CA: Harcourt Brace Jovanovich, 1976.

Strolonga, Polyxeni. "Aphrodite Against Athena, Artemis, and Hestia: A Contest of Erga." Illinois Classical Studies. 37. 2012. Gale Academic OneFile. 1-20.

Stuckey, Johanna. "Inanna and the 'Sacred Marriage.'" MatriFocus: Cross-Quarterly for the Goddess Woman. 2005. 4.2. www.matrifocus.com/IMB05/spotlight.htm.

"Swans in Ancient Egypt." At the Mummies Ball. March 8, 2018. https://www.atthemummiesball.com/swans-ancient-egypt/. January 17, 2020.

Sykes, Bryan. *The Seven Daughters of Eve: The Science That Reveals Our Genetic Ancestry*. NY: Norton, 2002.

Szarzynska, Krystyna. "Archaic Sumerian Standards." *Journal of Cuneiform Studies*. 48. 1996. 1-15.

-----. "Offerings for the Goddess Inana in Archaic Uruk." *Revue D'Assyriologie et D'Archeologie Orientale*. 87.1. 1993. 7-28.

Tanner, Nancy Makepeace. *On Becoming Human*. Cambridge: U of Cambridge P, 1981.

Taylor, Gary. *Castration: An Abbreviated History of Western Manhood*. Psychology Press, 2000.

Tedlock, Dennis, Trans. *Popol Vuh*, Rev. New York: Simon & Schuster, 1996.

Templo Mayor Museum. https://templomayor.inah.gob.mx/english Flier Describing Contents of Rooms 1-8. Image of Coyolxāuhqui Stone. https://www.templomayor.inah.gob.mx/images/avisos/2018/cudriptic oingles1.jpg

Tena-Colunga, Arturo. "The Meaning of the Word Mexico." December 1987. University of Illinois at Urbana-Champaign. themeaningofthewordmexico-vip.pdf.

"Textiles Made from Agave Fibres." Fibre2Fashion. April 2014. https://www.fibre2fashion.com/industry-article/7302/textiles-made-from-agave-fibres.

Thomson, A.M., F.E. Hytten, & A.E. Black. "Lactation and Reproduction." *Bull World Health Organization*.52.3 (1975): 227-349.

Tomkins, Peter. "Neolithic Antecedents." In *The Oxford Handbook of the Bronze Age Aegean*. Ed. Eric Cline. Oxford: Oxford UP, 2010. 31-49.

Tripp, Edward. *Classical Mythology*. New York: Penguin, 1974.

Uckelmann, Marion. "Land Transport in the Bronze Age." In *The Oxford Handbook of the European Bronze Age*. Eds. Harry Fokkens & Anthony Harding. Oxford UP, 2013. 398-413.

Ulfr23. "Making Copper the Ancient Way." YouTube. November 21, 2016. https://www.youtube.com/watch?v=OO747eWvGME.

Umberger, Emily. "Events Commemorated by Date Plaques at the Templo Mayor: Further Thoughts on the Solar Metaphor." *In The Aztec Templo Mayor: A Symposium at Dumbarton Oaks*. Ed. Elizabeth Hill Boone. 411-449. 1983.

University of Pennsylvania Museum. "Lilis [Instrument]." http://psd.museum.upenn.edu/epsd/e3338.html

Van Wees, Hans. "From Kings to Demigods: Epic Heroes and Social Change c. 750-600 BC." in *Ancient Greece: From the Mycenaean Palaces to the Age of Homer*. Eds. Sigrid Degev-Jalkotzy and Irene S. Lemos. Edinburgh UP, 2006. pp. 363-79.

-----. "Heroes, Knights and Knutters: Warrior Mentality in Homer." In *Battle in Antiquity*. Ed. Alan B. Lloyd. The Classical Press of Wales, 1996. Reprint 2009. Pp. 1-86.

"The Venus of Laussel." *World History Encyclopedia*. July 7, 2017. https://www.worldhistory.org/image/6866/the-venus-of-laussel/. September 5, 2021.

"Vulva." *Online Etymology Dictionary*. 2001-2021. https://www.etymonline.com/word/vulva

Wagner, Helmuth O. "Food and Feeding Habits of Mexican Hummingbirds." The Wilson Bulletin: A Quarterly Magazine of Ornithology. June 1946, Vol. 58, Number 2. Pages 69-132.

Watson, Sara Kiley. "Egypt Is Reclaiming Its Mummies and Its Past." *Popular Science*. December 3, 2019. https://www.popsci.com/story/science/egyptian-archaeology/

Weil, Simone. *The Iliad or the Poem of Force*. New York: Peter Lang Publishing, 2008.

Wengrow, David. *The Archaeology of Early Egypt: Social Transformation in North-East Africa, 10,000 to 2650 BC*. Cambridge: Cambridge UP, 2006.

-----. "The Voyages of Europa: Ritual and Trade in the Eastern Mediterranean circa 2300-1850 BC." In Archaic State Interaction: the Eastern Mediterranean in the Bronze Age. Eds. William A. Parkinson & Michael L. Galaty. Santa Fe: School for Advanced Research Press, 2009. Pp. 141-60.

Wessles, Leana. "An Analysis of the Extent to Which the Trickster Archetype Can Be Applied to the Goddess Inanna/Ishtar." *Semantics Scholar*. d8f718ee1df9483f99a73bf9d71aa820ea0d.pdf. 35-55.

West, Alan. "A Burial of Gold." Norwich Castle Museum and Art Gallery. July 30, 2020. https://norwichcastle.wordpress.com/2020/07/30/a-burial-of-gold-little-cressingham/

Weston, Phoebe. "Dozens of Black Granite Statues of a Fierce Lion-headed Goddess Known as the Lady of War Are Discovered in the Ancient Egyptian City of Luxor." Daily Mail. December 5, 2017.

https://www.dailymail.co.uk/sciencetech/article-5147121/Granite-statues-fierce-lion-goddess-discovered-Egypt.html

Whelan, Ed. "Proof Infants Sipped Animal Milk from Prehistoric Baby Bottles." Ancient Origins. September 26, 2019. https://www.ancient-origins.net/news-history-archaeology/baby-bottles-0012631?fbclid=IwAR058JK6KMg2KJ-SmnqwBJq29K7CYGP_gAOAu2s1FZ1lzMCy31S5v9lxRgU

Whitman, Cedric H. *Homer and the Heroic Tradition*. Cambridge: Harvard UP, 1958.

Whittaker, Helene. "Horns and Axes." In *Metaphysis: Ritual, Myth, and Symbolism in the Aegean Bronze Age*. Eds. Eva Alram-Stern, Fritz Blakolmer, Sigrid Deger-Jolkotzy, Robert Laffineur, and Jorg Weilhartner. Peeters Leuven-Liege, 2016. 109-114.

"Who Are the Luwians?" Luwian Studies. 2019. https://luwianstudies.org/who-are-the-luwians/

Wilkinson, Toby. *Genesis of the Pharaohs: Dramatic New Discoveries Rewrite the Origins of Ancient Egypt*. London: Thames & Hudson, 2003.

Wills, Matthew. "The Evolution of the Microscope." *JSTOR Daily*. March 27, 2018. https://daily.jstor.org/the-evolution-of-the-microscope/

Withrow, Brandon. "The US' 2000-year-old Mystery Mounds." BBC: Rediscovering America. December 5, 2022. https://www.bbc.com/travel/article/20221204-the-us-2000-year-old-mystery-mounds

Wolkstein, Diane & Samuel Noah Kramer. "Inanna's Descent." Deagona. N.D. descentofInanna.pdf

-----. *Inanna: Queen of Heaven and Earth, Her Stories and Hymns from Sumer*. New York: Harper & Row, 1983.

Woods, Shawn. "How to Make an Otzi the Iceman Copper Axe Blade. Ancient Bushcraft Survival Skills." YouTube. January 15, 2016. https://www.youtube.com/watch?v=aoNU4dEtKWE.

Wrangham, Richard. The Goodness Paradox: The Strange Relationship Between Virtue and Violence in Human Evolution. NY: Pantheon, 2019.

Wright, Rita P. "Sumerian and Akkadian Industries: Crafting Textiles." In *Sumerian World*, Ed. H.E.W. Crawford, Routledge Press, 2013. 395-417.

Xocoyotzin, Moctezuma. "Mythological Journey to the Aztec Underworld." WilderUtopia. October 10, 2018. https://www.wilderutopia.com/traditions/mythological-journey-to-the-aztec-lands-of-the-dead/

Younger, John G. & Paul Rehak. "The Material Culture of Neopalatial Crete." In *The Cambridge Companion to the Aegean Bronze Age*. Cynthia Shelmerdine, ed. Cambridge: UP, 2008. Pp. 140-64.

Zenger, Jack & Joseph Folkman. "Research: Women Are Better Leaders During a Crisis." *Harvard Business Review*. December 30, 2020. https://hbr.org/2020/12/research-women-are-better-leaders-during-a-crisis

Zimmerman, Kim Ann. "Aries Constellation: Facts About the Ram." Space.com. March 4, 2019. https://www.space.com/17052-aries-constellation.html

"Zion." Britannica. https://www.britannica.com/place/Zion-hill-Jerusalem

Index

2400 BCE 2, n. 22, 30, 51, 62, 70, 75-6, 79-80, 103, 105, 109, 111, 121, 128, 130, 151, 176, 184, 188-191, 198-9, 202, 207, 210, 218, 220-21, 225, 237, 291-2, 297, 300, 312, 320, 326, 336-7, 340, 344-5, 365, 369, 390, 392, 405, 409, 411, 453

10,000 Years Ago 5, 64, 68, 148, 152-3, 161, 173, 284, 369

30,000 Before Present 42, n. 192

Abiogenesis (see **parthenogenesis**) 53

Abrahamic religion(s) n. 61, n. 64, 97, 100, 107, 116, 119, 121, 125, 179-80, 202, n. 203, 208, 236, 266, 447, 452

Achilles 156-7, 367, n. 368, 371-2, n. 373, 376-88, 397-401, 414-15, 418-20, 422-7, 429, 435, 438-41

Achitometl 243

Administration 47

Aeneid n. 336, 434

Africa 8, 25, 38, 129, 148, 156, 159, 169, 207, 220, 248, 358, 412

African lions 25, 41-2, 148, 150, 158

Afterlife 8, 27, n. 77, 88, 90-92, n. 149, 184, 200, n. 204, 260, 272, 278, 286, 304, 387

After the Common Era (ACE) 14, 17, n. 18, n. 19, 26, 51, 53, 69, 76, 112, 129, 161, 180, 188, 199, n. 201, 211, 242, 257, n. 271, 350, 374, n. 390

Afterworld (see next world, afterlife) 82

Agave (also **maguey**) n. 270, 272-3

Ahuitzotl 267

Akan (Ghana) 102

Akkadian(s) n. 69, n. 77, 78, 84-5, n. 85, 136, 169, 180, n. 182, 199, n. 205, 207, 210, n. 210, 232, 234, 244-5, 260, n. 264, 282, 302, n. 318, 319, n. 320, n. 345, n. 349, 350, n. 391, 451

Alarm systems 69

Alcmene 433-4

Alligators 45, 252, 282-3

Alluvial (see also flood) 88

Amazon(s) 10, 24, 284, 424

Ama(r)-gi (Sumerian "return to mother") 34

American bison 25, 41, 52, n. 57, n. 103, 148, n. 370

Amphorae 130, n. 308, 314

Anal sex 72-3, 213

Ancestor(s) 2, 12, n. 18, 26, 35, 51, 53, 58, 61, 64, 68, n. 69, 73-4, 80, 84, 91, n. 92, 101, 104, 113, 115-6, 118-9, 126-7, 160, 163, 165, 168, 178, n. 178, 180-1, 183, 187, 191, 197, 199, 207, 214, 244, 289, 322, 337, 361, n. 368, n. 376, 377, n. 380, 386, n. 388, 395, n. 403, 405-7, 411, 445, n. 445

Ancestor worship 102, 376, n. 376, 380, n. 380, 386, n. 388, 405, 408-10, 413, 453

Ancient Architects 58

Androcentric/ism/ists 2-5, 7-8, 10-14, 16-7, 19-20, n. 20, 22-4, n. 26, 27, 29, 31,-32, 40, 41-4, 47-8, 75, 78, 80-1, 82-5, 89, 94, 97-8, 100-102, 104-5, 108-111, 113, 119, 121, 123, 127-8, 150, 156, 158, 160, n. 168, 173, n. 174, 176, 179-180, 182, 184-5, 189, n. 189, 193-196, 198-200, 207, n. 210, 212, 214, 217, 219, 223-224, 226-229, 230-34, 237, n. 237, 239, 241, 243, n. 244, n. 245, 246, 251-3, 266-7, 269, 288, 290-291, 293-4, n. 294, 297, 301, 306, 308-315, 317-9, n. 320, 324-325, 326-328, 330-31, 333, n. 335, 336, 339, 341-2, 344-345, 349-52, 354-56, 358, 360-363, 365-6, 369, 375-383, 385, n. 387, 388-92, 394-6, 400, 402, 404-405, 406, 408-9, 411-413, 417-419, 422-4, 431. 443-4, 447, 449, 450-3

n. 244, 246, n. 247, 252, 260, n. 261, 264-5, 268, 270, 281, 283-6, 288, 312, 319, n. 321, 335, 352-3, 366, n. 369, 395, 398, 402, n. 403, 410, 435-437, 451, 453
Birth canal 94, 319
Birth control 73, n. 168, n. 189
Birth defects 51, n. 66, 68, 202, n. 244, 286
Blackfoot 95
Blood (including **flesh and blood, blood bonds, blood bath,** and **blood red moon**) 1, 8-9, 12, 30, 32-3, 39, n. 39, n. 55, 56, n. 57, 59, 71, n. 73, 74-5, 78, 85, n. 85, 86, 88, 90, n. 92, 97, n. 101, 111, 113, 117-18, 124, 126, n. 126, 127, 136, n. 150, 159-60, 166, n. 162, 169, 176, n. 176, 183, 189, 197, 204, 233, 244-5, 247, 253, 260-1, n. 261, 263-70, n. 268, 274, 277-8, 282-9, n. 287, 292, 298, n. 299, n. 312, 319-20, n. 320, 327, 334-5, 348, 354, n. 354, 360, n. 369, n. 371, 375, n. 375, 378, 385, 383, n. 388, n. 390, 393, n. 403, 408, 410, 430, 431, 434, 436, 447, 450, 453
Blood guilt n. 354, n. 354
Blood rituals n. 85, 86
Blue Morpho butterfly 284
Bo (China) 93
Boat(s; including Boat Burials) n. 78, 82, 91-3, 95, 129, 132, 193, 205, 209-213, 308-9, n. 309, 310-11, n. 311, 313, 339, 394
Boat of Heaven n. 78, 193, 209-10, 308-9, n. 309, n. 311, 313, 339
Bond (**-ed and -ing,** including **house bond** and **pair bonding**) 6, 27-8, 36, 70, 79, 71, 77, 78, 103, 107, 116, 139, 159-61, 167-8, 169-170, 192, 203, 251, 341, 402
Bonobos 25, n. 79
Book of the Celestial Cow 110-11, 333-5
Book of Genesis 61, 321, n. 322, n. 402
Book of Isaiah 91
Book of Judges 21
Book of Matthew 112
Book of Ruth 21, 81, n. 402, 405
Book of the Dead (Egypt) 91, 183, 341
Breasts (including **Breast Feeding**) 62, n. 71, 117, 122-3, 160, n. 168, 170, n. 170, n. 184, n. 189, n. 202, 257, 262, 266, n. 270, 273, n. 293, n. 300, 328, 388, 402, 428
"Brides by capture" 106,
Bride gift (gynocentric term) 160, 347
Bride price (androcentric term) 160, 346-7
Briseis 231, 378, 386-7, 397, 420, 422
British 46, n. 46, 53, 93, 180, n. 201, 356
British Association for the Advancement of Science 52
Bronze (as weapons, as metal) 1, 131, 134, 136, 194, 324, 348, n. 367, 384, 420, 430, 431-2
Bronze Age 1, 5, 7, 13, 21, 28, 32, n. 37, n. 38, 42, 54, n. 55, 68, 103, 105-6, n. 105, 120, 122, 129-30, 132, 134, 177, 119, 126-7, 129, 131, 133-4, 136-7, 199, 204, 211-4, 217, 222, 228-30, 232, 239, 296-7, 318, 319, 362, n. 362, 381, 450-4
Buck (rabbit) 63
Bucrania v, 175, 178, 364, n. 364
Bull(s) 38, n. 149, n. 154, 156, 160, n. 161, 162, 169, 172-178, 195, 219, 234, n. 292, 303, n. 303, n. 304, 321-2, n. 333, 336, 348, 353, 356, 360, 364, n. 364, 384, 437, n. 437
Bull of Heaven 303, n. 303, 321
Bûr-Sin Palace 30
Bush Barrow 94
Cafer Höyük 160

242, 260, n. 264, 263-4, 267, 274, 281-2, 285, 296, 302, n. 311, 314, 319, n. 322, 332, 339, 342, 410-411, 447, 451

Creation of Patriarchy 13, 115, 447

Crete/Cretan(s) (aka Minoan) 46-48, n. 46, n. 55, 110, 119-20, 160, 172, 184, 198, 205, 220, 242, 259, 266, 318, 356-65, n. 356

Croatia 130

Crocodiles 31, 74, 252, n. 252, 285

Cronus 102, 246

Cross-dressing 29, n. 55, 258, 301

Cuauhnahuac 267

Culhuacan(s) 238-9, 243

Cult of the Hero 31, 238, 243, n. 367, 372, n. 373, 378, 381, 383-389, 403, 406, 414-16

Cuneiform n. 77, 84, 228, 228, 232, n. 245, n. 292, n. 293, n. 301, 308, 313, 345, 350

Cunnilingus 57, 65, 72, 99, 182, n. 298, n. 302, 307

Cunynghame 53

Custer, Libby Bacon n. 18

Cyrus 36

Daedalus 162

Daimona (Greek term for minor goddesses) 425-426, 434, 436-437

Dairy cattle 25, 42

Danc(e)ing 8, 58, 60, 150, n. 150, 164, n. 165, 195

Dark Earth (Hittite nether realm) 348

Dawn of Everything 33, 327

Death 3, 33, 45, 46, n. 61, 67, n. 73, 79, 88, 90-91, 93, 95-6, 106, 108, 127, n. 148, n. 149, n. 174, 177, 183, n. 183, 184-5, n. 190, 192, 200, 208, 220, 237, 242, 244, 247, n. 250, 260, 265, 270, 272, 276, 278-9, n. 284, n. 286, 288, 290, 301, 304, n. 306, n. 307, n. 317, 322, 324, 335, 340-1, 351, 372-4, n. 374, 375, 378, 382, 384, n. 384, 385, 387, 397, 399, 401, n. 406, 409, 411, 414-5, n. 418, 426, 428, 430, 436, 443, 454

"Debt peons" 35-6

Deconsecrated 61

Dehumanize 45-6

Deity(ies) 28, 52, n. 69, 86-8, 125, 168, 170, 179, 186, 197, 201, n. 204, 206, 242, 244, 252-3, 268, 275, 276, 280, 282, 291, 293, 297, 311, 323, 327, 331, 334-5, 350, 362, 363, 365, 382, 396

Delta (see **triangle**) 78, 169

Democracy(-tic) 11, 41, 234, 239, 249, 352, 381, n. 384, 412, 420

Desacralized 61

Dever, William 13, n. 344, n. 406

Diaz, Bernal del Castillo 280-1

Dichotomy(-ies) 46, 247, 249, 330, n. 343, n. 344

Dildo(s) 79, 99, 360

Diomedes n. 370, 396-7, n. 397, 430-4, 438

Dione 430, 433, 435

Disease(s) 53, n. 55, 67, n. 73, 105-6, 157, 161-162, 183, 202, 263, 360

Dismemberment 45-6, 89, n. 244, 245-7, 256, 260, 262-3, 266, 329

Divine ancestor 90, 368, 386, 415

Divine birth (see also **Virgin birth** and **parthenogenesis**) n. 72, n. 83, 99

Divine forces 55, 112, 203

Divine kingship (see also **god-king**) 111

DNA 7, 27, 67, 75, 94, 106, 120, 161, 204, 218, n. 222, 227, 230, 237, 253, 290, 301, 330, 392, n. 393, 409, n. 433

Doe (rabbit) 63
Dogs 63, 69, 143, 144, 146-7, 149, 153, 156, 158, 161, 173, 258
Dolphin 25
Dolní Věstonice 130
Domestication(-ated) 25, 40-1, 43, 64, 70-1, 73, 103, 108, 110, 146-158, 161, 168, 171, 173, n. 174, 176, 155, 158, 161-162, 164, 171, 174, 188, 214-216, 219-20, 223, 232-234, 250-251, 287, n. 291, 390, 407
Domination 3, 5, 10-11, 26, 33, 34-5, 43, 46, 49, 105, 124, 142, 185, 213, 247, n. 292, 300, 302-3, 312-313, 316, n. 317, 323-324, 326, n. 326, 331, 342, 344, 359, 412-413, 428, 450
Donkeys 153
Dresden Codex n. 271
Dual linear(-ality) (descent counted from both biological parents) 253, 257
Duat (Egypt) n. 77, 91, 334
Ducks 42, 63, n. 70, 169, n. 304
Dumul n. 244
Dumuzi 79, 133-134, 149, 193, 199, 221, 234, 288-289, n. 292, 293, 304-306, 312-14, n. 316, 317, n. 317, 335, 337, n. 337, 339, n. 390, 393, n. 395, 450
Duran 281
Dye(s)(d) n. 76, 245, n. 292, 334
Eanna Temple (Inanna's temple in Uruk) 300, n. 305, 307, 314, n. 317, 318-9, n. 336
Easter 76, n. 292, n. 304
Egalitarian(ism) 3, 5, 11, 13, 18, 23-24, 32-36, 39, 41, 43, 49-50, 81, 94, 115, 120, 124, 152, 163, 165, 176, 178, 186, 230, 234, 239, 244, 249, 288, n. 292, 296, 329, 357, 361-2, 369, 412
Eggs (aka **ova)** 68-71, 76, 74-76, 80, 104, 167, 170, 217, n. 304, 301, n. 390, 411
Egynovi(-ite; ition) 32-3, n. 32, n. 167, n. 172, 410, n. 412
Egypt(-ians) 8, n. 11, 13, n. 26, 28, 38-9, 59, 64, 70, n. 71, 77, n. 77, 79-80, 92-93, 103, 104, 113, 133, 149, 152, 168, n. 168, 169-170, 175, 193, 197-8, 201, 219, 223-226, 228, 233, n. 236, 237-8, 245, 250-251, 286, 293-294, 315, 326, 328-345, 351, 357, 364, 389, n. 389, n. 418
Ejaculate(tion) 58, 63, 99, 122, 164, 177, 190, 193-195, 197-9, 201, 219, 224, 235, 306, 317, 337, 340, 343, n. 343
Elect(-ing, -tions) 7, 11-12, 23, 122, 161, 186, 234, 276, 296, 316, 329, 351-352, 361, 364, 397, 412, 450
Elephants 24-25, 41, n. 103
Elizabethan 27, 104
Emasculate n. 32, n. 167, n. 170, 412, n. 412
Empathy 45
Eneolithic (Pre-Copper Age) 140
England 27, 93-94, 104, n. 149
Enheduanna 126, 235, 237, 259, 297-301, 317-18, 451
Enki n. 149, n. 182, n. 190, 199, 210, n. 210, 240, n. 244, 245, n. 245, 308-9, n. 308, n. 309, 310-1, 313, n. 314, n. 391, 394
Enkidu 60, 79, 125, 197, 231, 299, n. 303, 307, 315, 316, 322, n. 322, 318, n. 382, n. 390
Eos (Greek goddess of dawn) 426
Epic of Gilgamesh 60, 125, 197, n. 299, n. 303, 306, 316, 320-2, 346, n. 382, n. 390, n. 394, 418, 451
Equality (see **Egalitarian**) 33, 35, 37, 195, n. 288, 381, 446
Equinox(es) 89, n. 242, n. 306
Eridu 60, 198, 309
Ereshkigal 299, 303-4, n. 305, n. 317, 451

253, 259, 261, 266, 269, 271, 274-5, 281, 289, 307, 326-7, n. 327, 329-331, 344-5, 351, 356-7, 359-61, 363, 365, n. 371, 378, 381, 388, 401, 404-6, 408-10, 412, 417, 419, 444, 447, 452-3

Hamlet (community) 48
Hammurabi 51, 202, 318
Hammurabi's Code of Laws 51, 202, 318
Hannahanna 348-50, n. 348
Hantili 354-5
Harlot(s) 6, n. 6, 107, 207, 232, n. 245, 307, n. 317, n. 390
Hashadu 60, 79, 95, 107, 179, n. 182, 207, 235, 307, 315, 317, 342, n. 390
Hartland, Edwin Sidney 52
Hathor 38, n. 77, 79, 111, 175, 245, n. 245, n. 322, 327, n. 333, 331-7, 340-2, 364
Hatshepsut 111, 226, 227, 233, 235, 321, n. 329, 357-358
Hattuša 238, 246, 345, 351, 352, 354, n. 394
Hattuššili 354
Hawai'i 84, 95
Heaven(s) (see also **Sky Realm**) n. 78, 86-7, 108, 164, 176, 185, 193, 209-10, 266, 279, n. 292, 303, n. 303, n. 304, n. 305, 308-9, n. 309, 311, n. 311, 313, 321, 333, 339, 341, 376, 394, 442, 451
Hebat 353
Hebrew(s) 6, 22, 36, 148, 181, 241, n. 293, n. 312, 319, n. 322, 323, n. 326, 329, n. 344, 405
Heiros gamos n. 77, 237, 316, n. 313, 346
Hektor/Hector 231, 366, 370, n. 370, 376, 385, 397, 400, 414, 420, 421, 427-30
Helen 76, 106, 231, n. 231, n. 363, 392-4, n. 399, 401, 420, 431-2
Hens 69
Hepat 246
Hera 102, 248, 351, n. 372, 399, n. 372, 404, 425, n. 425, 433, 435-7
Hercules/Herakles/Heracles n. 372, 425-6, 430, 433, 436
Herders 19, 38, n. 65, 146, 149, 155-6, n. 292, 300
Hermes n. 77, 400, 435-6
Hermopolis (see also **Khmunu**) n. 77
Hero worship (see also Cult of the Hero) 104, n. 196, n. 373, 378, n. 378, 380, n. 380, 383, 386-7, n. 388, 395, 402, 405, 407-408, 415
Hestia 246, n. 246, 387, n. 387, 398, 438-440, n. 404, 441-3
Heterosexual (includes intercourse and pair bonding) 6, 30, 51-2, 57, 63-6, 68-70, 72-4, 76, n. 77, 78-79, 96, 98-99, 100-1, 113, 125, n. 126, 145, n. 148, 151, 156, 159, 162, 168, n. 168, 176-7, 182, n. 182, 188, 192-3, 195-6, 198, n. 199, 209, 216, 220-1, n. 220, 224, 233-4, 235-7, 241-2, n. 244, 245, 264, n. 291, 293, n. 293, n. 294, 299, n. 300, 316, 336, 345, n. 395, 397-9, 411, 446, 450
Hidden Place 77, 335, 341
Hierarchy(ies) 11, 21, 25-6, 36, n. 36, 39, 41-3, 46-7, 49, n. 49, 72, 94, 103, 112, 126, 129, 132, 152, 164, 176, 178, 181, 200, 202, 205, 210, 213, 227, 234-6, 238-9, 241, 244, 254, 257, 285, n. 287, 288, n. 290, 296, n. 326, 356, 359, n. 366, n. 369, 377, 380, 398, 401, 409-10, 412-3, 417, 420
Highway(s) 48
Hindu(s) n. 13, 76, 105, 193, n. 201, 218, 221
Hine-nui-te-poo (Maori death goddess) 95-6
Historians 1, 44, 46, 102, n. 104, 112, 212, 219, 227, 230, 238-9, 259, 278, 298, 301, 389, 403, 404, 447

Hittites 14, 81, 168, 170, 233, 238, 241, 246, 302, n. 306, 314, n. 318, 345-6, 348-9, 350-5, 369, 388

Holy of holies 77-8, 83, 85, 01, image 163, 191, 205, 319

Homemaker(s) 26, 30, 105, 227, 361, 412

Homer 1, 157, 185, 211, 218, 228, 241, 243, 367-93, 396-7, 399, 402, 403-5, 417-444, 454

Homosexual 57, 65, 69, 75, 79, 99, 110, 177, 182, 192, 219, 235, 318, 346, n. 398

Homunculi(-us) 75, n. 75, 81, 207, 391, 398

Hongshan culture (China) 87-8

Hopi 24, 115, 412

Horns (see also bucrania) 39, n. 57, 66, n. 69, n. 77, 148-9, 153-4, n. 153-4, 156, 165, n. 165, 169, 173-4, n. 174, 175-6, n. 176, 194, 203, 239, 332, n. 333, 334, 339, 343, 360, 363-5, n. 363

Horns of Consecration image, 175, 360

Horses 41, 63, 65, n. 80, 147, n. 154, 156-161, 192, 208, n. 239, 365, 426

Horus 169-70, 173-4, 212, 245, 327, 329, 335, 342-43, n. 342-3, 384

House (family run & owned by women) 22, 28, 30, 35-6, 38, n. 70, 79-80, 103-4, n. 105, 129, 135, 145, 159-61, 166, 170, 197, n. 200, 227, 233, 250-1, 253-4, 256-7, 268, 270, 274, 288, 297, 301, 310, 313-4, 319, n. 329, 347, 351, 354, 357, 359, 361, 399, 402, 405, 412-3, 449

Huanca (monolith stone at Caral) 89, n. 89, image 90

Huemac 276-7

Huey Tzompantili 280

Huitzilopochtli 85, 170, 242-243, 246-247, 257, 260-2, 268, 269, 276, 279, 281

Humility 38, 107

Hunahpu 278

Hunter-gatherers (see **Foragers**)

Hupasiya 346

Hurrian(s) 246, n. 322, 353-4

Husband(s) 6, n. 6, 22-4, n. 22, 79-80, n. 79, 91, n. 101, 103, 106, 107, 109, 112, 154, 159, 161-2, n. 200, 231-2, 237, 245, 257, 270, 287-8, 303, 306-7, n. 308, n. 322, 327, n. 332, 340, 342, 346-7, 351-2, 357, n. 373, n. 375, 391, 393, 428, n. 423, 429, 431

Huxley, Thomas Henry 53

Hymen 99, 102

Ibis(es) n. 70-1, 169

Ilhuicatl-Tonatiuh 284, 286

Iliad, The 1, 44, 106, 109, 155-157, 211-2, 214, 221, 230-1, 241, 246, 366-389, 393, 395-8, 401-402, n. 404, 405-6, 414-419, 421-2, 424, 426-8, 431, 433-5, 440, 443-4, 453

Illuyanka 346

Imam(s) 119

Inanna 10, n. 11, 59, 79, 89, 93, 112, 122-3, 125-6, 133-4, 146, n. 149, 169, 176, 179, 193, 199, 205, 209-10, 218-9, 221-2, n. 224, 228-9, 231-2, 234-5, 238, 241-2, 244-5, n. 245, 260, 290-325, 331-2, n. 331, 335, n. 336, 337, 339, 346, 390-391, n. 390, n. 391, 393-4, n. 394, n. 395, 431-2, 434, 450-1

Inara 346, 350

Inca 46

Incest 30, 51, 68, 108, n. 126, n. 199, 202, n. 241, 273, n. 273, 283, n. 283, 314-5, n. 317, 318, 326, 353

Individualism(ity) 37, 45, 377, 381

Indo-European language 211, n. 320, 345, n. 345, 349, 351

Industrial Revolution n. 37, 129, 137, 229

Indus Valley 152, 190, n. 290

Matrilineal descent (inheritance) 5, 10, 26-7, 36, 47, 56, 81, 116, 119, 159, 161, 227, 232-2, 239, 246, 253, 256-257, 259, 290, n. 288, 330-331, 343, 345, n. 348, 351-355, 357-61, n. 369, n. 380, 389, n. 391, 395, 401-2, 405, 409

Matrilocal practices (aka marriage, living arrangements, funeral practices) 11, 27, 120, 159, 227, 239, 253, 290, n. 290, 347, 355, 357, 359, 361, 402

Matrix, Mixed Model, or Circular Platform Model 24, 198, 451

Matron/matriarch 24-5, 27, 34, 41, 43, 101, 122, 125, 159, 410

Maui (aka Maaui) 95

Mauna Kea 84

Maya(ns) 85-86, 166, 171, n. 250, n. 251, 252-3, 255, 257, 264, 267-8, n. 270, n. 271, n. 272, 273, n. 273, 275, 278-9, 286, 295-6, n. 295

Mayahuel (woman of 400 breasts) n. 270, 273

Medicine wheels (aka sacred hoops) 61, 93, 95

Mediterranean (geographic area) 48, n. 65, 79, 83, 109-10, 112, 122, 129, 148, n. 148, 150, 152, n. 154, 155-7, 198, 205, 213, 217, 220, 227-8, 230, 233, 241, n. 245, 246, 296, 311, 318, n. 326, 331, 338, 342, 344-5, 351, n. 369, 389, 392, 394, n. 394, 402, 409, 452

Mediterranean Sea 48, 122, 129, 150, 156, 217, 296, 342, 409, 452

Mehet-Weret n. 77, 334

Menarche 9, 55, 101, 109-10, 175, n. 231, 263, n. 275, 312, n. 320, 335, n. 335, 407-8, 413, 421-2

Menopause(al) 9, 117, 175, 224, 263, 271, 275, 320, 335, n. 335

Men's halls/dorms (bachelor quarters/warrior house) 24-5, 256-7, 355, 359

Menses 8-9, n. 52, 118, n. 152, 203, n. 335, 361, n. 371

Menstrual(-ation) 4, 7-10, 21, 29-33, n. 39, 49, 55, n. 55, 59, 62, 67, 73-5, n. 73, 81, 85, n. 85, 95, 97, 101, n. 101, 107-11, 113, 117-9, 124, 126-7, n. 126, 134, 151, n. 152, n. 166, 169, 175, n. 176, 183, 188-9, 194, 201-3, n. 203, 247, 260, 261, 263-5, 267-8, 270, n. 275, 277, 283, 285, 289, n. 304, n. 311, n. 318, n. 319, n. 320, 327, 335, n. 335, 338, n. 354, 359-60, 363, n. 390, 408, 410, 413, 422-3, 446, 452-3

Menstrual fluids/blood 30, 59, 74-5, 113, 115, n. 126, 183, 189, 263, 265, 267-8, n. 354, 260, n. 390

Mesoamerica(n)(ns) 86, n. 86, 255, 257, 264, 266, 281, 282, 284, 286

Mesopotamia(n)(ns) 30, 36, 38, 60, 76, 134, 185, n. 203, 215, 217, 280, n. 290, n. 291, 297, 301, 311, 321, 340, n. 387, 451

Metal (production/metallurgy/weapons) 1, 30, n. 97, 106, 117, 130-1, 135, 140-143, 200, 211, n. 258, 384, 415

Metis (**Athena**'s mother)101-2, 395

Metztli (Nahuatl for "moon") 275

Mexica (see **Nahuatl** and **Aztec**) 242, 247, 250-254, 256, 258-9, 267, 271-272, n. 273, 275, 279-81, n. 279, 283, n. 283, 286, n. 286

Mictlán 284

Middle Ages (aka **Medieval** Europe) 206, 211

Middle Kingdom (Egypt) n. 77

"Might makes right" (aka bullying and oppression) 12, 34, 43, 64, 105, 179, 201, 242, 355, 379, 406, 450

Milk (see also **semen**) 42, 59, 75, 147, 151, 154, 158, n. 163, 173-4, n. 174, 191, 199, 221, 272-3, n. 292, n. 312, 332, 335-7, 341, n. 390, 441

Min (Egypt) 59, 79, 197, 327, 336-7

Minoans (see **Crete/Cretans**) 14, n. 46, 47-8, 79, 110, n. 153, n. 202, 241, n. 322, 329, n. 356, 357, 359-365, n. 360, n. 358, n. 406, 407, 409, 412, 414

Minos n. 46, 160, 356

Minotaur 117, 162, n. 304, 356

Pornography(ic) 72, 122-3
Poseidon 156, 161, n. 292, 363, n. 394, n. 398
Postpatriarchy(al) 3, 16
Pottery wheel 128-9
Pre-ceramic 89
Predynastic (Egypt) 37
Pregnancy(ies) (see also gestation) 6-7, 29, 51-53, 58-9, 62-66, 68-9, 71-4, 79, 95, 99-100, 107-8, 110-12, 120, 147, 149, 166, 174, 181-3, 187-189, 192, 198, 200, 209, 215, 217-221, 232-233, 235-6, 244, 247, 259-60, 262-3, 267-9, 271-2, 285, 292, 312, 314, 318, 329, 331, 334, 344, 364, 389, 397, 405, 419, 436, 445, 447
Pregnancy test 74
Preliterate cultures 145, 151, 178, 222
Premenarchal (prepubescent girls) 51, 100, 244, 333
Prepatriarchy(al) 16, 33, 187,
Pre-pottery 159
Priest(s) 36, 117, 123 (all genders), 134, 177, 203 (trans), 241, 248, n. 256, 268, 270, 272, 277, 283, 295, 308, 316 (trans), 357, 362, n. 403
Priestess(es) 79, 101, 118, 124, 131, 134, 180, 229, 233, n. 234, 246, 257, 266, 270, 295, 297, 298, 304, 314-16, n. 314, 336, 341, n. 341, 344, n. 388, 390, 402, 449
Primates 65, n. 215
"Principles of Statehood" 34
Procreate(ion) (discovery of) 14, n. 22, 26, 30, 35, n. 37, 51-3, 58, 62-71, 73-6, 78-80, 96, 101-5, 108-112, 118-9, 124, 126, 147, 149, 152, n. 180, 182, 187-9, 191, 196-7, 200, 216-225, 228, 231, 234, 242, 244, 252-4, 259, 262, 271, 284, 289-292, 295, 298, 306, 307, 310, 215-6, 322, 324, 327-8, 334, 340, 342, 350, 357, 360, n. 367, n. 369, 384, 387-390, n. 392, n. 396, 399, 403, 407-9, 412, n. 423, 444
Progenitor(s) 105, 384, 400, 407, 429
Prostitutes(tion) 6, n. 6, 205, n. 229, 230, 236, n. 314, n. 388, 448
Proto-dynastic (Egypt) 37, 90
Psychedelic (drugs) 74
Psychology 44, 328, 405-6, 449
Psychotropic (drugs) 74
Ptah-Hotep (Egyptian vizier) 70, 235
Puberty (see also menarche) 49, 57, 161, 406, n. 432
Public spaces 105, 340
Pudenda(um) (female; see Mons Venus, Mound of Creation) 8, 31, 76, n. 77, 78, n. 78, 81-3, 88, 94-6, 109, 182, 203, 208, n. 219, 240, 245, 257-8, 261, 263, 272, 283, 287, 337, 408, n. 409
Pulque (also **octli**) n. 268, 270-1
Puritans 53, 61, n. 61
Purulli Festival 344
Pyramid(s) 8-9, 32, 77-8, n. 77, 81-97, 109, 153, 182, 203, 208, 238, 240, 245, 257-8, 261, 263-4, 272, 283, 287, 337, 339, 342, 356, 408, 449
Quetzal birds n. 71, n. 74, 176, 196, n. 235, 253-4
Quetzalcoatl 266, 271, 281, 284
Quinceañeras n. 55
Qursu **ritual** n. 70, 77, n. 77, 90, 120, 124, 131, 132, 172, 205, 207, n. 208, n. 222, 230, 233, 235, 242, 257, 290, n. 290, n. 296, 301, 304, 306, n. 310, 312, 314-5, n. 314, n. 315, n. 321, 333, 337, 344, n. 388, 448
Ra (Egypt) n. 77, 110, 175, 233, 332-3
Raa (Māori) 94

Rabbis 117
Rabbits 63, 217, n. 302
Railroads 48
Rangi (Māori sky god) 94
Rape 52, 66, 76, 106-8, n. 180, n. 188, 197, 214, n. 393, 426, 442-3
Rebirth/reborn 8, 27, 30, n. 73, n. 77, 78, n. 78, 89, 90-1, 94-6, 109, 147, 182-3, n. 182, 196, 202-3, n. 202, 205, 219, 238, 240, 258, 264, n. 264, 269, 284, 286, 302, 332-3, 337, 339, n. 339, 380, 408-9, 449
Reed baskets 27, n. 220, 350
Reed boats 99, 133, 171, 207, 209, 309
Reed huts 312
Reed mats 157, 171, n. 447
Reed pipes/tubes 137, 140
Reincarnate(d) 73, 282, 302, 337, 408, 413
Religion(ous) 5-6, 12-13, n. 13, 17, n. 61, n. 64, 75, 80, 96, 99, 105-6, 114, 117, 119, 177-8, 183-4, 200, 206, 223, n. 244, 248, n. 248, 250, 252, 262, 266, 281, 311, n. 317, n. 342, 347, 349, 361, 363, 369, n. 374, n. 388, 423, 445, 450-1
Reproductive organs 52, n. 57, n. 69, 71, 77, 82-3, 115, n. 152, 163, 189-190, 193, n. 208, 317, n. 317, n. 362
Reptiles 71, 119, 172
Repudium (Latin) 81
Revelry 60
Ritual emblems/tokens 58
Roman(s) n. 6, 76, 108, 110, 190, 192-3, 202, 206, 208, 210, 292, 340, 435-6, n. 436, 442
Rondavel(s) (thatched roofs) 157
Roosters 69, 159
Rosette 94, n. 291, n. 293, n. 312
Royalty 28, 36, 51, 74, 133, 173, 202, 225, 231, 235, 252-3, 265, 284, 303, 308, 315, 326, 327, 341, 349-350, 352-3, 355, 387, 401, 406
Rule of Law 32
Rutting 34
Sacred hoops (see also Medicine Wheels) 92, 94
Sacred *me* 123, 180, 242, 305-6, 308, 314, n. 321, 333, 388-9, 392
Sacred meal 35,
Sacred ritual(s) 57, 68, n. 290, 451
Sacred Sex 57, 60, 79, 94, 114, 197, n. 202, 205, 230, 232-3, 242, 290, 304, 314-5, n. 380, n. 388, n. 416
Sacred space(s) 60, 77, 200, 202-3, 233, 305, 312-313
Sacrificial victims n. 6, 9, 33, 45-6, 84, 134, 141, 148, 153-4, 160, 181, 241, 244-5, 248, 251, 255-6, 258-9, 261, 263-5, 267-8, 272-8, 280-1, 283-4, 286, 303, 335, 345-7, 352, 354, 365, 372, 379, 382, 384, 386, 395, 402, 406, 413, 439-441
Sahagun, Bernardino de 248-9, 279, 285
San Andrés (Olmec city) 86
Saqqara 70
Sargon of Akkad 28, 124, 202-3, 233, 236, 256-7, 297-301, 304-5, 315-6, 319, 449
Saudi Arabia 23
Scapegoating (ritual) 146, n. 146, n. 242
Scientific method 30, 68
Seasonal sex cycles (see In heat, estrous) 64-6, n. 119, 149, n. 152, 252, 254, n. 290, 356
Sedentary 19, 116, 133, 142, 145, 256

Texas legislature 3
Tezcatlipoca 267, 282, 286
Thatch (roofs) 134, 158
Thetis 156, 376, 383, 388, 398, n. 398, 427, 442
Tiachoccalcatl 256
Tiamat 243, n. 243, 281
Tibet 24, 84
Tigris River 59, n. 319
Tipi(s) 104, 108, 158, 413
Tiyospaye (Lakota) 119
Tizaapan 241-2
Tlacaelel I 256-9, 265-67, 280
Tlaloc (Aztec god of rain) 85, 259, 267, n. 267, 284, 286
Tlalocán (Mansion of the Moon) 284
Tlaltecuhtli 281-2
Tlatoani (Aztec for Speaker) 242, 257, 266
Toltecs 45, 85, 275, 285, 287
Tonatiuh (also **Nanahuatzin**) 281, 283, 285
Totonac, Mexico 41
Town(s) 11, 47-8, 67, 87, 288, n. 326
Toxic masculinity 16, 106, 212, 354, n. 375, 408, 417
Transcendence(-ent) 33, 61-2, 300, 454
Transformative(-tional) 59-60, n. 70, 78-9, n. 85, 124, n. 124, 167, 181, 191, 196, 206, 208-9, 221, 284, n. 301, 305-6, n. 305, 316, 320, 333-4, n. 334, n. 398, 340-1
Transgendered (erers) 16, 124, n. 124, 187, n. 203, n. 257, 300
Transition (liminal movement) n. 57, 72, 104, 109, 158, 164, 183, 268, 330, 382, 389, 414
Travois 146, 152, 154
Triangle (aka **delta**) 78, n. 78, 94, 97, 122, 206, 337-338
Tribal cultures 19, 23, 73, 109, 146, 170, 174, n. 182, 219, n. 368, 450
Troy 1, 76, 109, 129, 155-6, 210, 249, 251, 265, 268, 280, 297-8, 302, 323, 333, 370, 372, 376-8, 387, 391-2, 395-7, 400-402, 428-429, 431, 435
Turkey (country) 147, 159, 193, 344
Turkeys 42
Two-spirited n. 23, 29, 327
Tzitzimeme 268-272, n. 268
Tzolkin n. 270. n. 294
Ubaid 121, 132, 210, 294, 296, 317
Uncle (matriarchal system and social role) 24, 56, 309, 330
Unclean 31, 33, 80, 107, 113, 118, n. 202, n. 311, n. 353, 454
United States 3, 24, 52, 114, n. 170, 274, 366
Uruk 39, 148, 181, 209, 231, 236, 298, n. 289, n. 296, n. 301, n. 302, 305, 307, 312, 317, 320, 322, n. 322, 328, 345
Uterus (see also **Holy of Holies**) 3, 53, 71, 76-7, 83, 96, 164, 190, 262, 317-8, 331-2
Vagina(al) 3, 31, 53, n. 57, 71, 77, n. 78, 83-4, 88, 91, 95-6, 99, 107, 124, n. 153, 164, 168, n. 181, 190-1, 208, 251, 263, 317, n. 317, 318, 323, 331, 366
Van Leewenhoek, Anthony n. 75
Venereal disease 53
Venus figurines 49, 115-7, 121, 126, 129
Venus (planet or Morning and Evening Star) n. 270, 294-5, 302, n. 303

Yolia (Nahuatl for "soul") 284
Yoni 78, 120, 166, 192
Yolk (of an egg) 75
Zacatapayolli 266
Zalpa 351
Zeus 76, 101, 111, 207, 218, 220, 245-6, 292, 363, 372, 377, 394-5, 399-400m 492, 423, 426, 428, 431, 434, 436-8, 443-4
Zephyrus 156
Ziggurat(s) 8, 77, 82-3, 97, 129, 204, 209, 237, 318, 343

About the Author

Ruth J. Heflin did not predict Jesus would appear on the Benny Hinn show, but she did survive growing up in the buckle of the bible belt with the name "Ruth." She still lives there with two cats, near her grown son, and with her monogamous (37 years and counting) husband, James P. Cooper, poet extraordinaire. After 34 years teaching as a college professor, she now spends her time researching things that fascinate her, writing other things, and spending most of her time editing for a small, private literary press.

Ruth also holds a PhD in American Ethnic Cultures and Literatures from Oklahoma State University. Her first scholarly book was, *I Remain Alive: The Sioux Literary Renaissance.*

In *Pitiless Bronze*, Ruth brings together her expertise in pre-literate cultures and her broad understanding of humanity to reread ancient symbols and texts.

Peer into her thoughts at *The Book of Ruth* at ruthjheflin.com.

Choeofpleirn Press, LLC

2023